FM 22-100

U.S. ARMY LEADERSHIP

U.S. DEPARTMENT OF THE ARMY

SILVER ROCK PUBLISHING

Published by Silver Rock Publishing

FM 22-100: The U.S. Army Leadership Field Manual
ISBN: 978-1-62654-429-1 (paperback)
978-1-62654-430-7 (casebound)
978-1-62654-431-4 (spiralbound)

FOREWORD

Soldiers represent what's best about our Army. Day in and day out, in the dark and in the mud and in faraway places, they execute tough missions whenever and wherever the Nation calls. They deserve our very best--leaders of character and competence who act to achieve excellence. That theme resounds throughout FM 22-100, Army Leadership, and echoes our time-honored principle of BE, KNOW, DO.

This leadership manual lays out a framework that applies to all Army leaders—officer and NCO, military and civilian, active and reserve component. At the core of our leadership doctrine are the same Army Values embedded in our force: loyalty, duty, respect, selfless service, honor, integrity, and personal courage (LDRSHIP). The framework also outlines physical, mental, and emotional attributes that together with values form *character*—what a leader must BE.

Being a person of character is fundamental to our Army. What makes Army leaders of *competence* are skills with people, ideas, things, and warfighting. We refer to those four sets of skills as interpersonal, conceptual, technical, and tactical. Many are common to leaders in all situations; some additional skills are required for those who gain increasing responsibility. Leaders of character and competence are those with the appropriate skills, leaders who KNOW their people, their equipment, and their profession.

All that is still not enough. We call on our leaders to translate character and competence into leader actions. Army leaders influence people—by providing purpose, direction, and motivation—while operating to accomplish the mission and improving the organization. Leaders inspire others toward common goals and never lose sight of the future even as they labor tirelessly for the demands of today. That is what we expect our leaders to DO.

Unlike the previous editions, this leadership manual covers three levels of leadership—direct, organizational, and strategic. While the skills and actions necessary for leadership success at the direct level continue to be important at higher levels, organizational and strategic leaders need additional skills to perform in their more complex roles.

I urge Army leaders to read this manual thoroughly, study it carefully, and teach it faithfully. But above all, I challenge you to be leaders of character and competence who lead others to excellence. Whether supporting, training, or fighting, America looks to you to BE, KNOW, and DO what is right.

ERIC K. SHINSEKI
General, United States Army
Chief of Staff

Field Manual
No. 22-100

Headquarters
Department of the Army
Washington, DC, 31 August 1999

Army Leadership

Contents

Page

PART ONE: THE LEADER, LEADERSHIP, AND THE HUMAN DIMENSION

DISTRIBUTION RESTRICTION: Approved for public release; distribution is unlimited.

*This publication supersedes FM 22-100, 31 July 1990; FM 22-101, 3 June 1985; FM 22-102, 2 March 1987; FM 22-103, 21 June 1987; DA Pam 600-80, 9 June 1987; and DA Form 4856, June 1985.

Examples

Preface

The Army consists of the active component, Army National Guard, Army Reserve, and Department of the Army (DA) civilians. It's the world's premier land combat force—a full-spectrum force trained and ready to answer the nation's call. The Army's foundation is confident and competent leaders of character. This manual is addressed to them and to those who train and develop them.

PURPOSE

FM 22-100 is a single-source reference for all Army leaders. Its purpose is threefold:

- To provide leadership doctrine for meeting mission requirements under all conditions.
- To establish a unified leadership theory for all Army leaders: military and civilian, active and reserve, officer and enlisted.
- To provide a comprehensive and adaptable leadership resource for the Army of the 21st century.

As the capstone leadership manual for the Army, FM 22-100 establishes the Army's leadership doctrine, the fundamental principles by which Army leaders act to accomplish the mission and take care of their people. The doctrine discusses how Army values form the basis of character. In addition, it links a suite of instruments, publications, and initiatives that the Army uses to develop leaders. Among these are—

- AR 600-100, which establishes the basis for leader development doctrine and training.
- DA Pam 350-58, which describes the Army's leader development model.
- DA Pam 600-3, which discusses qualification criteria and outlines development and career management programs for commissioned officers.
- DA Pam 600-11, which discusses qualification criteria and outlines development and career management programs for warrant officers.
- DA Pam 600-25, which discusses noncommissioned officer (NCO) career development.
- DA Pam 690-46, which discusses mentoring of DA civilians.
- The TRADOC Common Core, which lists tasks that military and DA civilian leaders must perform and establishes who is responsible for training leaders to perform them.
- Officer, NCO, and DA civilian evaluation reports.

FM 22-100 also serves as the basis for future leadership and leader development initiatives associated with the three pillars of the Army's leader development model. Specifically, FM 22-100 serves as—

- The basis for leadership assessment.

- The basis for developmental counseling and leader development.
- The basis for leadership evaluation.
- A reference for leadership development in operational assignments.
- A guide for institutional instruction at proponent schools.
- A resource for individual leaders' self-development goals and initiatives.

FM 22-100 directly supports the Army's keystone manuals, FM 100-1 and FM 100-5, which describe the Army and its missions. It contains principles all Army leaders use when they apply the doctrine, tactics, techniques, and procedures established in the following types of doctrinal publications:

- Combined arms publications, which describe the tactics and techniques of combined arms forces.
- Proponency publications, which describe doctrinal principles, tactics, techniques, and collective training tasks for branch-oriented or functional units.
- Employment procedure publications, which address the operation, employment, and maintenance of specific systems.
- Soldier publications, which address soldier duties.
- Reference publications, which focus on procedures (as opposed to doctrine, tactics, or techniques) for managing training, operating in special environments or against specific threats, providing leadership, and performing fundamental tasks.

This edition of FM 22-100 establishes a unified leadership theory for all Army leaders based on the Army leadership framework and three leadership levels. Specifically, it—

- Defines and discusses Army values and leader attributes.
- Discusses character-based leadership.
- Establishes leader attributes as part of character.
- Focuses on improving people and organizations for the long term.
- Outlines three levels of leadership—direct, organizational, and strategic.
- Identifies four skill domains that apply at all levels.
- Specifies leadership actions for each level.

The Army leadership framework brings together many existing leadership concepts by establishing leadership dimensions and showing how they relate to each other. Solidly based on BE, KNOW, DO—that is, character, competence, and action—the Army leadership framework provides a single instrument for leader development. Individuals can use it for self-development. Leaders can use it to develop subordinates. Commanders can use it to focus their programs. By establishing leadership dimensions grouped under the skill domains of values, attributes, skills, and actions, the Army leadership framework provides a simple way to think about and discuss leadership.

The Army is a values-based institution. FM 22-100 establishes and clarifies those values. Army leaders must set high standards, lead by example, do what is legally and morally right, and influence other people to do the same. They must establish and sustain a climate that ensures people are treated with dignity and respect and create an environment in which people are challenged and motivated to be all they can be. FM 22-100 discusses these aspects of leadership and how they contribute to developing leaders of character and competence. These are the leaders who make the Army a trained and ready force prepared to fight and win the nation's wars.

The three leadership levels—direct, organizational, and strategic—reflect the different challenges facing leaders as they move into positions of increasing responsibility. Direct leaders lead face to face: they are the Army's first-line leaders. Organizational leaders lead large organizations, usually brigade-sized and larger. Strategic leaders are the Army's most senior leaders. They lead at the major command and national levels.

Unlike previous editions of FM 22-100—which focused exclusively on leadership by uniformed leaders at battalion level and below—this edition addresses leadership at all levels and is addressed to all Army leaders, military and DA civilian. It supersedes four publications—FM 22-101, *Leadership Counseling*; FM 22-102, *Soldier Team Development*; FM 22-103, *Leadership and Command at Senior Levels*; and DA Pam 600-80 *Executive Leadership*—as well as the previous edition of FM 22-100. A comprehensive reference, this manual shows how leader skills, actions, and concerns at the different levels are linked and allows direct leaders to read about issues that affect organizational and strategic leaders. This information can assist leaders serving in positions supporting organizational and strategic leaders and to other leaders who must work with members of organizational- and strategic-level staffs.

FM 22-100 emphasizes self-development and development of subordinates. It includes performance indicators to help leaders assess the values, attributes, skills, and actions that the rest of the manual discusses. It discusses developmental counseling, a skill all Army leaders must perfect so they can mentor their subordinates and leave their organization and people better than they found them. FM 22-100 prescribes DA Form 4856-E (Developmental Counseling Form), which supersedes DA Form 4856 (General Counseling Form). DA Form 4856-E is designed to support leader development. Its format follows the counseling steps outlined in Appendix C.

FM 22-100 offers a framework for how to lead and provides points for Army leaders to consider when assessing and developing themselves, their people, and their organizations. It doesn't presume to tell Army leaders exactly how they should lead every step of the way. They must be themselves and apply this leadership doctrine as appropriate to the situations they face.

SCOPE

FM 22-100 is divided into three parts. Part I (Chapters 1, 2, and 3) discusses leadership aspects common to all Army leaders. Part II (Chapters 4 and 5) addresses the skills and actions required of direct leaders. Part III (Chapters 6 and 7) discusses the skills and actions required of organizational and strategic leaders. The manual also includes six appendixes.

Chapter 1 defines Army leadership, establishes the Army leadership framework, and describes the three Army leadership levels. It addresses the characteristics of an Army leader (BE, KNOW, DO), the importance of being a good subordinate, and how all Army leaders lead other leaders. Chapter 1 concludes with a discussion of moral and collective excellence.

Chapter 2 examines character, competence, and leadership—what an Army leader must BE, KNOW, and DO. The chapter addresses character in terms of Army values and leader attributes. In addition, it describes character development and how character is related to ethics, orders—to include illegal orders—and beliefs. Chapter 2 concludes by introducing the categories of leader skills—interpersonal, conceptual, technical, and tactical—and the categories of leader actions—influencing, operating, and improving.

Chapter 3 covers the human dimension of leadership. The chapter begins by discussing discipline, morale, and care of subordinates. It then addresses stress, both combat- and change-related. Discussions of organizational climate, institutional culture, and leadership styles follow. Chapter 3 concludes by examining intended and unintended consequences of decisions and leader actions.

Chapters 4 and 5 discuss the skills and actions required of direct leaders. The skills and actions are grouped under the categories introduced at the end of Chapter 2.

Chapters 6 and 7 provide an overview of the skills and actions required of organizational and strategic leaders. These chapters introduce direct leaders to the concerns faced by leaders and staffs operating at the organizational and strategic levels. Like Chapters 4 and 5, Chapters 6 and 7 group skills and actions under the categories introduced in Chapter 2.

Appendix A outlines the roles and relationships of commissioned, warrant, and noncommissioned officers. It includes discussions of authority, responsibility, the chain of command, the NCO support channel, and DA civilian support.

Appendix B lists performance indicators for Army values and leader attributes, skills, and actions. It provides general examples of what Army leaders must BE, KNOW, and DO.

Appendix C addresses developmental counseling in detail. It begins with a discussion of the characteristics of a good counselor, the skills a counselor requires, and the limitations leaders face when they counsel subordinates. The appendix then examines the types of developmental counseling, counseling approaches, and counseling techniques. Appendix C concludes by describing the counseling process and explaining how to use DA Form 4856-E, the Developmental Counseling Form.

Appendix D explains how to prepare a leader plan of action and provides an example of a direct leader preparing a leader plan of action based on information gathered using an ethical climate assessment survey (ECAS). The example explains how to conduct an ECAS.

Appendix E discusses how Army values contribute to character development and the importance of developing the character of subordinates.

Appendix F contains a copy of the Constitution of the United States. All members of the Army take an oath to "support and defend the Constitution of the United States." It is included so it will be immediately available for Army leaders.

APPLICABILITY

FM 22-100's primary audience is direct leaders, military leaders serving at battalion level and below and DA civilian leaders in comparable organizations. However, FM 22-100 contains doctrine applicable at all leadership levels, to all military and DA civilian leaders of the Army.

The proponent of this publication is Headquarters (HQ), TRADOC. Send comments and recommendations on DA Form 2028 directly to Commander, US Army Combined Arms Center and Fort Leavenworth, Center for Army Leadership, ATTN: ATZL-SWC, Fort Leavenworth, KS 66027-2300.

Unless this publication states otherwise, masculine nouns and pronouns do not refer exclusively to men.

This publication contains copyrighted material.

ACKNOWLEDGMENTS

The copyright owners listed here have granted permission to reproduce material from their works. Other sources of quotations and material used in examples are listed in the source notes.

The quotations in Chapters 1 and 2 from Geoffrey C. Ward, _The Civil War: An Illustrated History_ (New York: Knopf, 1990) are reprinted with permission of the publisher.

Portions of the example Task Force Kingston are reprinted from _ARMY Magazine_, April 1964. Copyright © 1964 by the Association of the United States Army and reproduced by permission.

The quotation by Thomas J. Jackson in Chapter 2 is reprinted from Robert D. Heinl, _Dictionary of Military and Naval Quotations_ (Annapolis: US Naval Institute Press, 1988).

The quotation by Dandridge M. Malone in Chapter 2 is reproduced from Dandridge M. Malone, _Small Unit Leadership: A Commonsense Approach_ (Novato, Calif.: Presidio Press, 1983).

The quotations by Theodore Roosevelt in Chapter 5 and Dwight D. Eisenhower in Chapter 7 are reprinted from John C. Maxwell, _Leadership 101—Inspirational Quotes and Insights for Leaders_ (Tulsa, Okla.: Honor Books, 1994).

The quotation by Arleigh Burke in Chapter 7 is reproduced from Karel Montor and others, _Naval Leadership: Voices of Experiences_ (Annapolis: US Naval Institute Press, 1987).

Excerpts from WAR AS I KNEW IT by General George S. Patton. Copyright © 1947 by Beatrice Patton Walters, Ruth Patton Totten, and George Smith Totten. Copyright © renewed 1975 by Major General George Patton, Ruth Patton Totten, John K. Waters Jr., and George P. Waters. Reprinted by permission of Houghton Mifflin Co. All rights reserved.

The quotation by George C. Marshall in Chapter 2 is reprinted from Forrest C. Pogue, _George C. Marshall: Ordeal and Hope 1939-1942_ (New York: Viking Press, 1966). Copyright © 1965, 1966 by George C. Marshall Research Foundation.

This manual is a tool to help you answer these questions, to begin or continue becoming a leader of character and competence, an Army leader. Chapter 1 starts with an overview of what the Army requires of you as an Army leader. This is the Army leadership framework; it forms the structure of the Army's leadership doctrine. Chapter 1 also discusses the three levels of Army leadership: direct, organizational, and strategic. Chapter 2 discusses character, competence, and leadership—what you must BE, KNOW, and DO as an Army leader. Chapter 3 talks about the human dimension, the many factors that affect the people and teams that you lead and the institution of which you and they are a part.

CHAPTER 1

The Army Leadership Framework

Just as the diamond requires three properties for its formation—carbon, heat, and pressure—successful leaders require the interaction of three properties—character, knowledge, and application. Like carbon to the diamond, character is the basic quality of the leader....But as carbon alone does not create a diamond, neither can character alone create a leader. The diamond needs heat. Man needs knowledge, study, and preparation....The third property, pressure—acting in conjunction with carbon and heat—forms the diamond. Similarly, one's character, attended by knowledge, blooms through application to produce a leader.

General Edward C. Meyer
Former Army Chief of Staff

1-1. The Army's ultimate responsibility is to win the nation's wars. For you as an Army leader, leadership in combat is your primary mission and most important challenge. To meet this challenge, you must develop character and competence while achieving excellence. This manual is about leadership. It focuses on character, competence, and excellence. It's about accomplishing the mission and taking care of people. It's about living up to your ultimate responsibility, leading your soldiers in combat and winning our nation's wars.

1-2. Figure 1-1 shows the Army leadership framework. The top of the figure shows the four categories of things leaders must BE, KNOW, and DO. The bottom of the figure lists dimensions of Army leadership, grouped under these four categories. The dimensions consist of Army values and subcategories under attributes, skills, and actions.

1-3. Leadership starts at the top, with the character of the leader, with your character. In order to lead others, you must first make sure your own house is in order. For example, the first line of *The Creed of the Noncommissioned Officer* states, "No one is more professional than I." But it takes a remarkable person to move from memorizing a creed to actually living that creed; a true leader _is_ that remarkable person.

1-4. Army leadership begins with what the leader must BE, the values and attributes that shape a leader's character. It may be helpful to think of these as internal qualities: you possess them all the time, alone and with others. They define who you are; they give you a solid footing. These values and attributes are the same for all leaders, regardless of position, although

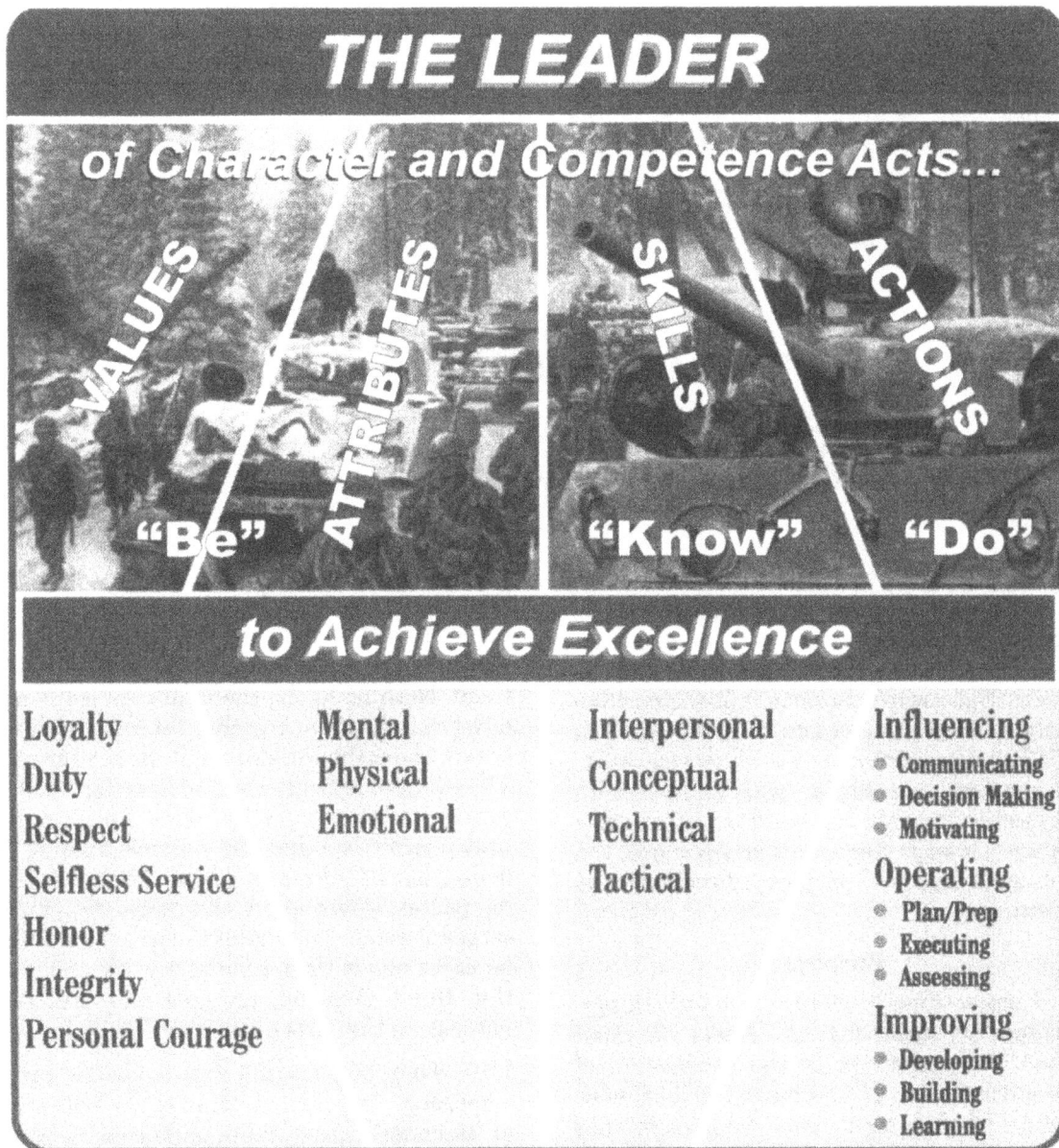

THE LEADER

of Character and Competence Acts...

VALUES   ATTRIBUTES   SKILLS   ACTIONS

"Be"   **"Know"**   **"Do"**

to Achieve Excellence

Loyalty	Mental	Interpersonal	Influencing
Duty	Physical	Conceptual	• Communicating
Respect	Emotional	Technical	• Decision Making
Selfless Service		Tactical	• Motivating
Honor			Operating
Integrity			• Plan/Prep
Personal Courage			• Executing
			• Assessing
			Improving
			• Developing
			• Building
			• Learning

Leaders of character and competence act to achieve excellence by developing a force that can fight and win the nation's wars and serve the common defense of the United States.

Figure 1-1. The Army Leadership Framework

you certainly refine your understanding of them as you become more experienced and assume positions of greater responsibility. For example, a sergeant major with combat experience has a deeper understanding of selfless service and personal courage than a new soldier does.

1-5. Your skills are those things you KNOW how to do, your competence in everything from the technical side of your job to the people skills a leader requires. The skill categories of the Army leadership framework apply to all leaders. However, as you assume positions of greater responsibility, you must master additional skills in each category. Army leadership positions fall into one of three levels: direct, organizational, and strategic. These levels are described later in this chapter. Chapters 4, 6, and 7 describe the skills leaders at each level require.

1-6. But character and knowledge—while absolutely necessary—are not enough. You cannot be effective, you cannot be a leader, until you *apply* what you know, until you act and DO what you must. As with skills, you will learn more leadership actions as you serve in different positions. Because actions are the essence of leadership, the discussion begins with them.

LEADERSHIP DEFINED

> Leadership is **influencing** people—by providing purpose, direction, and motivation—while **operating** to accomplish the mission and **improving** the organization.

INFLUENCING

1-7. Influencing means getting people to do what you want them to do. It is the means or method to achieve two ends: operating and improving. But there's more to influencing than simply passing along orders. The example you set is just as important as the words you speak. And you set an example—good or bad—with every action you take and word you utter, on or off duty. Through your words and example, you must communicate purpose, direction, and motivation.

Purpose

1-8. Purpose gives people a reason to do things. This does not mean that as a leader you must explain every decision to the satisfaction of your subordinates. It does mean you must earn their trust: they must know from experience that you care about them and would not ask them to do something—particularly something dangerous—unless there was a good reason, unless the task was essential to mission accomplishment.

1-9. Look, for example, at a battalion maintenance section. Its motor sergeant always takes the time—and has the patience—to explain to the mechanics what is required of them. Nothing fancy; the motor sergeant usually just calls them together for a few minutes to talk about the workload and the time crunch. The soldiers may get tired of hearing "And, of course, unless we get the work finished, this unit doesn't roll and the mission doesn't get done," but they know it's true. And every time he passes information this way, the motor sergeant sends this signal to the soldiers: that he cares about their time and work and what they think, that they are members of a team, not cogs in the "green machine."

1-10. Then one day the unit is alerted for an emergency deployment. Things are happening at breakneck speed; there is no time to pause, and everything and everyone is under stress. The motor sergeant cannot stop to explain things, pat people on the back, or talk them up. But the soldiers will work themselves to exhaustion, if need be, because the motor sergeant has earned their trust. They know and appreciate their leader's normal way of operating, and they will assume there is a good reason the leader is doing things differently this time. And should the deployment lead to a combat mission, the team will be better prepared to accomplish their mission under fire. Trust is a basic bond of leadership, and it must be developed over time.

Direction

1-11. When providing direction, you communicate the way you want the mission accomplished. You prioritize tasks, assign responsibility for completing them (delegating authority when necessary), and make sure your people understand the standard. In short, you figure out how to get the work done right with the available people, time, and other resources; then you communicate that information to your subordinates: "We'll do these things first. You people work here; you people work there." As you think the job through, you can better aim your effort and resources at the right targets.

1-12. People want direction. They want to be given challenging tasks, training in how to accomplish them, and the resources necessary to do them well. Then they want to be left alone to do the job.

Motivation

1-13. Motivation gives subordinates the will to do everything they can to accomplish a mission. It results in their acting on their own initiative when they see something needs to be done.

1-14. To motivate your people, give them missions that challenge them. After all, they did not join the Army to be bored. Get to know your people and their capabilities; that way you can tell just how far to push each one. Give them as much responsibility as they can handle; then let them do the work without looking over their shoulders and nagging them. When they succeed, praise them. When they fall short, give them credit for what they have done and coach or counsel them on how to do better next time.

1-15. People who are trained this way will accomplish the mission, even when no one is watching. They will work harder than they thought they could. And when their leader

notices and gives them credit (with something more than the offhand comment "good job"), they will be ready to take on even more next time.

1-16. But Army leaders motivate their people by more than words. The example you set is at least as important as what you say and how well you manage the work. As the unit prepares for the rollout, the motor sergeant you just read about is in the motor pool with the mechanics on Friday night and Saturday morning. If his people are working in the rain, the NCO's uniform will be wet too. If they have missed breakfast, the leader's stomach will be growling just as loudly. The best leaders lead from the front. Don't underestimate the importance of being where the action is.

OPERATING

1-17. Actions taken to influence others serve to accomplish operating actions, those actions you take to achieve the short-term goal of accomplishing the mission. The motor sergeant will make sure the vehicles roll out, on time and combat ready, through planning and preparing (laying out the work and making the necessary arrangements), executing (doing the job), and assessing (learning how to work smarter next time). The motor sergeant provides an example of how direct leaders perform operating actions. All leaders execute these operating actions, which become more complex as they assume positions of increasing responsibility.

IMPROVING

1-18. The motor sergeant's job is not complete when the last vehicle clears the gate. While getting the job done is key, the Army also expects him to do far more than just accomplish the day's work. Army leaders also strive to improve everything entrusted to them: their people, facilities, equipment, training, and resources. There will be a new mission, of course, but part of finishing the old one is improving the organization.

1-19. After checking to be sure the tools are repaired, cleaned, accounted for, and put away, the motor sergeant conducts an informal after-action review (AAR) with the section. (An AAR is a professional discussion of an event, focused on performance standards, that allows participants to discover for themselves what happened, why it happened, and how to sustain strengths and improve on weaknesses. Chapter 5 discusses AARs.) The motor sergeant is self-confident enough to ask subordinates for their ideas on how to make things work better (always a key goal). He then acts based on his own and team members' observations. The motor sergeant looks for strong areas to sustain and praises team members as appropriate; however if the motor sergeant saw the team members spend too much time on some tasks and not enough on others, he changes the section standing operating procedures (SOP) or counsels the people involved. (Developmental counseling is not an adverse action; it is a skill you use to help your subordinates become better team members, improve performance, and prepare for the future. Counseling should address strong areas as well as weak ones and successes as well as failures. Appendix C discusses developmental counseling.) If the motor sergeant discovers gaps in individual or collective skills, he plans and conducts the training necessary to fill them. If something the motor sergeant did or a decision he made didn't turn out quite right, he will not make the same error again. More than that, the motor sergeant lets his people know what went wrong, finds out their impressions of why it happened, and determines how they will make it work next time.

1-20. By doing these things, the motor sergeant is creating a better organization, one that will work smarter the next time. His example sends an important message. The soldiers see their leader look at his own and the organization's performance, evaluate it, identify strong areas to sustain as well as mistakes and shortcomings, and commit to a better way of doing things. These actions are more powerful than any lecture on leadership.

BE, KNOW, DO

1-21. BE, KNOW, DO clearly and concisely state the characteristics of an Army leader. You have just read about leader actions, the DO of BE, KNOW, DO. Leadership is about taking action, but there's more to being a leader than just what you do. Character and competence, the BE and the KNOW, underlie everything a leader does. So becoming a leader involves developing all aspects of yourself. This includes adopting and living Army values. It means developing the attributes and learning the skills of an Army leader. Only by this self-development will you become a confident and competent leader of character. Being an Army leader is not easy. There are no cookie-cutter solutions to leadership challenges, and there are no shortcuts to success. However, the tools are available to every leader. It is up to you to master and use them.

BE

1-22. Character describes a person's inner strength, the BE of BE, KNOW, DO. Your character helps you know what is right; more than that, it links that knowledge to action. Character gives you the courage to do what is right regardless of the circumstances or the consequences. (Appendix E discusses character development.)

1-23. You demonstrate character through your behavior. One of your key responsibilities as a leader is to teach Army values to your subordinates. The old saying that actions speak louder than words has never been more true than here. Leaders who talk about honor, loyalty, and selfless service but do not live these values—both on and off duty—send the wrong message, that this "values stuff" is all just talk.

1-24. Understanding Army values and leader attributes (which Chapter 2 discusses) is only the first step. You also must embrace Army values and develop leader attributes, living them until they become habit. You must teach Army values to your subordinates through action and example and help them develop leader attributes in themselves.

KNOW

1-25. A leader must have a certain level of knowledge to be competent. That knowledge is spread across four skill domains. You must develop **interpersonal skills**, knowledge of your people and how to work with them. You must have **conceptual skills**, the ability to understand and apply the doctrine and other ideas required to do your job. You must learn **technical skills**, how to use your equipment. Finally, warrior leaders must master **tactical skills**, the ability to make the right decisions concerning employment of units in combat. Tactical skills include mastery of the art of tactics appropriate to the leader's level of responsibility and unit type. They're amplified by the other skills—interpersonal, conceptual, and technical—and are the most important skills for warfighters. (FM 100-40 discusses the art of tactics.)

1-26. Mastery of different skills in these domains is essential to the Army's success in peace and war. But a true leader is not satisfied with knowing only how to do what will get the organization through today; you must also be concerned about what it will need tomorrow. You must strive to master your job and prepare to take over your boss's job. In addition, as you move to jobs of increasing responsibility, you'll face new equipment, new ideas, and new ways of thinking and doing things. You must learn to apply all these to accomplish your mission.

1-27. Army schools teach you basic job skills, but they are only part of the learning picture. You'll learn even more on the job. Good leaders add to their knowledge and skills every day. True leaders seek out opportunities; they're always looking for ways to increase their professional knowledge and skills. Dedicated squad leaders jump at the chance to fill in as acting platoon sergeant, not because they've mastered the platoon sergeant's job but because they know the best place to learn about it is in the thick of the action. Those squad leaders challenge

themselves and will learn through doing; what's more, with coaching, they'll learn as much from their mistakes as from their successes.

DO

1-28. You read about leader actions, the DO of Army leadership doctrine, at the beginning of this chapter. Leader actions include—

- **Influencing:** making decisions, communicating those decisions, and motivating people.
- **Operating:** the things you do to accomplish your organization's immediate mission.
- **Improving:** the things you do to increase the organization's capability to accomplish current or future missions.

1-29. Earlier in this chapter, you read about a motor sergeant who lives Army values, has developed leader attributes, and routinely performs leader actions. But that was an example, and a garrison example at that. What about reality? What about combat? Trained soldiers know what they are supposed to do, but under stress, their instincts might tell them to do something different. The exhausted, hungry, cold, wet, disoriented, and frightened soldier is more likely to do the wrong thing—stop moving, lie down, retreat—than one not under that kind of stress. This is when the leader must step in—when things are falling apart, when there seems to be no hope—and get the job done.

1-30. The fight between the 20th Regiment of Maine Volunteers and the 15th and 47th Regiments of Alabama Infantry during the Civil War illustrates what can happen when a leader acts decisively. It shows how the actions of one leader, in a situation that looked hopeless, not only saved his unit, but allowed the entire Union Army to maintain its position and defeat the Confederate invasion of Pennsylvania. The story's hero is a colonel—but it could have been a captain, or a sergeant, or a corporal. At other times and in other places it has been.

COL Chamberlain at Gettysburg

In late June 1863 GEN Robert E. Lee's Army of Northern Virginia passed through western Maryland and invaded Pennsylvania. For five days, the Army of the Potomac hurried to get between the Confederates and the national capital. On 1 July the 20th Maine received word to press on to Gettysburg. The Union Army had engaged the Confederates there, and Union commanders were hurrying all available forces to the hills south of the little town.

The 20th Maine arrived at Gettysburg near midday on 2 July, after marching more than one hundred miles in five days. They had had only two hours sleep and no hot food during the previous 24 hours. The regiment was preparing to go into a defensive position as part of the brigade commanded by COL Strong Vincent when a staff officer rode up to COL Vincent and began gesturing towards a little hill at the extreme southern end of the Union line. The hill, Little Round Top, dominated the Union position and, at that moment, was unoccupied. If the Confederates placed artillery on it, they could force the entire Union Army to withdraw. The hill had been left unprotected through a series of mistakes—wrong assumptions, the failure to communicate clearly, and the failure to check—and the situation was critical.

Realizing the danger, COL Vincent ordered his brigade to occupy Little Round Top. He positioned the 20th Maine, commanded by COL Joshua L. Chamberlain, on his brigade's left flank, the extreme left of the Union line. COL Vincent told COL Chamberlain to "hold at all hazards."

On Little Round Top, COL Chamberlain told his company commanders the purpose and importance of their mission. He ordered the right flank company to tie in with the 83d Pennsylvania and the left flank company to anchor on a large boulder. His thoughts turned to his left flank. There was nothing there except a small hollow and the rising slope of Big Round Top. The 20th Maine was literally at the end of the line.

COL Chamberlain then showed a skill common to good tactical leaders. He imagined threats to his unit, did what he could to guard against them, and considered what he would do to meet other possible threats. Since his left flank was open, COL Chamberlain sent B Company, commanded by CPT Walter G. Morrill, off to guard it and "act as the necessities of battle required." The captain positioned his men behind a stone wall that would face the flank of any Confederate advance. There, fourteen soldiers from the 2d US Sharpshooters, who had been separated from their unit, joined them.

The 20th Maine had been in position only a few minutes when the soldiers of the 15th and 47th Alabama attacked. The Confederates had also marched all night and were tired and thirsty. Even so, they attacked ferociously.

The Maine men held their ground, but then one of COL Chamberlain's officers reported seeing a large body of Confederate soldiers moving laterally behind the attacking force. COL Chamberlain climbed on a rock—exposing himself to enemy fire—and saw a Confederate unit moving around his exposed left flank. If they outflanked him, his unit would be pushed off its position and destroyed. He would have failed his mission.

COL Chamberlain had to think fast. The tactical manuals he had so diligently studied called for a maneuver that would not work on this terrain. The colonel had to create a new maneuver, one that his soldiers could execute, and execute now.

The 20th Maine was in a defensive line, two ranks deep. It was threatened by an attack around its left flank. So the colonel ordered his company commanders to stretch the line to the left and bend it back to form an angle, concealing the maneuver by keeping up a steady rate of fire. The corner of the angle would be the large boulder he had pointed out earlier. The sidestep maneuver was tricky, but it was a combination of other battle drills his soldiers knew. In spite of the terrible noise that made voice commands useless, in spite of the blinding smoke, the cries of the wounded, and the continuing Confederate attack, the Maine men were able to pull it off. Now COL Chamberlain's thin line was only

COL Chamberlain at Gettysburg (continued)

one rank deep. His units, covering twice their normal frontage, were bent back into an L shape. Minutes after COL Chamberlain repositioned his force, the Confederate infantry, moving up what they thought was an open flank, were thrown back by the redeployed left wing of the 20th Maine. Surprised and angry, they nonetheless attacked again.

The Maine men rallied and held; the Confederates regrouped and attacked. "The Alabamians drove the Maine men from their positions five times. Five times they fought their way back again. At some places, the muzzles of the opposing guns almost touched." After these assaults, the Maine men were down to one or two rounds per man, and the determined Confederates were regrouping for another try. COL Chamberlain saw that he could not stay where he was and could not withdraw. So he decided to counterattack. His men would have the advantage of attacking down the steep hill, he reasoned, and the Confederates would not be expecting it. Clearly he was risking his entire unit, but the fate of the Union Army depended on his men.

The decision left COL Chamberlain with another problem: there was nothing in the tactics book about how to get his unit from their L-shaped position into a line of advance. Under tremendous fire and in the midst of the battle, COL Chamberlain again called his commanders together. He explained that the regiment's left wing would swing around "like a barn door on a hinge" until it was even with the right wing. Then the entire regiment, bayonets fixed, would charge downhill, staying anchored to the 83d Pennsylvania on its right. The explanation was clear and the situation clearly desperate.

When COL Chamberlain gave the order, 1LT Holman Melcher of F Company leaped forward and led the left wing downhill toward the surprised Confederates. COL Chamberlain had positioned himself at the boulder at the center of the L. When the left wing was abreast of the right wing, he jumped off the rock and led the right wing down the hill. The entire regiment was now charging on line, swinging like a great barn door—just as its commander had intended.

The Alabama soldiers, stunned at the sight of the charging Union troops, fell back on the positions behind them. There the 20th Maine's charge might have failed if not for a surprise resulting from COL Chamberlain's foresight. Just then CPT Morrill's B Company and the sharpshooters opened fire on the Confederate flank and rear. The exhausted and shattered Alabama regiments thought they were surrounded. They broke and ran, not realizing that one more attack would have carried the hill.

The slopes of Little Round Top were littered with bodies. Saplings halfway up the hill had been sawed in half by weapons fire. A third of the 20th Maine had fallen, 130 men out of 386. Nonetheless, the farmers, woodsmen, and fishermen from Maine—under the command of a brave and creative leader who had anticipated enemy actions, improvised under fire, and applied disciplined initiative in the heat of battle—had fought through to victory.

1-31. COL Joshua Chamberlain was awarded the Medal of Honor for his actions on 2 July 1863. After surviving terrible wounds at Petersburg, Virginia, he and his command were chosen to receive the surrender of Confederate units at Appomattox in April 1865. His actions there contributed to national reconciliation and are described in Chapter 7.

PUTTING IT TOGETHER

1-32. Study the Army leadership framework; it is the Army's common basis for thinking about leadership. With all the day-to-day tasks you must do, it's easy to get lost in particulars. The Army leadership framework is a tool that allows you to step back and think about leadership as a whole. It is a canopy that covers the hundreds of things you do every day. The Army leadership framework gives you the big picture and can help you put your job, your people, and your organization in perspective.

1-33. The dimensions of the Army leadership framework shown in Figure 1-1—the values, attributes, skills, and actions that support BE, KNOW, and DO—each contain components. All are interrelated; none stands alone. For example, *will* is very important, as you saw in the case of COL Chamberlain. It's discussed in Chapter 2 under mental attributes. Yet will cannot stand by itself. Left unchecked and without moral boundaries, will can be dangerous. The case of Adolf Hitler shows this fact. Will misapplied can also produce disastrous results. Early in World War I, French forces attacked German machine gun positions across open fields, believing their élan (unit morale and will to win) would overcome a technologically advanced weapon. The cost in lives was catastrophic. Nevertheless, the will of leaders of character and competence—like the small unit leaders at Normandy that you'll read about later in this chapter—can make the difference between victory and defeat.

1-34. This is how you should think about the Army leadership framework: all its pieces work in combination to produce something bigger and better than the sum of the parts. BE the leader of character: embrace Army values and demonstrate leader attributes. Study and practice so that you have the skills to KNOW your job. Then act, DO what's right to achieve excellence.

1-35. The Army leadership framework applies to all Army leaders. However, as you assume positions of increasing responsibility, you'll need to develop additional attributes and master more skills and actions. Part of this knowledge includes understanding what your bosses are doing—the factors that affect their decisions and the environment in which they work. To help you do this, Army leadership positions are divided into three levels—direct, organizational, and strategic.

LEVELS OF LEADERSHIP

NCOs like to make a decision right away and move on to the next thing...so the higher up the flagpole you go, the more you have to learn a very different style of leadership.

Command Sergeant Major Douglas E. Murray
United States Army Reserve

Figure 1-2. Army Leadership Levels

1-36. Figure 1-2 shows the perspectives of the three levels of Army leadership: direct, organizational, and strategic. Factors that determine a position's leadership level can include the position's span of control, its headquarters level, and the extent of the influence the leader holding the position exerts. Other factors include the size of the unit or organization, the type of operations it conducts, the number of people assigned, and its planning horizon.

1-37. Sometimes the rank or grade of the leader holding a position does not indicate the position's leadership level. That's why Figure 1-2 does not show rank. A sergeant first class serving as a platoon sergeant works at the direct leadership level. If the same NCO holds a headquarters job dealing with issues and policy affecting a brigade-sized or larger organization, the NCO works at the organizational leadership level. However, if the NCO's primary duty is running a staff section that supports the leaders who run the organization, the NCO is a direct leader. In fact, most leadership positions are direct leadership positions, and every leader at every level acts as a direct leader when dealing with immediate subordinates.

1-38. The headquarters echelon alone doesn't determine a position's leadership level. Soldiers and DA civilians of all ranks and grades serve in strategic-level headquarters, but they are not all strategic-level leaders. The responsibilities of a duty position, together with the other factors paragraph 1-36 lists, determine its leadership level. For example, a DA civilian at a training area range control with a dozen subordinates works at the direct leadership level while a DA civilian deputy garrison commander with a span of influence over several thousand people works at the organizational leadership level. Most NCOs, company grade officers, field grade officers, and DA civilian leaders serve at the direct leadership level. Some senior NCOs, field grade officers, and higher-grade DA civilians serve at the organizational leadership level. Most general officers and equivalent Senior Executive Service DA civilians serve at the organizational or strategic leadership levels.

DIRECT LEADERSHIP

1-39. Direct leadership is face-to-face, first-line leadership. It takes place in those organizations where subordinates are used to seeing their leaders all the time: teams and squads, sections and platoons, companies, batteries, and troops—even squadrons and battalions. The direct leader's span of influence, those whose lives he can reach out and touch, may range from a handful to several hundred people.

1-40. Direct leaders develop their subordinates one-on-one; however, they also influence their organization through their subordinates. For instance, a cavalry squadron commander is close enough to his soldiers to have a direct influence on them. They're used to seeing him regularly, even if it is only once a week in garrison; they expect to see him from time to time in the field. Still, during daily operations, the commander guides the organization primarily through his subordinate officers and NCOs.

1-41. For direct leaders there is more certainty and less complexity than for organizational and strategic leaders. Direct leaders are close enough to see—very quickly—how things work, how things don't work, and how to address any problems. (Chapter 4 discusses direct leader skills. Chapter 5 discusses direct leader actions.)

ORGANIZATIONAL LEADERSHIP

1-42. Organizational leaders influence several hundred to several thousand people. They do this indirectly, generally through more levels of subordinates than do direct leaders. The additional levels of subordinates can make it more difficult for them to see results. Organizational leaders have staffs to help them lead their people and manage their organizations' resources. They establish policies and the organizational climate that support their subordinate leaders. (Chapter 3 introduces climate and culture and explains the role of direct leaders in setting the organizational climate. Chapters 6 and 7 discuss the roles of organizational and strategic leaders in establishing and maintaining the organizational climate and institutional culture.)

1-43. Organizational leadership skills differ from direct leadership skills in degree, but not in kind. That is, the skill domains are the same, but organizational leaders must deal with more complexity, more people, greater uncertainty, and a greater number of unintended consequences. They find themselves influencing people more through policymaking and systems integration than through face-to-face contact.

1-44. Organizational leaders include military leaders at the brigade through corps levels, military and DA civilian leaders at directorate through installation levels, and DA civilians at the assistant through undersecretary of the Army levels. They focus on planning and mission accomplishment over the next two to ten years.

1-45. Getting out of their offices and visiting the parts of their organizations where the work is done is especially important for organizational leaders. They must make time to get to the field to compare the reports their staff gives them with the actual conditions their people face and the perceptions of the organization and mission they hold. Because of their less-frequent presence among their soldiers and DA civilians, organizational leaders must use those visits they are able to make to assess how well the commander's intent is understood and to reinforce the organization's priorities.

STRATEGIC LEADERSHIP

1-46. Strategic leaders include military and DA civilian leaders at the major command through Department of Defense levels. Strategic leaders are responsible for large organizations and influence several thousand to hundreds of thousands of people. They establish force structure, allocate resources, communicate strategic vision, and prepare their commands and the Army as a whole for their future roles.

1-47. Strategic leaders work in an uncertain environment on highly complex problems that affect and are affected by events and organizations outside the Army. Actions of a theater commander in chief (CINC), for example, may even have an impact on global politics. (CINCs command combatant commands, very large, joint organizations assigned broad, continuing missions. Theater CINCs are assigned responsibilities for a geographic area (a theater); for example, the CINC of the US Central Command is responsible for most of southwestern Asia and part of eastern Africa. Functional CINCs are assigned responsibilities not bounded by geography; for example, the CINC of the US Transportation Command is responsible for providing integrated land, sea, and air

transportation to all services. (JP 0-2, JP 3-0, and FM 100-7 discuss combatant commands.) Although civilian leaders make national policy, decisions a CINC makes while carrying out that policy may affect whether or not a national objective is achieved. Strategic leaders apply many of the same leadership skills and actions they mastered as direct and organizational leaders; however, strategic leadership requires others that are more complex and indirectly applied.

1-48. Strategic leaders concern themselves with the total environment in which the Army functions; their decisions take into account such things as congressional hearings, Army budgetary constraints, new systems acquisition, civilian programs, research, development, and interservice cooperation—just to name a few.

1-49. Strategic leaders, like direct and organizational leaders, process information quickly, assess alternatives based on incomplete data, make decisions, and generate support. However, strategic leaders' decisions affect more people, commit more resources, and have wider-ranging consequences in both space and time than do decisions of organizational and direct leaders.

1-50. Strategic leaders often do not see their ideas come to fruition during their "watch"; their initiatives may take years to plan, prepare, and execute. In-process reviews (IPRs) might not even begin until after the leader has left the job. This has important implications for long-range planning. On the other hand, some strategic decisions may become a front-page headline of the next morning's newspaper. Strategic leaders have very few opportunities to visit the lowest-level organizations of their commands; thus, their sense of when and where to visit is crucial. Because they exert influence primarily through subordinates, strategic leaders must develop strong skills in picking and developing good ones. This is an important improving skill, which Chapter 7 discusses.

LEADERS OF LEADERS

More than anything else, I had confidence in my soldiers, junior leaders, and staff. They were trained, and I knew they would carry the fight to the enemy. I trusted them, and they knew I trusted them. I think in Just Cause, which was a company commander's war, being a decentralized commander paid big dividends because I wasn't in the knickers of my company commanders all the time. I gave them the mission and let them do it. I couldn't do it for them.

A Battalion Commander, Operation Just Cause
Panama, 1989

1-51. At any level, anyone responsible for supervising people or accomplishing a mission that involves other people is a leader. Anyone who influences others, motivating them to action or influencing their thinking or decision making, is a leader. It's not a function only of position; it's also a function of role. In addition, everyone in the Army—including every leader—fits somewhere in a chain of command. Everyone in the Army is also a follower or subordinate. There are, obviously, many leaders in an organization, and it's important to understand that you don't just lead subordinates—you lead other leaders. Even at the lowest level, you are a leader of leaders.

1-52. For example, a rifle company has four leadership levels: the company commander leads through platoon leaders, the platoon leaders through squad leaders, and the squad leaders through team leaders. At each level, the leader must let subordinate leaders do their jobs. Practicing this kind of decentralized execution based on mission orders in peacetime trains subordinates who will, in battle, exercise disciplined initiative in the absence of orders. They'll continue to fight when the radios are jammed, when the plan falls apart, when the enemy does something unexpected. (Appendix A discusses leader roles and relationships. FM 100-34 discusses mission orders and initiative.)

1-53. This decentralization does not mean that a commander never steps in and takes direct control. There will be times when a leader has to stop leading through subordinates, step forward, and say, "Follow me!" A situation like this may occur in combat, when things are falling apart and, like BG Thomas J. Jackson, you'll need to "stand like a stone wall" and save victory. (You'll read about BG Jackson in Chapter 2.) Or it may occur during training, when a subordinate is about to make a mistake that could result in serious injury or death and you must act to prevent disaster.

1-54. More often, however, you should empower your subordinate leaders: give them a task, delegate the necessary authority, and let them do the work. Of course you need to check periodically. How else will you be able to critique, coach, and evaluate them? But the point is to "power down without powering off." Give your subordinate leaders the authority they need to get the job done. Then check on them frequently enough to keep track of what is going on but not so often that you get in their way. You can develop this skill through experience.

1-55. It takes personal courage to operate this way. But a leader must let subordinate leaders learn by doing. Is there a risk that, for instance, a squad leader—especially an inexperienced one—will make mistakes? Of course there is. But if your subordinate leaders are to grow, you must let them take risks. This means you must let go of some control and let your subordinate leaders do things on their own—within bounds established by mission orders and your expressed intent.

1-56. A company commander who routinely steps in and gives orders directly to squad leaders weakens the whole chain of command, denies squad leaders valuable learning experiences, and sends a signal to the whole company that the chain of command and NCO support channel can be bypassed at any time. On the other hand, successful accomplishment of specified and implied missions results from subordinate leaders at all levels exercising disciplined initiative within the commander's intent. Effective leaders strive to create an environment of trust and understanding that encourages their subordinates to seize the initiative and act. (Appendix A discusses authority, the chain of command, and the NCO support channel. FM 100-34 contains information about building trust up and down the chain of command.)

1-57. Weak leaders who have not trained their subordinates sometimes say, "My organization can't do it without me." Many people, used to being at the center of the action, begin to feel as if they're indispensable. You have heard them: "I can't take a day off. I have to be here all the time. I must watch my subordinates' every move, or who knows what will happen?" But no one is irreplaceable. The Army is not going to stop functioning because one leader—no matter how senior, no matter how central—steps aside. In combat, the loss of a leader is a shock to a unit, but the unit must continue its mission. If leaders train their subordinates properly, one of them will take charge.

1-58. Strong commanders—those with personal courage—realize their subordinate leaders need room to work. This doesn't mean that you should let your subordinates make the same mistake over and over. Part of your responsibility as a leader is to help your subordinates succeed. You can achieve this through empowering and coaching. Train your subordinates to plan, prepare, execute, and assess well enough to operate independently. Provide sufficient purpose, direction, and motivation for them to operate in support of the overall plan.

1-59. Finally, check and make corrections. Take time to help your subordinates sort out what happened and why. Conduct AARs so your people don't just make mistakes, but learn from them. There is not a soldier out there, from private to general, who has not slipped up from time to time. Good soldiers, and especially good leaders, learn from those mistakes. Good leaders help their subordinates grow by teaching, coaching, and counseling.

LEADERSHIP AND COMMAND

When you are commanding, leading [soldiers] under conditions where physical exhaustion and privations must be ignored, where the lives of [soldiers] may be sacrificed, then, the efficiency of your leadership will depend only to a minor degree on your tactical ability. It will primarily be determined by your character, your reputation, not much for courage—which will be accepted as a matter of course—but by the previous reputation you have established for fairness, for that high-minded patriotic purpose, that quality of unswerving determination to carry through any military task assigned to you.

General of the Army George C. Marshall
Speaking to officer candidates in September, 1941

1-60. Command is a specific and legal position unique to the military. It's where the buck stops. Like all leaders, commanders are responsible for the success of their organizations, but commanders have special accountability to their superiors, the institution, and the nation. Commanders must think deeply and creatively, for their concerns encompass

yesterday's heritage, today's mission, and tomorrow's force. To maintain their balance among all the demands on them, they must exemplify Army values. The nation, as well as the members of the Army, hold commanders accountable for accomplishing the mission, keeping the institution sound, and caring for its people.

1-61. Command is a sacred trust. The legal and moral responsibilities of commanders exceed those of any other leader of similar position or authority. Nowhere else does a boss have to answer for how subordinates live and what they do after work. Our society and the institution look to commanders to make sure that missions succeed, that people receive the proper training and care, that values survive. On the one hand, the nation grants commanders special authority to be good stewards of its most precious resources: freedom and people. On the other hand, those citizens serving in the Army also trust their commanders to lead them well. NCOs probably have a more immediate impact on their people, but commanders set the policies that reward superior performance and personally punish misconduct. It's no wonder that organizations take on the personal stamp of their commanders. Those selected to command offer something beyond their formal authority: their personal example and public actions have tremendous moral force. Because of that powerful aspect of their position, people inside and outside the Army see a commander as the human face of "the system"—the person who embodies the commitment of the Army to operational readiness and care of its people.

SUBORDINATES

To our subordinates we owe everything we are or hope to be. For it is our subordinates, not our superiors, who raise us to the dizziest of professional heights, and it is our subordinates who can and will, if we deserve it, bury us in the deepest mire of disgrace. When the chips are down and our subordinates have accepted us as their leader, we don't need any superior to tell us; we see it in their eyes and in their faces, in the barracks, on the field, and on the battle line. And on that final day when we must be ruthlessly demanding, cruel and heartless, they will rise as one to do our bidding, knowing full well that it may be their last act in this life.

Colonel Albert G. Jenkins, CSA
8th Virginia Cavalry

1-62. No one is only a leader; each of you is also a subordinate, and all members of the Army are part of a team. A technical supervisor leading a team of DA civilian specialists, for instance, isn't just the leader of that group. The team chief also works for someone else, and the team has a place in a larger organization.

1-63. Part of being a good subordinate is supporting your chain of command. And it's your responsibility to make sure your team supports the larger organization. Consider a leader whose team is responsible for handling the pay administration of a large organization. The chief knows that when the team makes a mistake or falls behind in its work, its customers—soldiers and DA civilians—pay the price in terms of late pay actions. One day a message from the boss introducing a new computer system for handling payroll changes arrives. The team chief looks hard at the new system and decides it will not work as well as the old one. The team will spend a lot of time installing the new system, all the while keeping up with their regular workload. Then they'll have to spend more time undoing the work once the new system fails. And the team chief believes it will fail—all his experience points to that.

1-64. But the team chief cannot simply say, "We'll let these actions pile up; that'll send a signal to the commander about just how bad the new system is and how important we are down here." The team does not exist in a vacuum; it's part of a larger organization that serves soldiers and DA civilians. For the good of the organization and the people in it, the team chief must make sure the job gets done.

1-65. Since the team chief disagrees with the boss's order and it affects both the team's mission and the welfare of its members, the team chief must tell the boss; he must have the moral courage to make his opinions known. Of course, the team chief must also have the right attitude; disagreement doesn't mean it's okay to be disrespectful. He must choose the time and place—usually in private—to explain his concerns to the boss fully and clearly. In addition, the team chief must go into the meeting knowing that, at some point, the discussion will be over and he must execute the boss's decision, whatever it is.

1-66. Once the boss has listened to all the arguments and made a decision, the team chief must support that decision as if it were his own. If he goes to the team and says, "I still don't think this is a good idea, but we're going to do it anyway," the team chief undermines the chain of command and teaches his people a bad lesson. Imagine what it would do to an organization's effectiveness if subordinates chose which orders to pursue vigorously and which ones to half step.

1-67. Such an action would also damage the team chief himself: in the future the team may treat his orders as he treated the boss's. And there is no great leap between people thinking their leader is disloyal to the boss to the same people thinking their leader will be disloyal to them as well. The good leader executes the boss's decision with energy and enthusiasm; looking at their leader, subordinates will believe the leader thinks it's absolutely the best possible solution. The only exception to this involves your duty to disobey obviously illegal orders. This is not a privilege you can claim, but a duty you must perform. (Chapter 2 discusses character and illegal orders. Chapter 4 discusses ethical reasoning.)

1-68. Loyalty to superiors and subordinates does more than ensure smooth-running peacetime organizations. It prepares units for combat by building soldiers' trust in leaders and leaders' faith in soldiers. The success of the airborne assault prior to the 1944 Normandy invasion is one example of how well-trained subordinate leaders can make the difference between victory and defeat.

Small Unit Leaders' Initiative in Normandy

The amphibious landings in Normandy on D-Day, 1944, were preceded by a corps-sized, night parachute assault by American and British airborne units. Many of the thousands of aircraft that delivered the 82d and 101st (US) Airborne Divisions to Normandy on the night of 5-6 June 1944 were blown off course. Some wound up in the wrong place because of enemy fire; others were simply lost. Thousands of paratroopers, the spearhead of the Allied invasion of Western Europe, found themselves scattered across unfamiliar countryside, many of them miles from their drop zones. They wandered about in the night, searching for their units, their buddies, their leaders, and their objectives. In those first few hours, the fate of the invasion hung in the balance; if the airborne forces did not cut the roads leading to the beaches, the Germans could counterattack the landing forces at the water's edge, crushing the invasion before it even began.

Fortunately for the Allies and the soldiers in the landing craft, the leaders in these airborne forces had trained their subordinate leaders well, encouraging their initiative, allowing them to do their jobs. Small unit leaders scattered around the darkened, unfamiliar countryside knew they were part of a larger effort, and they knew its success was up to them. They had been trained to act instead of waiting to be told what to do; they knew that if the invasion was to succeed, their small units had to accomplish their individual missions.

Among these leaders were men like CPT Sam Gibbons of the 505th Parachute Infantry Regiment. He gathered a group of 12 soldiers—from different commands—and liberated a tiny village—which turned out to be outside the division area of operations—before heading south toward his original objective, the Douve River bridges. CPT Gibbons set off with a dozen people he had never seen before and no demolition equipment to destroy a bridge nearly 15 kilometers away. Later, he remarked,

---

Small Unit Leaders' Initiative in Normandy (continued)

"This certainly wasn't the way I had thought the invasion would go, nor had we ever rehearsed it in this manner." But he was moving out to accomplish the mission. Throughout the Cotentin Peninsula, small unit leaders from both divisions were doing the same.

This was the payoff for hard training and leaders who valued soldiers, communicated the importance of the mission, and trusted their subordinate leaders to accomplish it. As they trained their commands for the invasion, organizational leaders focused downward as well as upward. They took care of their soldiers' needs while providing the most realistic training possible. This freed their subordinate leaders to focus upward as well as downward. Because they knew their units were well-trained and their leaders would do everything in their power to support them, small unit leaders were able to focus on the force's overall mission. They knew and understood the commander's intent. They believed that if they exercised disciplined initiative within that intent, things would turn out right. Events proved them correct.

1-69. You read earlier about how COL Joshua Chamberlain accomplished his mission and took care of his soldiers at Little Round Top. Empower subordinates to take initiative and be the subordinate leader who stands up and makes a difference. That lesson applies in peace and in combat, from the smallest organization to the largest. Consider the words of GEN Edward C. Meyer, former Army Chief of Staff:

When I became chief of staff, I set two personal goals for myself. The first was to ensure that the Army was continually prepared to go to war, and the second was to create a climate in which each member could find personal meaning and fulfillment. It is my belief that only by attainment of the second goal will we ensure the first.

1-70. GEN Meyer's words and COL Chamberlain's actions both say the same thing: leaders must accomplish the mission and take care of their soldiers. For COL Chamberlain, this

meant he had to personally lead his men in a bayonet charge and show he believed they could do what he asked of them. For GEN Meyer the challenge was on a larger scale: his task was to make sure the entire Army was ready to fight and win. He knew—and he tells us—that the only way to accomplish such a huge goal is to pay attention to the smallest parts of the machine, the individual soldiers and DA civilians. Through his subordinate leaders, GEN Meyer offered challenges and guidance and set the example so that every member of the Army felt a part of the team and knew that the team was doing important work.

1-71. Both leaders understood the path to excellence: disciplined leaders with strong values produce disciplined soldiers with strong values. Together they become disciplined, cohesive units that train hard, fight honorably, and win decisively.

THE PAYOFF: EXCELLENCE

Leaders of character and competence act to achieve excellence by developing a force that can fight and win the nation's wars and serve the common defense of the United States.

1-72. You achieve excellence when your people are disciplined and committed to Army values. Individuals and organizations pursue excellence to improve, to get better and better. The Army is led by leaders of character who are good role models, consistently set the example, and accomplish the mission while improving their units. It is a cohesive organization of high-performing units characterized by the warrior ethos.

1-73. Army leaders get the job done. Sometimes it's on a large scale, such as GEN Meyer's role in making sure the Army was ready to fight. Other times it may be amid the terror of combat, as with COL Chamberlain at Gettysburg. However, most of you will not become Army Chief of Staff. Not all of you will face the challenge of combat. So it would be a mistake to think that the only time mission accomplishment and leadership are important is with the obvious examples—the general officer, the combat leader. The Army cannot accomplish its mission unless all Army leaders, soldiers, and DA civilians accomplish theirs—whether that means filling out a status report, repairing a vehicle, planning a budget, packing a parachute, maintaining pay records, or walking guard duty. The Army isn't a single general or a handful of combat heroes; it's hundreds of thousands of soldiers and DA civilians, tens of thousands of leaders, all striving to do the right things. Every soldier, every DA civilian, is important to the success of the Army.

MORAL EXCELLENCE: ACCOMPLISHING THE MISSION WITH CHARACTER

To the brave men and women who wear the uniform of the United States of America —thank you. Your calling is a high one—to be the defenders of freedom and the guarantors of liberty.

George Bush
41st President of the United States

1-74 The ultimate end of war, at least as America fights it, is to restore peace. For this reason the Army must accomplish its mission honorably. The Army fights to win, but with one eye on the kind of peace that will follow the war. The actions of Ulysses S. Grant, general in chief of the Union Army at the end of the Civil War, provide an example of balancing fighting to win with restoring the peace.

1-75. In combat GEN Grant had been a relentless and determined commander. During the final days of campaigning in Virginia, he hounded his exhausted foes and pushed his own troops on forced marches of 30 and 40 miles to end the war quickly. GEN Grant's approach to war was best summed up by President Lincoln, who said simply, "He fights."

1-76. Yet even before the surrender was signed, GEN Grant had shifted his focus to the peace. Although some of his subordinates wanted the Confederates to submit to the humiliation of an unconditional surrender, GEN Grant treated his former enemies with respect and considered the long-term effects of his decisions. Rather than demanding an unconditional surrender, GEN Grant negotiated terms with GEN Lee. One of those was allowing his former enemies to keep their horses because they needed them for spring plowing. GEN Grant reasoned that peace would best be served if the Southerners got back to a normal existence as quickly as possible. GEN Grant's decisions and actions sent a message to every man in the Union Army: that it was time to move on, to get back to peacetime concerns.

1-77. At the same time, the Union commander insisted on a formal surrender. He realized that for a true peace to prevail, the Confederates had to publicly acknowledge that organized hostility to the Union had ended. GEN Grant knew that true peace would come about only if both victors and vanquished put the war behind them—a timeless lesson.

1-78. The Army must accomplish its mission honorably. FM 27-10 discusses the law of war and reminds you of the importance of restoring peace. The Army minimizes collateral damage and avoids harming noncombatants for practical as well as honorable reasons. No matter what, though, soldiers fight to win, to live or die with honor for the benefit of our country and its citizens.

1-79. Army leaders often make decisions amid uncertainty, without guidance or precedent, in situations dominated by fear and risk, and sometimes under the threat of sudden, violent death. At those times leaders fall back on their values and Army values and ask, What is right? The question is simple; the answer, often, is not. Having made the decision, the leader depends on self-discipline to see it through.

ACHIEVING COLLECTIVE EXCELLENCE

1-80. Some examples of excellence are obvious: COL Chamberlain's imaginative defense of Little Round Top, GA Dwight Eisenhower drafting his D-Day message (you'll read about it in Chapter 2), MSG Gary Gordon and SFC Randall Shughart putting their lives on the line to save other soldiers in Somalia (their story is in Chapter 3). Those examples of excellence shine, and good leaders teach these stories; soldiers must know they are part of a long tradition of excellence and valor.

1-81. But good leaders see excellence wherever and whenever it happens. Excellent leaders make certain all subordinates know the important roles they play. Look for everyday examples that occur under ordinary circumstances: the way a soldier digs a fighting position, prepares for guard duty, fixes a radio, lays an artillery battery; the way a DA civilian handles an action, takes care of customers, meets a deadline on short notice. Good leaders know that each of these people is contributing in a small but important way to the business of the Army. An excellent Army is the collection of small tasks done to standard, day in and day out. At the end of the day, at the end of a career, those leaders, soldiers and DA civilians—the ones whose excellent work created an excellent Army—can look back confidently. Whether they commanded an invasion armada of thousands of soldiers or supervised a technical section of three people, they know they did the job well and made a difference.

1-82. Excellence in leadership does not mean perfection; on the contrary, an excellent leader allows subordinates room to learn from their mistakes as well as their successes. In such a climate, people work to improve and take the risks necessary to learn. They know that when they fall short—as they will—their leader will pick them up, give them new or more detailed instructions, and send them on their way again. This is the only way to improve the force, the only way to train leaders.

1-83. A leader who sets a standard of "zero defects, no mistakes" is also saying "Don't take any chances. Don't try anything you can't already do perfectly, and for heaven's sake, don't try anything new." That organization will not improve; in fact, its ability to perform the mission will deteriorate rapidly. Accomplishing the Army's mission requires leaders who are imaginative, flexible, and daring. Improving the Army for future missions requires leaders who are thoughtful and reflective. These qualities are incompatible with a "zero-defects" attitude.

1-84. Competent, confident leaders tolerate honest mistakes that do not result from negligence. The pursuit of excellence is not a game to achieve perfection; it involves trying, learning, trying again, and getting better each time. This in no way justifies or excuses failure. Even the best efforts and good intentions cannot take away an individual's responsibility for his actions.

SUMMARY

1-85. Leadership in combat is your primary and most important challenge. It requires you to accept a set of values that contributes to a core of motivation and will. If you fail to accept and live these Army values, your soldiers may die unnecessarily. Army leaders of character and competence act to achieve excellence by developing a force that can fight and win the nation's wars and serve the common defense of the United States. The Army leadership framework identifies the dimensions of Army leadership: what the Army expects you, as one of its leaders, to BE, KNOW, and DO.

1-86. Leadership positions fall into one of three leadership levels: direct, organizational, and strategic. The perspective and focus of leaders change and the actions they must DO become more complex with greater consequences as they assume positions of greater responsibility. Nonetheless, they must still live Army values and possess leader attributes.

1-87. Being a good subordinate is part of being a good leader. Everyone is part of a team, and all members have responsibilities that go with belonging to that team. But every soldier and DA civilian who is responsible for supervising people or accomplishing a mission that involves other people is a leader. All soldiers and DA civilians at one time or another must act as leaders.

1-88. Values and attributes make up a leader's character, the BE of Army leadership. Character embodies self-discipline and the will to win, among other things. It contributes to the motivation to persevere. From this motivation comes the lifelong work of mastering the skills that define competence, the KNOW of Army leadership. As you reflect on Army values and leadership attributes and develop the skills your position and experience require, you become a leader of character and competence, one who can act to achieve excellence, who can DO what is necessary to accomplish the mission and take care of your people. That is leadership—influencing people by providing purpose, direction, and motivation while operating to accomplish the mission and improving the organization. That is what makes a successful leader, one who lives the principles of BE, KNOW, DO.

CHAPTER 2

The Leader and Leadership:
What the Leader Must Be, Know, and Do

I do solemnly swear (or affirm) that I will support and defend the Constitution of the United States against all enemies, foreign and domestic; that I will bear true faith and allegiance to the same; and that I will obey the orders of the President of the United States and the orders of the officers appointed over me, according to regulations and the Uniform Code of Military Justice. So help me God.

Oath of Enlistment

I [full name], having been appointed a [rank] in the United States Army, do solemnly swear (or affirm) that I will support and defend the Constitution of the United States against all enemies, foreign and domestic; that I will bear true faith and allegiance to the same; that I take this obligation freely, without any mental reservation or purpose of evasion, and that I will well and faithfully discharge the duties of the office upon which I am about to enter. So help me God.

Oath of office taken by commissioned officers and DA civilians

2-1. Beneath the Army leadership framework shown in Figure 1-1, 30 words spell out your job as a leader: **Leaders of character and competence act to achieve excellence by developing a force that can fight and win the nation's wars and serve the common defense of the United States.** There's a lot in that sentence. This chapter looks at it in detail.

2-2. Army leadership doctrine addresses what makes leaders of character and competence and what makes leadership. Figure 2-1 highlights these values and attributes. Remember from Chapter 1 that character describes what leaders must BE; competence refers to what leaders must KNOW; and action is what leaders must DO. Although this chapter discusses these concepts one at a time, they don't stand alone; they are closely connected and together make up who you seek to be (a leader of character and competence) and what you need to do (leadership).

SECTION I

CHARACTER: WHAT A LEADER MUST BE

Everywhere you look—on the fields of athletic competition, in combat training, operations, and in civilian communities—soldiers are doing what is right.

Former Sergeant Major of the Army
Julius W. Gates

2-3. Character—who you are—contributes significantly to how you act. Character helps you know what's right and do what's right, all the time and at whatever the cost. Character is made up of two interacting parts: values and attributes. Stephen Ambrose, speaking about the Civil War, says that "at the pivotal point in the war it was always the character of individuals that made the difference." Army leaders must be those critical individuals of character themselves and in turn develop character in those they lead. (Appendix E discusses character development.)

ARMY VALUES

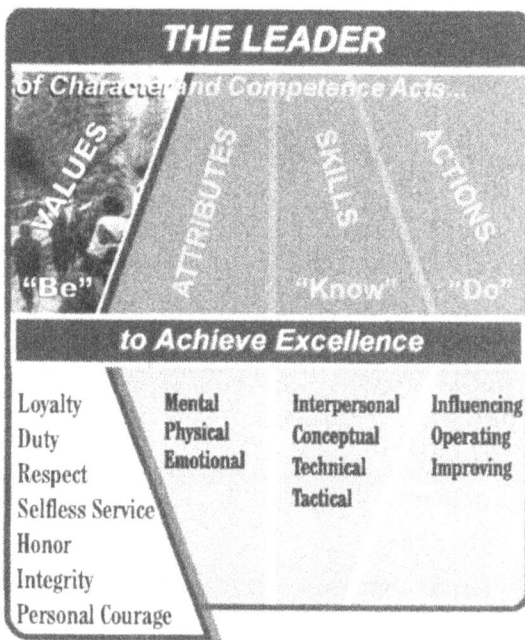

Figure 2-1. Army Values

2-4. Your attitudes about the worth of people, concepts, and other things describe your values. Everything begins there. Your subordinates enter the Army with their own values, developed in childhood and nurtured through experience. All people are all shaped by what they've seen, what they've learned, and whom they've met.

But when soldiers and DA civilians take the oath, they enter an institution guided by Army values. These are more than a system of rules. They're not just a code tucked away in a drawer or a list in a dusty book. These values tell you what you need to be, every day, in every action you take. Army values form the very identity of the Army, the solid rock upon which everything else stands, especially in combat. They are the glue that binds together the members of a noble profession. As a result, the whole is much greater than the sum of its parts. Army values are nonnegotiable: they apply to everyone and in every situation throughout the Army.

2-5. Army values remind us and tell the rest of the world—the civilian government we serve, the nation we protect, even our enemies—who we are and what we stand for. The trust soldiers and DA civilians have for each other and the trust the American people have in us depends on how well we live up to Army values. They are the fundamental building blocks that enable us to discern right from wrong in any situation. Army values are consistent; they support one another. You can't follow one value and ignore another.

2-6. Here are the Army values that guide you, the leader, and the rest of the Army. They form the acronym LDRSHIP:

Loyalty
 Duty
  Respect
   Selfless Service
    Honor
     Integrity
      Personal Courage

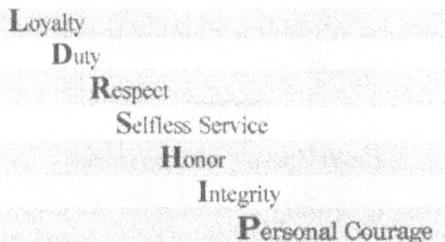

2-7. The following discussions can help you understand Army values, but understanding is only the first step. As a leader, you must not only understand them; you must believe in them, model them in your own actions, and teach others to accept and live by them.

LOYALTY

Bear true faith and allegiance to the US Constitution, the Army, your unit, and other soldiers.

Loyalty is the big thing, the greatest battle asset of all. But no man ever wins the loyalty of troops by preaching loyalty. It is given to him as he proves his possession of the other virtues.

Brigadier General S. L. A. Marshall
Men Against Fire

2-8. Since before the founding of the republic, the Army has respected its subordination to its civilian political leaders. This subordination is fundamental to preserving the liberty of all Americans. You began your Army career by swearing allegiance to the Constitution, the basis of our government and laws. If you've never read it or if it has been a while, the Constitution is in Appendix F. Pay particular attention to Article I, Section 8, which outlines congressional responsibilities regarding the armed forces, and Article II, Section 2, which designates the president as commander in chief. Beyond your allegiance to the Constitution, you have an obligation to be faithful to the Army—the institution and its people—and to your unit or organization. Few examples illustrate loyalty to country and institution as well as the example of GEN George Washington in 1782.

2-9. GEN Washington's example shows how the obligation to subordinates and peers fits in the context of loyalty to the chain of command and the institution at large. As commander of the Continental Army, GEN Washington was obligated to see that his soldiers were taken care of. However, he also was obligated to ensure that the new nation remained secure and that the Continental Army remained able to fight if necessary. If the Continental Army had marched on the seat of government, it may well have destroyed the nation by undermining the law that held it together. It also would have destroyed the Army as an institution by destroying the basis for the authority under which it served. GEN Washington realized these things and acted based on his knowledge. Had he done nothing else, this single act would have been enough to establish GEN George Washington as the father of his country.

GEN Washington at Newburgh

Following its victory at Yorktown in 1781, the Continental Army set up camp at Newburgh, New York, to wait for peace with Great Britain. The central government formed under the Articles of Confederation proved weak and unwilling to supply the Army properly or even pay the soldiers who had won the war for independence. After months of waiting many officers, angry and impatient, suggested that the Army march on the seat of government in Philadelphia, Pennsylvania, and force Congress to meet the Army's demands. One colonel even suggested that GEN Washington become King George I.

Upon hearing this, GEN Washington assembled his officers and publicly and emphatically rejected the suggestion. He believed that seizing power by force would have destroyed everything for which the Revolutionary War had been fought. By this action, GEN Washington firmly established an enduring precedent: America's armed forces are subordinate to civilian authority and serve the democratic principles that are now enshrined in the Constitution. GEN Washington's action demonstrated the loyalty to country that the Army must maintain in order to protect the freedom enjoyed by all Americans.

2-10. Loyalty is a two-way street: you should not expect loyalty without being prepared to give it as well. Leaders can neither demand loyalty nor win it from their people by talking about it. The loyalty of your people is a gift they give you when, and only when, you deserve it—when you train them well, treat them fairly, and live by the concepts you talk about. Leaders who are loyal to their subordinates never let them be misused.

2-11. Soldiers fight for each other—loyalty is commitment. Some of you will encounter the most important way of earning this loyalty: leading your soldiers well in combat. There's no loyalty fiercer than that of soldiers who trust their leader to take them through the dangers of combat. However, loyalty extends to all members of an organization—to your superiors and subordinates, as well as your peers.

2-12. Loyalty extends to all members of all components of the Army. The reserve components—Army National Guard and Army Reserve—play an increasingly active role in the Army's mission. Most DA civilians will not be called upon to serve in combat theaters, but their contributions to mission accomplishment are nonetheless vital. As an Army leader, you'll serve throughout your career with soldiers of the active and reserve components as well as

DA civilians. All are members of the same team, loyal to one another.

DUTY

Fulfill your obligations.

The essence of duty is acting in the absence of orders or direction from others, based on an inner sense of what is morally and professionally right....

General John A. Wickham Jr.
Former Army Chief of Staff

2-13. Duty begins with everything required of you by law, regulation, and orders; but it includes much more than that. Professionals do their work not just to the minimum standard, but to the very best of their ability. Soldiers and DA civilians commit to excellence in all aspects of their professional responsibility so that when the job is done they can look back and say, "I couldn't have given any more."

2-14. Army leaders take the initiative, figuring out what needs to be done before being told what to do. What's more, they take full responsibility for their actions and those of their subordinates. Army leaders never shade the truth to make the unit look good—or even to make their subordinates feel good. Instead, they follow their higher duty to the Army and the nation.

Duty in Korea

CPT Viola B. McConnell was the only Army nurse on duty in Korea in July of 1950. When hostilities broke out, she escorted nearly 700 American evacuees from Seoul to Japan aboard a freighter designed to accommodate only 12 passengers. CPT McConnell assessed priorities for care of the evacuees and worked exhaustively with a medical team to care for them. Once in Japan, she requested reassignment back to Korea. After all she had already done, CPT McConnell returned to Taejon to care for and evacuate wounded soldiers of the 24th Infantry Division.

2-15. CPT McConnell understood and fulfilled her duty to the Army and to the soldiers she supported in ways that went beyond her medical training. A leader's duty is to take charge, even in unfamiliar circumstances. But duty isn't reserved for special occasions. When a platoon sergeant tells a squad leader to inspect weapons, the squad leader has fulfilled his

minimum obligation when he has checked the weapons. He's done what he was told to do. But if the squad leader finds weapons that are not clean or serviced, his sense of duty tells him to go beyond the platoon sergeant's instructions. The squad leader does his duty when he corrects the problem and ensures the weapons are up to standard.

2-16. In extremely rare cases, you may receive an illegal order. Duty requires that you refuse to obey it. You have no choice but to do what's ethically and legally correct. Paragraphs 2-97 through 2-99 discuss illegal orders.

RESPECT

Treat people as they should be treated.

The discipline which makes the soldiers of a free country reliable in battle is not to be gained by harsh or tyrannical treatment. On the contrary, such treatment is far more likely to destroy than to make an army. It is possible to impart instruction and to give commands in such manner and such a tone of voice to inspire in the soldier no feeling but an intense desire to obey, while the opposite manner and tone of voice cannot fail to excite strong resentment and a desire to disobey. The one mode or the other of dealing with subordinates springs from a corresponding spirit in the breast of the commander. He who feels the respect which is due to others cannot fail to inspire in them regard for himself, while he who feels, and hence manifests, disrespect toward others, especially his inferiors, cannot fail to inspire hatred against himself.

Major General John M. Schofield
Address to the United States Corps of Cadets
11 August 1879

2-17. Respect for the individual forms the basis for the rule of law, the very essence of what makes America. In the Army, respect means recognizing and appreciating the inherent dignity and worth of all people. This value reminds you that your people are your greatest resource. Army leaders honor everyone's individual worth by treating all people with dignity and respect.

2-18. As America becomes more culturally diverse, Army leaders must be aware that they will deal with people from a wider range of ethnic, racial, and religious backgrounds. Effective leaders are tolerant of beliefs different from their own as long as those beliefs don't conflict with Army values, are not illegal, and are not unethical. As an Army leader, you need to avoid misunderstandings arising from cultural differences. Actively seeking to learn about people and cultures different from your own can help you do this. Being sensitive to other cultures can also aid you in counseling your people more effectively. You show respect when you seek to understand your people's background, see things from their perspective, and appreciate what's important to them.

2-19. As an Army leader, you must also foster a climate in which everyone is treated with dignity and respect regardless of race, gender, creed, or religious belief. Fostering this climate begins with your example: how you live Army values shows your people how they should live them. However, values training is another major contributor. Effective training helps create a common understanding of Army values and the standards you expect. When you conduct it as part of your regular routine—such as during developmental counseling sessions—you reinforce the message that respect for others is part of the character of every soldier and DA civilian. Combined with your example, such training creates an organizational climate that promotes consideration for others, fairness in all dealings, and equal opportunity. In essence, Army leaders treat others as they wish to be treated.

2-20. As part of this consideration, leaders create an environment in which subordinates are challenged, where they can reach their full potential and be all they can be. Providing tough training doesn't demean subordinates; in fact, building their capabilities and showing faith in their potential is the essence of respect. Effective leaders take the time to learn what their subordinates want to accomplish. They advise their people on how they can grow, personally and professionally. Not all of your subordinates will succeed equally, but they all deserve respect.

2-21. Respect is also an essential component for the development of disciplined, cohesive, and effective warfighting teams. In the deadly confusion of combat, soldiers often overcome incredible odds to accomplish the mission and protect the lives of their comrades. This spirit of selfless service and duty is built on a soldier's personal trust and regard for fellow soldiers. A leader's willingness to tolerate discrimination or harassment on any basis, or a failure to cultivate a climate of respect, eats away at this trust and erodes unit cohesion. But respect goes beyond issues of discrimination and harassment; it includes the broader issue of civility, the way people treat each other and those they come in contact with. It involves being sensitive to diversity and one's own behaviors that others may find insensitive, offensive, or abusive. Soldiers and DA civilians, like their leaders, treat everyone with dignity and respect.

SELFLESS SERVICE

Put the welfare of the nation, the Army, and subordinates before your own.

The nation today needs men who think in terms of service to their country and not in terms of their country's debt to them.

General of the Army Omar N. Bradley

2-22. You have often heard the military referred to as "the service." As a member of the Army, you serve the United States. Selfless service means doing what's right for the nation, the Army, your organization, and your people—and putting these responsibilities above your own interests. The needs of the Army and the nation come first. This doesn't mean that you neglect your family or yourself; in fact, such neglect weakens a leader and can cause the Army more harm than good. Selfless service doesn't mean that you can't have a strong ego, high self-esteem, or even healthy ambition. Rather, selfless service means that you don't make decisions or take actions that help your image or your career but hurt others or sabotage the mission. The selfish superior claims credit for work his subordinates do; the selfless leader gives credit to those who earned it. The Army can't function except as a team, and for a team to work, the individual has to give up self-interest for the good of the whole.

2-23. Soldiers are not the only members of the Army who display selfless service. DA civilians display this value as well. Then Army Chief of Staff, Gordon R. Sullivan assessed the DA civilian contribution to Operation Desert Storm this way:

Not surprisingly, most of the civilians deployed to Southwest Asia volunteered to serve there. But the civilian presence in the Gulf region meant more than moral support and filling in for soldiers. Gulf War veterans say that many of the combat soldiers could owe their lives to the DA civilians who helped maintain equipment by speeding up the process of getting parts and other support from 60 logistics agencies Army-wide.

2-24. As GEN Sullivan's comment indicates, selfless service is an essential component of teamwork. Team members give of themselves so that the team may succeed. In combat some soldiers give themselves completely so that their comrades can live and the mission can be accomplished. But the need for selflessness isn't limited to combat situations. Requirements for individuals to place their own needs below those of their organization can occur during peacetime as well. And the requirement for selflessness doesn't decrease with one's rank; it increases. Consider this example of a soldier of long service and high rank who demonstrated the value of selfless service.

GA Marshall Continues to Serve

GA George C. Marshall served as Army Chief of Staff from 1939 until 1945. He led the Army through the buildup, deployment, and worldwide operations of World War II. Chapter 7 outlines some of his contributions to the Allied victory. In November 1945 he retired to a well-deserved rest at his home in Leesburg, Virginia. Just six days later President Harry S Truman called on him to serve as Special Ambassador to China. From the White House President Truman telephoned GA Marshall at his home: "General, I want you to go to China for me," the president said. "Yes, Mr. President," GA Marshall replied. He then hung up the telephone, informed his wife of the president's request and his reply, and prepared to return to government service.

> ### GA Marshall Continues to Serve (continued)
>
> President Truman didn't appoint GA Marshall a special ambassador to reward his faithful service; he appointed GA Marshall because there was a tough job in China that needed to be done. The Chinese communists under Mao Tse-tung were battling the Nationalists under Chiang Kai-shek, who had been America's ally against the Japanese; GA Marshall's job was to mediate peace between them. In the end, he was unsuccessful in spite of a year of frustrating work; the scale of the problem was more than any one person could handle. However, in January 1947 President Truman appointed GA Marshall Secretary of State. The Cold War had begun and the president needed a leader Americans trusted. GA Marshall's reputation made him the one; his selflessness led him to continue to serve.

2-25. When faced with a request to solve a difficult problem in an overseas theater after six years of demanding work, GA Marshall didn't say, "I've been in uniform for over thirty years, we just won a world war, and I think I've done enough." Instead, he responded to his commander in chief the only way a professional could. He said yes, took care of his family, and prepared to accomplish the mission. After a year overseas, when faced with a similar question, he gave the same answer. GA Marshall always placed his country's interests first and his own second. Army leaders who follow his example do the same.

HONOR

Live up to all the Army values.

What is life without honor? Degradation is worse than death.

Lieutenant General Thomas J. "Stonewall" Jackson

2-26. Honor provides the "moral compass" for character and personal conduct in the Army. Though many people struggle to define the term, most recognize instinctively those with a keen sense of right and wrong, those who live such that their words and deeds are above reproach. The expression "honorable person," therefore, refers to both the character traits an individual actually possesses and the fact that the community recognizes and respects them.

2-27. Honor holds Army values together while at the same time being a value itself. Together, Army values describe the foundation essential to develop leaders of character. Honor means demonstrating an understanding of what's right and taking pride in the community's

acknowledgment of that reputation. Military ceremonies recognizing individual and unit achievement demonstrate and reinforce the importance the Army places on honor.

2-28. For you as an Army leader, demonstrating an understanding of what's right and taking pride in that reputation means this: **Live up to all the Army values**. Implicitly, that's what you promised when you took your oath of office or enlistment. You made this promise publicly, and the standards—Army values—are also public. To be an honorable person, you must be true to your oath and live Army values in all you do. Living honorably strengthens Army values, not only for yourself but for others as well: all members of an organization contribute to the organization's climate (which you'll read about in Chapter 3). By what they do, people living out Army values contribute to a climate that encourages all members of the Army to do the same.

2-29. How you conduct yourself and meet your obligations defines who you are as a person; how the Army meets the nation's commitments defines the Army as an institution. For you as an Army leader, honor means putting Army values above self-interest, above career and comfort. For all soldiers, it means putting Army values above self-preservation as well. This honor is essential for creating a bond of trust among members of the Army and between the Army and the nation it serves. Army leaders have the strength of will to live according to Army values, even though the temptations to do otherwise are strong, especially in the face of personal danger. The military's highest award is the Medal of Honor. Its recipients didn't do just what was required of them; they went beyond the expected, above and beyond the call of duty. Some gave their own lives so that others could live. It's fitting that the word we use to describe their achievements is "honor."

MSG Gordon and SFC Shughart in Somalia

During a raid in Mogadishu in October 1993, MSG Gary Gordon and SFC Randall Shughart, leader and member of a sniper team with Task Force Ranger in Somalia, were providing precision and suppressive fires from helicopters above two helicopter crash sites. Learning that no ground forces were available to rescue one of the downed aircrews and aware that a growing number of enemy were closing in on the site, MSG Gordon and SFC Shughart volunteered to be inserted to protect their critically wounded comrades. Their initial request was turned down because of the danger of the situation. They asked a second time; permission was denied. Only after their third request were they inserted.

MSG Gordon and SFC Shughart were inserted one hundred meters south of the downed chopper. Armed only with their personal weapons, the two NCOs fought their way to the downed fliers through intense small arms fire, a maze of shanties and shacks, and the enemy converging on the site. After MSG Gordon and SFC Shughart pulled the wounded from the wreckage, they established a perimeter, put themselves in the most dangerous position, and fought off a series of attacks. The two NCOs continued to protect their comrades until they had depleted their ammunition and were themselves fatally wounded. Their actions saved the life of an Army pilot.

2-30. No one will ever know what was running through the minds of MSG Gordon and SFC Shughart as they left the comparative safety of their helicopter to go to the aid of the downed aircrew. The two NCOs knew there was no ground rescue force available, and they certainly knew there was no going back to their helicopter. They may have suspected that things would turn out as they did; nonetheless, they did what they believed to be the right thing. They acted based on Army values, which they had clearly made their own: *loyalty* to their fellow soldiers; the *duty* to stand by them, regardless of the circumstances; the *personal courage* to act, even in the face of great danger; *selfless service*, the willingness to give their all. MSG Gary I. Gordon and SFC Randall D. Shughart lived Army values to the end; they were posthumously awarded Medals of Honor.

INTEGRITY

Do what's right—legally and morally.

The American people rightly look to their military leaders not only to be skilled in the technical aspects of the profession of arms, but also to be men of integrity.

General J. Lawton Collins
Former Army Chief of Staff

2-31. People of integrity consistently act according to principles—not just what might work at the moment. Leaders of integrity make their principles known and consistently act in accordance with them. The Army requires leaders of integrity who possess high moral standards and are honest in word and deed. Being honest means being truthful and upright all the time, despite pressures to do otherwise. Having integrity means being both morally complete and true to yourself. As an Army leader, you're honest to yourself by committing to and consistently living Army values; you're honest to others by not presenting yourself or your actions as anything other than what they are. Army leaders say what they mean and do what they say. If you can't accomplish a mission, inform your chain of command. If you inadvertently pass on bad information, correct it as soon as you find out it's wrong. People of integrity do the right thing not because it's convenient or because they have no choice. They choose the right thing because their character permits no less. Conducting yourself with integrity has three parts:

* Separating what's right from what's wrong.

* Always acting according to what you know to be right, even at personal cost.

* Saying openly that you're acting on your understanding of right versus wrong.

2-32. Leaders can't hide what they do: that's why you must carefully decide how you act. As an Army leader, you're always on display. If you want to instill Army values in others, you must internalize and demonstrate them yourself. Your personal values may and probably do extend beyond the Army values, to include such things as political, cultural, or religious beliefs. However, if you're to be an Army leader *and* a person of integrity, these values must reinforce, not contradict, Army values.

2-33. Any conflict between your personal values and Army values must be resolved before you can become a morally complete Army leader. You may need to consult with someone whose values and judgment you respect. You would not be the first person to face this issue, and as a leader, you can expect others to come to you, too. Chapter 5 contains the story of how SGT Alvin York and his leaders confronted and resolved a conflict between SGT York's personal values and Army values. Read it and reflect on it. If one of your subordinates asks you to help resolve a similar conflict, you must be prepared by being sure your own values align with Army values. Resolving such conflicts is necessary to become a leader of integrity.

PERSONAL COURAGE

Face fear, danger, or adversity (physical or moral).

The concept of professional courage does not always mean being as tough as nails either. It also suggests a willingness to listen to the soldiers' problems, to go to bat for them in a tough situation, and it means knowing just how far they can go. It also means being willing to tell the boss when he's wrong.

Former Sergeant Major of the Army William Connelly

2-34. Personal courage isn't the absence of fear; rather, it's the ability to put fear aside and do what's necessary. It takes two forms, physical and moral. Good leaders demonstrate both.

2-35. Physical courage means overcoming fears of bodily harm and doing your duty. It's the bravery that allows a soldier to take risks in combat in spite of the fear of wounds or death. Physical courage is what gets the soldier at Airborne School out the aircraft door. It's what allows an infantryman to assault a bunker to save his buddies.

2-36. In contrast, moral courage is the willingness to stand firm on your values, principles, and convictions—even when threatened. It enables leaders to stand up for what they believe is right, regardless of the consequences. Leaders who take responsibility for their decisions and actions, even when things go wrong, display moral courage. Courageous leaders are willing to look critically inside themselves, consider new ideas, and change what needs changing.

2-37. Moral courage is sometimes overlooked, both in discussions of personal courage and in the everyday rush of business. A DA civilian at a meeting heard *courage* mentioned several times in the context of combat. The DA civilian pointed out that consistent moral courage is every bit as important as momentary physical courage. Situations requiring physical courage are rare; situations requiring moral courage can occur frequently. Moral courage is essential to living the Army values of integrity and honor every day.

2-38. Moral courage often expresses itself as candor. Candor means being frank, honest, and sincere with others while keeping your words free from bias, prejudice, or malice. Candor means calling things as you see them, even when it's uncomfortable or you think it might be better for you to just keep quiet. It means not allowing your feelings to affect what you say about a person or situation. A candid company commander calmly points out the first sergeant's mistake. Likewise, the candid first sergeant respectfully points out when the company commander's pet project isn't working and they need to do something different. For trust to exist between leaders and subordinates, candor is essential. Without it, subordinates won't know if they've met the standard and leaders won't know what's going on.

2-39. In combat physical and moral courage may blend together. The right thing to do may not only be unpopular, but dangerous as well. Situations of that sort reveal who's a leader of character and who's not. Consider this example.

WO1 Thompson at My Lai

Personal courage—whether physical, moral, or a combination of the two—may be manifested in a variety of ways, both on and off the battlefield. On March 16, 1968 Warrant Officer (WO1) Hugh C. Thompson Jr. and his two-man crew were on a reconnaissance mission over the village of My Lai, Republic of Vietnam. WO1 Thompson watched in horror as he saw an American soldier shoot an injured Vietnamese child. Minutes later, when he observed American soldiers advancing on a number of civilians in a ditch, WO1 Thompson landed his helicopter and questioned a young officer about what was happening on the ground. Told that the ground action was none of his business, WO1 Thompson took off and continued to circle the area.

When it became apparent that the American soldiers were now firing on civilians, WO1 Thompson landed his helicopter between the soldiers and a group of 10 villagers who were headed for a homemade bomb shelter. He ordered his gunner to train his weapon on the approaching American soldiers and to fire if necessary. Then he personally coaxed the civilians out of the shelter and airlifted them to safety. WO1 Thompson's radio reports of what was happening were instrumental in bringing about the cease-fire order that saved the lives of more civilians. His willingness to place himself in physical danger in order to do the morally right thing is a sterling example of personal courage.

LEADER ATTRIBUTES

Leadership is not a natural trait, something inherited like the color of eyes or hair...Leadership is a skill that can be studied, learned, and perfected by practice.

The Noncom's Guide, 1962

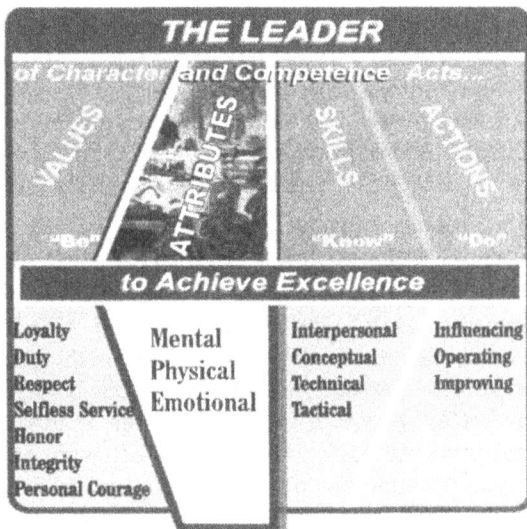

Figure 2-2. Leader Attributes

2-40. Values tell us part of what the leader must BE; the other side of what a leader must BE are the attributes listed in Figure 2-2. Leader attributes influence leader actions; leader actions, in turn, always influence the unit or organization. As an example, if you're physically fit, you're more likely to inspire your subordinates to be physically fit.

2-41. Attributes are a person's fundamental qualities and characteristics. People are born with some attributes; for instance, a person's genetic code determines eye, hair, and skin color. However, other attributes—including leader attributes—are learned and can be changed. Leader attributes can be characterized as mental, physical, and emotional. Successful leaders work to improve those attributes.

MENTAL ATTRIBUTES

2-42. The mental attributes of an Army leader include will, self-discipline, initiative, judgment, self-confidence, intelligence, and cultural awareness.

Will

The will of soldiers is three times more important than their weapons.

Colonel Dandridge M. "Mike" Malone
Small Unit Leadership: A Commonsense Approach

2-43. Will is the inner drive that compels soldiers and leaders to keep going when they are exhausted, hungry, afraid, cold, and wet—when it would be easier to quit. Will enables soldiers to press the fight to its conclusion. Yet will without competence is useless. It's not enough that soldiers are willing, or even eager, to fight; they must know how to fight. Likewise, soldiers who have competence but no will don't fight. The leader's task is to develop a winning spirit by building their subordinates' will as well as their skill. That begins with hard, realistic training.

2-44. Will is an attribute essential to all members of the Army. Work conditions vary among branches and components, between those deployed and those closer to home. In the Army, personal attitude must prevail over any adverse external conditions. All members of the Army—active, reserve, and DA civilian—will experience situations when it would be easier to quit rather than finish the task at hand. At those times, everyone needs that inner drive to press on to mission completion.

2-45. It's easy to talk about will when things go well. But the test of your will comes when things go badly— when events seem to be out of control, when you think your bosses have forgotten you, when the plan doesn't seem to work and it looks like you're going to lose. It's then that you must draw on your inner reserves to persevere—to do your job until there's nothing left to do it with and then to remain faithful to your people, your organization, and your country. The story of the American and Filipino stand on the Bataan Peninsula and their subsequent captivity is one of individuals, leaders, and units deciding to remain true to the end—and living and dying by that decision.

The Will to Persevere

On 8 December 1941, hours after the attack on Pearl Harbor, Japanese forces attacked the American and Filipino forces defending the Philippines. With insufficient combat power to launch a counterattack, GEN Douglas MacArthur, the American commander, ordered his force to consolidate on the Bataan Peninsula and hold as long as possible. Among his units was the 12th Quartermaster (QM) Regiment, which had the mission of supporting the force.

Completely cut off from outside support, the Allies held against an overwhelming Japanese army for the next three and a half months. Soldiers of the 12th QM Regiment worked in the debris of warehouses and repair shops under merciless shelling and bombing, fighting to make the meager supplies last. They slaughtered water buffaloes for meat, caught fish with traps they built themselves, and distilled salt from sea water. In coffeepots made from oil drums they boiled and reboiled the tiny coffee supply until the grounds were white. As long as an ounce of food existed, it was used. In the last desperate days, they resorted to killing horses and pack mules. More important, these supporters delivered rations to the foxholes on the front lines—fighting their way in when necessary. After Bataan and Corregidor fell, members of the 12th QM Regiment were prominent among the 7,000 Americans and Filipinos who died on the infamous Bataan Death March.

Though captured, the soldiers of the 12th QM Regiment maintained their will to resist. 1LT Beulah Greenwalt, a nurse assigned to the 12th QM Regiment, personified this will. Realizing the regimental colors represent the soul of a regiment and that they could serve as a symbol for resistance, 1LT Greenwalt assumed the mission of protecting the colors from the Japanese. She carried the colors to the prisoner of war (PW) camp in Manila by wrapping them around her

---

The Will to Persevere (continued)

shoulders and convincing her Japanese captors that they were "only a shawl." For the next 33 months 1LT Greenwalt and the remains of the regiment remained PWs, living on starvation diets and denied all comforts. But through it all, 1LT Greenwalt held onto the flag. The regimental colors were safeguarded: the soul of the regiment remained with the regiment, and its soldiers continued to resist.

When the war ended in 1945 and the surviving PWs were released, 1LT Greenwalt presented the colors to the regimental commander. She and her fellow PWs had persevered. They had resisted on Bataan until they had no more means to resist. They continued to resist through three long years of captivity. They decided on Bataan to carry on, and they renewed that decision daily until they were liberated. The 12th QM Regiment—and the other units that had fought and resisted with them—remained true to themselves, the Army, and their country. Their will allowed them to see events through to the end.

Self-Discipline

The core of a soldier is moral discipline. It is intertwined with the discipline of physical and mental achievement. Total discipline overcomes adversity, and physical stamina draws on an inner strength that says "drive on."

Former Sergeant Major of the Army
William G. Bainbridge

2-46. Self-disciplined people are masters of their impulses. This mastery comes from the habit of doing the right thing. Self-discipline allows Army leaders to do the right thing regardless of the consequences for them or their subordinates. Under the extreme stress of combat, you and your team might be cut off and alone, fearing for your lives, and having to act without guidance or knowledge of what's going on around you. Still, you—the leader—must think clearly and act reasonably. Self-discipline is the key to this kind of behavior.

2-47. In peacetime, self-discipline gets the unit out for the hard training. Self-discipline makes the tank commander demand another run-through of a battle drill if the performance doesn't meet the standard—even though everyone is long past ready to quit. Self-discipline doesn't mean that you never get tired or discouraged—after all, you're only human. It does mean that you do what needs to be done regardless of your feelings.

Initiative

The leader must be an aggressive thinker—always anticipating and analyzing.

He must be able to make good assessments and solid tactical judgments.

Brigadier General John. T. Nelson II

2-48. Initiative is the ability to be a self-starter—to act when there are no clear instructions, to act when the situation changes or when the plan falls apart. In the operational context, it means setting and dictating the terms of action throughout the battle or operation. An individual leader with initiative is willing to decide and initiate independent actions when the concept of operations no longer applies or when an unanticipated opportunity leading to accomplishment of the commander's intent presents itself. Initiative drives the Army leader to seek a better method, anticipate what must be done, and perform without waiting for instructions. Balanced with good judgment, it becomes *disciplined* initiative, an essential leader attribute. (FM 100-5 discusses initiative as it relates to military actions at the operational level. FM 100-34 discusses the relationship of initiative to command and control. FM 100-40 discusses the place of initiative in the art of tactics.)

2-49. As an Army leader, you can't just give orders: you must make clear the intent of those orders, the final goal of the mission. In combat, it's critically important for subordinates to understand their commander's intent. When they are cut off or enemy actions derail the original plan, well-trained soldiers who understand the commander's intent will apply disciplined initiative to accomplish the mission.

2-50. Disciplined initiative doesn't just appear; you must develop it within your subordinates. Your leadership style and the organizational climate you establish can either encourage or discourage initiative: you can instill initiative in your subordinates or you can drive it out. If you underwrite honest mistakes, your subordinates will be more likely to develop initiative. If you set a "zero defects" standard, you risk strangling initiative in its cradle, the hearts of your subordinates. (Chapter 5 discusses "zero defects" and learning.)

The Quick Reaction Platoon

On 26 December 1994 a group of armed and disgruntled members of the Haitian Army entered the Haitian Army Headquarters in Port-au-Prince demanding back pay. A gunfight ensued less than 150 meters from the grounds of the Haitian Palace, seat of the new government. American soldiers from C Company, 1-22 Infantry, who had deployed to Haiti as part of Operation Uphold Democracy, were guarding the palace grounds. The quick reaction platoon leader deployed and immediately maneuvered his platoon towards the gunfire. The platoon attacked, inflicting at least four casualties and causing the rest of the hostile soldiers to flee. The platoon quelled a potentially explosive situation by responding correctly and aggressively to the orders of their leader, who knew his mission and the commander's intent.

Judgment

I learned that good judgment comes from experience and that experience grows out of mistakes.

General of the Army Omar N. Bradley

2-51. Leaders must often juggle hard facts, questionable data, and gut-level intuition to arrive at a decision. Good judgment means making the best decision for the situation. It's a key attribute of the art of command and the transformation of knowledge into understanding. (FM 100-34 discusses how leaders convert data and information into knowledge and understanding.)

2-52. Good judgment is the ability to size up a situation quickly, determine what's important, and decide what needs to be done. Given a problem, you should consider a range of alternatives before you act. You need to think through the consequences of what you're about to do before you do it. In addition to considering the consequences, you should also think methodically. Some sources that aid judgment are the boss's intent, the desired goal, rules, laws, regulations, experience, and values. Good judgment also includes the ability to size up subordinates, peers, and the enemy for strengths, weaknesses, and potential actions. It's a critical part of problem solving and decision making. (Chapter 5 discusses problem solving and decision making).

2-53. Judgment and initiative go hand in hand. As an Army leader, you must weigh what you know and make decisions in situations where others do nothing. There will be times when you'll have to make decisions under severe time constraints. In all cases, however, you must take responsibility for your actions. In addition, you must encourage disciplined initiative in, and teach good judgment to, your subordinates. Help your subordinates learn from mistakes by coaching and mentoring them along the way. (Chapter 5 discusses mentoring.)

Self-Confidence

2-54. Self-confidence is the faith that you'll act correctly and properly in any situation, even one in which you're under stress and don't have all the information you want. Self-confidence comes from competence: it's based on mastering skills, which takes hard work and dedication. Leaders who know their own capabilities and believe in themselves are self-confident. Don't mistake bluster—loudmouthed bragging or self-promotion—for self-confidence. Truly self-confident leaders don't need to advertise; their actions say it all.

2-55. Self-confidence is important for leaders and teams. People want self-confident leaders, leaders who understand the situation, know what needs to be done, and demonstrate that understanding and knowledge. Self-confident leaders instill self-confidence in their people. In combat, self-confidence helps soldiers control doubt and reduce anxiety. Together with will and self-discipline, self-confidence helps leaders act—do what must be done in circumstances where it would be easier to do nothing—and to convince their people to act as well.

Intelligence

2-56. Intelligent leaders think, learn, and reflect; then they apply what they learn. Intelligence is more than knowledge, and the ability to think isn't the same as book learning. All people have some intellectual ability that, when developed, allows them to analyze and understand a situation. And although some people are smarter than others, all people can develop the capabilities they have. Napoleon himself observed how a leader's intellectual development applies directly to battlefield success:

It is not genius which reveals to me suddenly and secretly what I should do in circumstances unexpected by others; it is thought and meditation.

2-57. Knowledge is only part of the equation. Smart decisions result when you combine professional skills (which you learn through study) with experience (which you gain on the job) and your ability to reason through a problem based on the information available. Reflection is also important. From time to time, you find yourself carefully and thoughtfully considering how leadership, values, and other military principles apply to you and your job. When things don't go quite the way they intended, intelligent leaders are confident enough to step back and ask, "Why did things turn out that way?" Then they are smart enough to build on their strengths and avoid making the same mistake again.

2-58. Reflection also contributes to your originality (the ability to innovate, rather than only adopt others' methods) and intuition (direct, immediate insight or understanding of important factors without apparent rational thought or inference). Remember COL Chamberlain at Little Round Top. To his soldiers, it sometimes appeared that he could "see through forests and hills and know what was coming." But this was no magical ability. Through study and reflection, the colonel had learned how to analyze terrain and imagine how the enemy might attempt to use it to his advantage. He had applied his intelligence and developed his intellectual capabilities. Good leaders follow COL Chamberlain's example.

Cultural Awareness

2-59. Culture is a group's shared set of beliefs, values, and assumptions about what's important. As an Army leader, you must be aware of cultural factors in three contexts:

- You must be sensitive to the different backgrounds of your people.
- You must be aware of the culture of the country in which your organization is operating.
- You must take into account your partners' customs and traditions when you're working with forces of another nation.

2-60. Within the Army, people come from widely different backgrounds: they are shaped by their schooling, race, gender, and religion as well as a host of other influences. Although they share Army values, an African-American man from rural Texas may look at many things differently from, say, a third-generation Irish-American man who grew up in Philadelphia or a Native American woman from the Pacific Northwest. But be aware that perspectives vary within groups as well. That's why you should try to understand individuals based on their own ideas, qualifications, and contributions and not jump to conclusions based on stereotypes.

2-61. Army values are part of the Army's institutional culture, a starting point for how you as a member of the Army should think and act. Beyond that, Army leaders not only recognize that people are different; they value them because of their differences, because they are people. Your job as a leader isn't to make everyone the same.

Instead, your job is to take advantage of the fact that everyone is different and build a cohesive team. (Chapter 7 discusses the role strategic leaders play in establishing and maintaining the Army's institutional culture.)

2-62. There's great diversity in the Army—religious, ethnic, and social—and people of different backgrounds bring different talents to the table. By joining the Army, these people have agreed to adopt the Army culture. Army leaders make this easier by embracing and making use of everyone's talents. What's more, they create a team where subordinates know they are valuable and their talents are important.

2-63. You never know how the talents of an individual or group will contribute to mission accomplishment. For example, during World War II US Marines from the Navajo nation formed a group of radio communications specialists dubbed the Navajo Code Talkers. The code talkers used their native language—a unique talent—to handle command radio traffic. Not even the best Japanese code breakers could decipher what was being said.

2-64. Understanding the culture of your adversaries and of the country in which your organization is operating is just as important as understanding the culture of your own country and organization. This aspect of cultural awareness has always been important, but today's operational environment of frequent deployments—often conducted by small units under constant media coverage—makes it even more so. As an Army leader, you need to remain aware of current events—particularly those in areas where America has national interests. You may have to deal with people who live in those areas, either as partners, neutrals, or adversaries. The more you know about them, the better prepared you'll be.

2-65. You may think that understanding other cultures applies mostly to stability operations and support operations. However, it's critical to planning offensive and defensive operations as well. For example, you may employ different tactics against an adversary who considers surrender a dishonor worse than death than against those for whom surrender is an honorable option. Likewise, if your organization is operating as part of a multinational team, how well you understand your partners will affect how well the team accomplishes its mission.

2-66. Cultural awareness is crucial to the success of multinational operations. In such situations Army leaders take the time to learn the customs and traditions of the partners' cultures. They learn how and why others think and act as they do. In multinational forces, effective leaders create a "third culture," which is the bridge or the compromise among partners. This is what GA Eisenhower did in the following example.

GA Eisenhower Forms SHAEF

During World War II, one of GA Eisenhower's duties as Supreme Allied Commander in the European Theater of Operations (ETO) was to form his theater headquarters, the Supreme Headquarters, Allied Expeditionary Force (SHAEF). GA Eisenhower had to create an environment in this multinational headquarters in which staff members from the different Allied armies could work together harmoniously. It was one of GA Eisenhower's toughest jobs.

The forces under his command—American, British, French, Canadian, and Polish—brought not only different languages, but different ways of thinking, different ideas about what was important, and different strategies. GA Eisenhower could have tried to bend everyone to his will and his way of thinking; he was the boss, after all. But it's doubtful the Allies would have fought as well for a bullying commander or that a bullying commander would have survived politically. Instead, he created a positive organizational climate that made best use of the various capabilities of his subordinates. This kind of work takes tact, patience, and trust. It doesn't destroy existing cultures but creates a new one. (Chapter 7 discusses how building this coalition contributed to the Allied victory in the ETO.)

PHYSICAL ATTRIBUTES

2-67. Physical attributes—health fitness, physical fitness, and military and professional bearing—can be developed. Army leaders maintain the appropriate level of physical fitness and military bearing.

Health Fitness

Disease was the chief killer in the [American Civil] war. Two soldiers died of it for every one killed in battle...In one year, 995 of every thousand men in the Union army contracted diarrhea and dysentery.

Geoffrey C. Ward
The Civil War

2-68. Health fitness is everything you do to maintain good health, things such as undergoing routine physical exams, practicing good dental hygiene, maintaining deployability standards, and even personal grooming and cleanliness. A soldier unable to fight because of dysentery is as much a loss as one who's wounded. Healthy soldiers can perform under extremes in temperature, humidity, and other conditions better than unhealthy ones. Health fitness also includes avoiding things that degrade your health, such as substance abuse, obesity, and smoking.

Physical Fitness

Fatigue makes cowards of us all.

General George S. Patton Jr.
Commanding General, Third Army, World War II

2-69. Unit readiness begins with physically fit soldiers and leaders. Combat drains soldiers physically, mentally, and emotionally. To minimize those effects, Army leaders are physically fit, and they make sure their subordinates are fit as well. Physically fit soldiers perform better in all areas, and physically fit leaders are better able to think, decide, and act appropriately under pressure. Physical readiness provides a foundation for combat readiness, and it's up to you, the leader, to get your soldiers ready.

2-70. Although physical fitness is a crucial element of success in battle, it's not just for frontline soldiers. Wherever they are, people who are physically fit feel more competent and confident. That attitude reassures and inspires those around them. Physically fit soldiers and DA civilians can handle stress better, work longer and harder, and recover faster than ones who are not fit. These payoffs are valuable in both peace and war.

2-71. The physical demands of leadership positions, prolonged deployments, and continuous operations can erode more than just physical attributes. Soldiers must show up ready for deprivations because it's difficult to maintain high levels of fitness during deployments and demanding operations. Trying to get fit under those conditions is even harder. If a person isn't physically fit, the effects of additional stress snowball until their mental and emotional fitness are compromised as well. Army leaders' physical fitness has significance beyond their personal performance and well-being. Since leaders' decisions affect their organizations' combat effectiveness, health, and safety and not just their own, maintaining physical fitness is an ethical as well as a practical imperative.

2-72. The Army Physical Fitness Test (APFT) measures a baseline level of physical fitness. As an Army leader, you're required to develop a physical fitness program that enhances your soldiers' ability to complete soldier and leader tasks that support the unit's mission essential task list (METL). (FM 25-101 discusses METL-based integration of soldier, leader, and collective training.) Fitness programs that emphasize training specifically for the APFT are boring and don't prepare soldiers for the varied stresses of combat. Make every effort to design a physical fitness program that prepares your people for what you expect them to do in combat. Readiness should be your program's primary focus; preparation for the APFT itself is secondary. (FM 21-20 is your primary physical fitness resource.)

You have to lead men in war by requiring more from the individual than he thinks he can do. You have to [bring] them along to endure and to display qualities of fortitude that are beyond the average man's thought of what he should be expected to do. You have to inspire them when they are hungry and exhausted and desperately uncomfortable and in great danger; and only a man of positive characteristics of leadership, with the physical stamina [fitness] that goes with it, can function under those conditions.

General of the Army George C. Marshall
Army Chief of Staff, World War II

Military and Professional Bearing

Our...soldiers should look as good as they are.

Sergeant Major of the Army Julius W. Gates

2-73. As an Army leader, you're expected to look like a soldier. Know how to wear the uniform and wear it with pride at all times. Meet height and weight standards. By the way you carry yourself and through your military courtesy and appearance, you send a signal: I am proud of my uniform, my unit, and myself. Skillful use of your professional bearing—fitness, courtesy, and military appearance—can often help you manage difficult situations. A professional—DA civilian or soldier—presents a professional appearance, but there's more to being an Army professional than looking good. Professionals are competent as well; the Army requires you to both *look* good and *be* good.

EMOTIONAL ATTRIBUTES

Anyone can become angry—that is easy. But to be angry with the right person, to the right degree, at the right time, for the right purpose, and in the right way—that is not easy.

Aristotle
Greek philosopher and tutor to Alexander the Great

2-74. As an Army leader, your emotional attributes—self-control, balance, and stability—contribute to how you feel and therefore to how you interact with others. Your people are human beings with hopes, fears, concerns, and dreams. When you understand that will and endurance come from emotional energy, you possess a powerful leadership tool. The feedback you give can help your subordinates use their emotional energy to accomplish amazing feats in tough times.

Self-Control in Combat

An American infantry company in Vietnam had been taking a lot of casualties from booby traps. The soldiers were frustrated because they could not fight back. One night, snipers ambushed the company near a village, killing two soldiers. The rest of the company—scared, anguished, and frustrated—wanted to enter the village, but the commander—who was just as angry—knew that the snipers were long gone. Further, he knew that there was a danger his soldiers would let their emotions get the upper hand, that they might injure or kill some villagers out of a desire to strike back at something. Besides being criminal, such killings would drive more villagers to the Viet Cong. The commander maintained control of his emotions, and the company avoided the village.

2-75. Self-control, balance, and stability also help you make the right ethical choices. Chapter 4 discusses the steps of ethical reasoning. However, in order to follow those steps, you must remain in control of yourself; you can't be at the mercy of your impulses. You must remain calm under pressure, "watch your lane," and expend energy on things you can fix. Inform your boss of things you can't fix and don't worry about things you can't affect.

2-76. Leaders who are emotionally mature also have a better awareness of their own strengths and weaknesses. Mature leaders spend their energy on self-improvement; immature leaders spend their energy denying there's anything wrong. Mature, less defensive leaders benefit from constructive criticism in ways that immature people cannot.

Self-Control

Sure I was scared, but under the circumstances, I'd have been crazy not to be scared....There's nothing wrong with fear. Without fear, you can't have acts of courage.

Sergeant Theresa Kristek
Operation Just Cause, Panama

2-77. Leaders control their emotions. No one wants to work for a hysterical leader who might lose control in a tough situation. This doesn't mean you never show emotion. Instead, you must display the proper amount of emotion and passion—somewhere between too much and too little—required to tap into your subordinates' emotions. Maintaining self-control inspires calm confidence in subordinates, the coolness under fire so essential to a successful unit. It also encourages feedback from your subordinates that can expand your sense of what's really going on.

Balance

An officer or noncommissioned officer who loses his temper and flies into a tantrum has failed to obtain his first triumph in discipline.

Noncommissioned Officer's Manual, 1917

2-78. Emotionally balanced leaders display the right emotion for the situation and can also read others' emotional state. They draw on their experience and provide their subordinates the proper perspective on events. They have a range of attitudes—from relaxed to intense—with which to approach situations and can choose the one appropriate to the circumstances. Such leaders know when it's time to send a message that things are urgent and how to do that without throwing the organization into chaos. They also know how to encourage people at the toughest moments and keep them driving on.

Stability

Never let yourself be driven by impatience or anger. One always regrets having followed the first dictates of his emotions.

Marshal de Belle-Isle
French Minister of War, 1757-1760

2-79. Effective leaders are steady, levelheaded under pressure and fatigue, and calm in the face of danger. These characteristics calm their subordinates, who are always looking to their leader's example. Display the emotions you want your people to display; don't give in to the temptation to do what feels good for you. If you're under great stress, it might feel better to vent—scream, throw things, kick furniture—but that will not help the organization. If you want your subordinates to be calm and rational under pressure, you must be also.

BG Jackson at First Bull Run

At a crucial juncture in the First Battle of Bull Run, the Confederate line was being beaten back from Matthews Hill by Union forces. Confederate BG Thomas J. Jackson and his 2,000-man brigade of Virginians, hearing the sounds of battle to the left of their position, pressed on to the action. Despite a painful shrapnel wound, BG Jackson calmly placed his men in a defensive position on Henry Hill and assured them that all was well.

As men of the broken regiments flowed past, one of their officers, BG Barnard E. Bee, exclaimed to BG Jackson, "General, they are driving us!" Looking toward the direction of the enemy, BG Jackson replied, "Sir, we will give them the bayonet." Impressed by BG Jackson's confidence and self-control, BG Bee rode off towards what was left of the officers and men of his brigade. As he rode into the throng he gestured with his sword toward Henry Hill and shouted, "Look, men! There is Jackson standing like a stone wall! Let us determine to die here, and we will conquer! Follow me!"

BG Bee would later be mortally wounded, but the Confederate line stiffened and the nickname he gave to BG Jackson would live on in American military history. This example shows how one leader's self-control under fire can turn the tide of battle by influencing not only the leader's own soldiers, but the leaders and soldiers of other units as well.

FOCUS ON CHARACTER

Just as fire tempers iron into fine steel, so does adversity temper one's character into firmness, tolerance, and determination.

Margaret Chase Smith
Lieutenant Colonel, US Air Force Reserve
and United States Senator

2-80. Earlier in this chapter, you read how character is made up of two interacting sets of characteristics: values and attributes. People enter the Army with values and attributes they've developed over the course of a lifetime, but those are just the starting points for further character development. Army leaders continuously develop in themselves and their subordinates the Army values and leader attributes that this chapter discusses and Figure 1-1 shows. This isn't just an academic exercise, another mandatory training topic to address once a year. Your character shows through in your actions—on and off duty.

2-81. Character helps you determine what's right and motivates you to do it, regardless of the circumstances or the consequences. What's more, an informed ethical conscience consistent with Army values steels you for making the right choices when faced with tough questions. Since Army leaders seek to do what's right and inspire others to do the same, you must be concerned with character development. Examine the actions in this example, taken from the report of a platoon sergeant during Operation Desert Storm. Consider the aspects of character that contributed to them.

Character and Prisoners

The morning of [28 February 1991], about a half-hour prior to the cease-fire, we had a T-55 tank in front of us and we were getting ready [to engage it with a TOW]. We had the TOW up and we were tracking him and my wingman saw him just stop and a head pop up out of it. And Neil started calling me saying, "Don't shoot, don't shoot, I think they're getting off the tank." And they did. Three of them jumped off the tank and ran around a sand dune. I told my wingman, "I'll cover the tank, you go on down and check around the back side and see what's down there." He went down there and found about 150 PWs....

[T]he only way we could handle that many was just to line them up and run them through...a little gauntlet...[W]e had to check them for weapons and stuff and we lined them up and called for the PW handlers to pick them up. It was just amazing.

We had to blow the tank up. My instructions were to destroy the tank, so I told them to go ahead and move it around the back side of the berm a little bit to safeguard us, so we wouldn't catch any shrapnel or ammunition coming off. When the tank blew up, these guys started yelling and screaming at my soldiers, "Don't shoot us, don't shoot us," and one of my soldiers said, "Hey, we're from America; we don't shoot our prisoners." That sort of stuck with me.

2-82. The soldier's comment at the end of this story captures the essence of character. He said, "We're from America..." He defined, in a very simple way, the connection between who you are—your character—and what you do. This example illustrates character—shared values and attributes—telling soldiers what to do and what not to do. However, it's interesting for other reasons. Read it again: You can almost feel the soldiers' surprise when they realized what the Iraqi PWs were afraid of. You can picture the young soldier, nervous, hands on his weapon, but still managing to be a bit amused. The right thing, the ethical choice, was so deeply ingrained in those soldiers that it never occurred to them to do anything other than safeguard the PWs.

The Battle of the Bulge

In December 1944 the German Army launched its last major offensive on the Western Front of the ETO, sending massive infantry and armor formations into a lightly-held sector of the Allied line in Belgium. American units were overrun. Thousands of green troops, sent to that sector because it was quiet, were captured. For two desperate weeks the Allies fought to check the enemy advance. The 101st Airborne Division was sent to the town of Bastogne. The Germans needed to control the crossroads there to move equipment to the front; the 101st was there to stop them.

Outnumbered, surrounded, low on ammunition, out of medical supplies, and with wounded piling up, the 101st, elements of the 9th and 10th Armored Divisions, and a tank destroyer battalion fought off repeated attacks through some of the coldest weather Europe had seen in 50 years. Wounded men froze to death in their foxholes. Paratroopers fought tanks. Nonetheless, when the German commander demanded American surrender, BG Anthony C. McAuliffe, acting division commander, sent a one-word reply: "Nuts."

The Americans held. By the time the Allies regained control of the area and pushed the Germans back, Hitler's "Thousand Year Reich" had fewer than four months remaining.

2-83. BG McAuliffe spoke based on what he knew his soldiers were capable of, even in the most extreme circumstances. This kind of courage and toughness didn't develop overnight. Every Allied soldier brought a lifetime's worth of character to that battle; that character was the foundation for everything else that made them successful.

GA Eisenhower's Message

On 5 June 1944, the day before the D-Day invasion, with his hundreds of thousands of soldiers, sailors and airmen poised to invade France, GA Dwight D. Eisenhower took a few minutes to draft a message he hoped he would never deliver. It was a "statement he wrote out to have ready when the invasion was repulsed, his troops torn apart for nothing, his planes ripped and smashed to no end, his warships sunk, his reputation blasted."

In his handwritten statement, GA Eisenhower began, "Our landings in the Cherbourg-Havre area have failed to gain a satisfactory foothold and I have withdrawn the troops." Originally he had written, the "troops have been withdrawn," a use of the passive voice that conceals the actor. But he changed the wording to reflect his acceptance of full personal accountability.

GA Eisenhower went on, "My decision to attack at this time and place was based on the best information available." And after recognizing the courage and sacrifice of the troops he concluded, "If any blame or fault attaches to this attempt, it is mine alone."

2-84. GA Eisenhower, in command of the largest invasion force ever assembled and poised on the eve of a battle that would decide the fate of millions of people, was guided by the same values and attributes that shaped the actions of the soldiers in the Desert Storm example. His character allowed for nothing less than acceptance of total personal responsibility. If things went badly, he was ready to take the blame. When things went well, he gave credit to his subordinates. The Army values GA Eisenhower personified provide a powerful example for all members of the Army.

CHARACTER AND THE WARRIOR ETHOS

2-85. The _warrior ethos_ refers to the professional attitudes and beliefs that characterize the American soldier. At its core, the warrior ethos grounds itself on the refusal to accept failure. The Army has forged the warrior ethos on training grounds from Valley Forge to the CTCs and honed it in battle from Bunker Hill to San Juan Hill, from the Meuse-Argonne to Omaha Beach, from Pork Chop Hill to the Ia Drang Valley, from Salinas Airfield to the Battle of 73 Easting. It derives from the unique realities of battle. It echoes through the precepts in the Code of Conduct. Developed through discipline, commitment to Army values, and knowledge of the Army's proud heritage, the warrior ethos makes clear that military service is much more than just another job: the purpose of winning the nation's wars calls for total commitment.

2-86. America has a proud tradition of winning. The ability to forge victory out of the chaos of battle includes overcoming fear, hunger, deprivation, and fatigue. The Army wins because it fights hard; it fights hard because it trains hard; and it trains hard because that's the way to _win_. Thus, the warrior ethos is about more than persevering under the worst of conditions; it fuels the fire to fight through those conditions to victory no matter how long it takes, no matter how much effort is required. It's one thing to make a snap decision to risk your life for a brief period of time. It's quite another to sustain the will to win when the situation looks hopeless and doesn't show any indications of getting better, when being away from home and family is a profound hardship. The soldier who jumps on a grenade to save his comrades is courageous, without question. That action requires great physical courage, but pursuing victory over time also requires a deep moral courage that concentrates on the mission.

2-87. The warrior ethos concerns character, shaping who you are and what you do. In that sense, it's clearly linked to Army values such as _personal courage, loyalty to comrades, and dedication to duty._ Both loyalty and duty involve putting your life on the line, even when there's little chance of survival, for the good of a cause larger than yourself. That's the clearest example of _selfless service._ American soldiers never give up on their fellow soldiers, and they never compromise on doing their duty. _Integrity_ underlies the character of the Army as well. The warrior ethos requires unrelenting and consistent determination to do what is right and to do it with pride, both in war and military operations other than war. Understanding what is right requires _respect_ for both your comrades and other people involved in such complex arenas as peace operations and nation assistance. In such ambiguous situations, decisions to use lethal or nonlethal force severely test judgment and discipline. In whatever conditions Army leaders find themselves, they turn the personal warrior ethos into a collective commitment to win with _honor._

2-88. The warrior ethos is crucial—and perishable—so the Army must continually affirm, develop, and sustain it. Its martial ethic connects American warriors today with those whose sacrifices have allowed our very existence. The Army's continuing drive to be the best, to triumph over all adversity, and to remain focused on mission accomplishment does more than preserve the Army's institutional culture; it sustains the nation.

2-89. Actions that safeguard the nation occur everywhere you find soldiers. The warrior ethos spurs the lead tank driver across a line of departure into uncertainty. It drives the bone-tired medic continually to put others first. It pushes the sweat-soaked gunner near muscle failure to keep up the fire. It drives the heavily loaded infantry soldier into an icy wind, steadily uphill to the objective. It presses the signaler through fatigue to provide communications. And the warrior ethos urges the truck driver across frozen roads bounded by minefields because fellow soldiers at an isolated outpost need supplies. Such tireless motivation comes in part from the comradeship that springs from the warrior ethos. Soldiers fight for each other; they would rather die than let their buddies down. That loyalty runs front to rear as well as left to right: mutual support marks Army culture regardless of who you are, where you are, or what you are doing.

2-90. That tight fabric of loyalty to one another and to collective victory reflects perhaps the noblest aspect of our American warrior ethos: the military's subordinate relationship to civilian authority. That subordination began in 1775, was reconfirmed at Newburgh, New York, in 1782, and continues to this day. It's established in the Constitution and makes possible the freedom all Americans enjoy. The Army sets out to achieve national objectives, not its own, for *selfless service* is an institutional as well as an individual value. And in the end, the Army returns its people back to the nation. America's sons and daughters return with their experience as part of a winning team and share that spirit as citizens. The traditions and values of the service derive from a commitment to excellent performance and operational success. They also point to the Army's unwavering commitment to the society we serve. Those characteristics serve America and its citizens—both in and out of uniform—well.

CHARACTER DEVELOPMENT

2-91. People come to the Army with a character formed by their background, religious or philosophical beliefs, education, and experience. Your job as an Army leader would be a great deal easier if you could check the values of a new DA civilian or soldier the way medics check teeth or run a blood test. You could figure out what values were missing by a quick glance at Figure 1-1 and administer the right combination, maybe with an injection or magic pill.

2-92. But character development is a complex, lifelong process. No scientist can point to a person and say, "This is when it all happens." However, there are a few things you can count on. You build character in subordinates by creating organizations in which Army values are not just words in a book but precepts for what their members do. You help build subordinates' character by acting the way you want them to act. You teach by example, and coach along the way. (Appendix E contains additional information on character development.) When you hold yourself and your subordinates to the highest standards, you reinforce the values those standards embody. They spread throughout the team, unit, or organization—throughout the Army—like the waves from a pebble dropped into a pond.

CHARACTER AND ETHICS

2-93. When you talk about character, you help your people answer the question, What kind of person should I be? You must not only embrace Army values and leader attributes but also use them to think, reason, and—after reflection—act. Acting in a situation that tests your character requires moral courage. Consider this example.

The Qualification Report

A battalion in a newly activated division had just spent a great deal of time and effort on weapons qualification. When the companies reported results, the battalion commander could not understand why B and C Companies had reported all machine gunners fully qualified while A Company had not. The A Company Commander said that he could not report his gunners qualified because they had only fired on the 10-meter range and the manual for qualification clearly stated that the gunners had to fire on the transition range as well. The battalion commander responded that since the transition range was not built yet, the gunners should be reported as qualified: "They fired on the only range we have. And besides, that's how we did it at Fort Braxton."

Some of the A Company NCOs, who had also been at Fort Braxton, tried to tell their company commander the same thing. But the captain insisted the A Company gunners were not fully qualified, and that's how the report went to the brigade commander.

> ### The Qualification Report (continued)
>
> The brigade commander asked for an explanation of the qualification scores. After hearing the A Company Commander's story, he agreed that the brigade would be doing itself no favors by reporting partially qualified gunners as fully qualified. The incident also sent a message to division: get that transition range built.
>
> The A Company Commander's choice was not between loyalty to his battalion commander and honesty; doing the right thing here meant being loyal and honest. And the company commander had the moral courage to be both honest and loyal—loyal to the Army, loyal to his unit, and loyal to his soldiers.

2-94. The A Company Commander made his decision and submitted his report without knowing how it would turn out. He didn't know the brigade commander would back him up, but he reported his company's status relative to the published Army standard anyway. He insisted on reporting the truth—which took character—because it was the right thing to do.

2-95. Character is important in living a consistent and moral life, but character doesn't always provide the final answer to the specific question, What should I do now? Finding that answer can be called ethical reasoning. Chapter 4 outlines a process for ethical reasoning. When you read it, keep in mind that the process is much more complex than the steps indicate and that you must apply your own values, critical reasoning skills, and imagination to the situation. There are no formulas that will serve every time; sometimes you may not even come up with an answer that completely satisfies you. But if you embrace Army values and let them govern your actions, if you learn from your experiences and develop your skills over time, you're as prepared as you can be to face the tough calls.

2-96. Some people try to set different Army values against one another, saying a problem is about loyalty versus honesty or duty versus respect. Leadership is more complicated than that; the world isn't always black and white. If it were, leadership would be easy and anybody could do it. However, in the vast majority of cases, Army values are perfectly compatible; in fact, they reinforce each other.

CHARACTER AND ORDERS

2-97. Making the right choice and acting on it when faced with an ethical question can be difficult. Sometimes it means standing your ground. Sometimes it means telling your boss you think the boss is wrong, like the finance supervisor in Chapter 1 did. Situations like these test your character. But a situation in which you think you've received an illegal order can be even more difficult.

2-98. In Chapter 1 you read that a good leader executes the boss's decision with energy and enthusiasm. The only exception to this principle is your duty to disobey illegal orders. This isn't a privilege you can conveniently claim, but a duty you must perform. If you think an order is illegal, first be sure that you understand both the details of the order and its original intent. Seek clarification from the person who gave the order. This takes moral courage, but the question will be straightforward: Did you really mean for me to...steal the part...submit a false report...shoot the prisoners? If the question is complex or time permits, always seek legal counsel. However, if you must decide immediately—as may happen in the heat of combat—make the best judgment possible based on Army values, your experience, and your previous study and reflection. You take a risk when you disobey what you believe to be an illegal order. It may be the most difficult decision you'll ever make, but that's what leaders do.

2-99. While you'll never be completely prepared for such a situation, spending time reflecting on Army values and leader attributes may help. Talk to your superiors, particularly those who have done what you aspire to do or what you think you'll be called on to do; providing counsel of this sort is an important part of mentoring (which Chapter 5 discusses). Obviously, you need to make time to do this before you're faced with a tough call. When you're in the middle of a firefight, you don't have time to reflect.

CHARACTER AND BELIEFS

2-100. What role do beliefs play in ethical matters? Beliefs are convictions people hold as true; they are based on their upbringing, culture, heritage, families, and traditions. As a result, different moral beliefs have been and will continue to be shaped by diverse religious and philosophical traditions. You serve a nation that takes very seriously the notion that people are free to choose their own beliefs and the basis for those beliefs. In fact, America's strength comes from that diversity. The Army respects different moral backgrounds and personal convictions—as long as they don't conflict with Army values.

2-101. Beliefs matter because they are the way people make sense of what they experience. Beliefs also provide the basis for personal values; values are moral beliefs that shape a person's behavior. Effective leaders are careful not to require their people to violate their beliefs by ordering or encouraging any illegal or unethical action.

2-102. The Constitution reflects our deepest national values. One of these values is the guarantee of freedom of religion. While religious beliefs and practices are left to individual conscience, Army leaders are responsible for ensuring their soldiers' right to freely practice their religion. Title 10 of the United States Code states, "Each commanding officer shall furnish facilities, including necessary transportation, to any chaplain assigned to his command, to assist the chaplain in performing his duties." What does this mean for Army leaders? The commander delegates staff responsibility to the chaplain for programs to enhance spiritual fitness since many people draw moral fortitude and inner strength from a spiritual foundation. At the same time, no leader may apply undue influence or coerce others in matters of religion—whether to practice or not to practice specific religious beliefs. (The first ten amendments to the Constitution are called the Bill of Rights. Freedom of religion is guaranteed by the First Amendment, an indication of how important the Founders considered it. You can read the Bill of Rights in Appendix F.)

2-103. Army leaders also recognize the role beliefs play in preparing soldiers for battle. Soldiers often fight and win over tremendous odds when they are convinced of the ideals (beliefs) for which they are fighting. Commitment to such beliefs as justice, liberty, freedom, and not letting down your fellow soldier can be essential ingredients in creating and sustaining the will to fight and prevail. A common theme expressed by American PWs during the Vietnam Conflict was the importance of values instilled by a common American culture. Those values helped them to withstand torture and the hardships of captivity.

SECTION II

COMPETENCE: WHAT A LEADER MUST KNOW

The American soldier...demands professional competence in his leaders. In battle, he wants to know that the job is going to be done right, with no unnecessary casualties. The noncommissioned officer wearing the chevron is supposed to be the best soldier in the platoon and he is supposed to know how to perform all the duties expected of him. The American soldier expects his sergeant to be able to teach him how to do his job. And he expects even more from his officers.

General of the Army Omar N. Bradley

2-104. Army values and leader attributes form the foundation of the character of soldiers and DA civilians. Character, in turn, serves as the basis of knowing (competence) and doing (leadership). The self-discipline that leads to teamwork is rooted in character. In the Army, teamwork depends on the actions of competent leaders of proven character who know their profession and act to improve their organizations. The best Army leaders constantly strive to improve, to get better at what they do. Their self-discipline focuses on learning more about their profession and continually getting the team to perform better. They build competence in themselves and their subordinates. Leader skills increase in scope and complexity as one moves from direct leader positions to organizational and strategic leader positions. Chapters 4, 6, and 7 discuss in detail the different skills direct, organizational, and strategic leaders require.

2-105. Competence results from hard, realistic training. That's why Basic Training starts with simple skills, such as drill and marksmanship. Soldiers who master these skills have a couple of victories under their belts. The message from the drill sergeants—explicit or not—is, "You've learned how to do those things; now you're ready to take on something tougher." When you lead people through progressively more complex tasks this way, they develop the confidence and will—the inner drive—to take on the next, more difficult challenge.

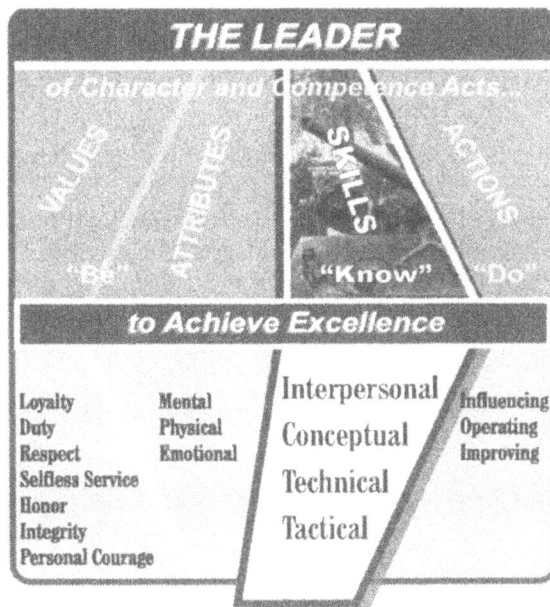

Figure 2-3. Leader Skills

2-106. For you as an Army leader, competence means much more than being well-trained. Competence links character (knowing the right thing to do) and leadership (doing or influencing your people to do the right thing). Leaders are responsible for being personally competent, but even that isn't enough: as a leader, you're responsible for your subordinates' competence as well.

2-107. Figure 2-3 highlights the four categories containing skills an Army leader must KNOW:

- **Interpersonal skills** affect how you deal with people. They include coaching, teaching, counseling, motivating, and empowering.

- **Conceptual skills** enable you to handle ideas. They require sound judgment as well as the ability to think creatively and reason analytically, critically, and ethically.

- **Technical skills** are job-related abilities. They include basic soldier skills. As an Army leader, you must possess the expertise necessary to accomplish all tasks and functions you're assigned.

- **Tactical skills** apply to solving tactical problems, that is, problems concerning employment of units in combat. You enhance tactical skills when you combine them with interpersonal, conceptual, and technical skills to accomplish a mission.

2-108. Leaders in combat combine interpersonal, conceptual, technical, and tactical skills to accomplish the mission. They use their interpersonal skills to communicate their intent effectively and motivate their soldiers. They apply their conceptual skills to determine viable concepts of operations, make the right decisions, and execute the tactics the operational environment requires. They capitalize on their technical skills to properly employ the techniques, procedures, fieldcraft, and equipment that fit the situation. Finally, combat leaders employ tactical skill, combining skills from the other skill categories with knowledge of the art of tactics appropriate to their level of responsibility and unit type to accomplish the mission. When plans go wrong and leadership must turn the tide, it is tactical skill, combined with character, that enables an Army leader to seize control of the situation and lead the unit to mission accomplishment.

2-109. The Army leadership framework draws a distinction between developing skills and performing actions. Army leaders who take their units to a combat training center (CTC) improve their skills by performing actions—by doing their jobs on the ground in the midst of intense simulated combat. But they don't wait until they arrive at the CTC to develop their skills; they practice ahead of time in command post exercises, in combat drills, on firing ranges, and even on the physical training (PT) field.

2-110. Your leader skills will improve as your experience broadens. A platoon sergeant gains valuable experience on the job that will help him be a better first sergeant. Army leaders take advantage of every chance to improve: they look for new learning opportunities, ask questions, seek training opportunities, and request performance critiques.

SECTION III

LEADERSHIP: WHAT A LEADER MUST DO

He gets his men to go along with him because they want to do it for him and they believe in him.

General of the Army Dwight D. Eisenhower

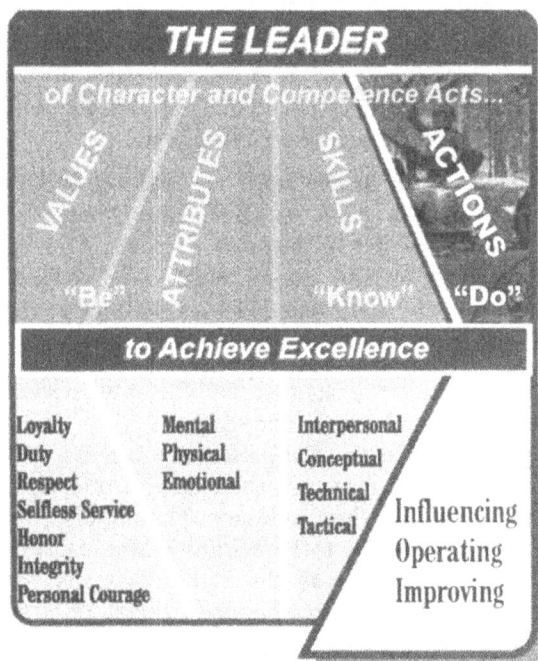

THE LEADER of Character and Competence Acts...

VALUES | ATTRIBUTES | SKILLS | ACTIONS
"Be" | "Know" | "Do"

to Achieve Excellence

Loyalty	Mental	Interpersonal	
Duty	Physical	Conceptual	
Respect	Emotional	Technical	Influencing
Selfless Service		Tactical	Operating
Honor			Improving
Integrity			
Personal Courage			

Figure 2-4. Leader Actions

2-111. Leaders act. They bring together everything they are, everything they believe, and everything they know how to do to provide purpose, direction, and motivation. Army leaders work to influence people, operate to accomplish the mission, and act to improve their organization. This section introduces leader actions. Chapters 5, 6, and 7 discuss them more fully. As with leader skills, leader actions increase in scope and complexity as you move from direct leader positions to organizational and strategic leader positions.

2-112. Developing the right values, attributes, and skills is only preparation to lead. Leadership doesn't begin until you act. Leaders who live up to Army values, who display leader attributes, who are competent, who act at all times as they would have their people act, will succeed. Leaders who talk a good game but can't back their words with actions will fail in the long run.

INFLUENCING

2-113. Army leaders use interpersonal skills to guide others toward a goal. Direct leaders most often influence subordinates face to face—such as when a team leader gives instructions, recognizes achievement, and encourages hard work. Organizational and strategic leaders also influence their immediate subordinates and staff face to face; however, they guide their organizations primarily by indirect influence. Squad leaders, for example, know what their division commander wants, not because the general has briefed each one personally, but because his intent is passed through the chain of command. Influencing actions fall into these categories:

- **Communicating** involves displaying good oral, written, and listening skills for individuals and groups.
- **Decision making** involves selecting the line of action intended to be followed as the one most favorable to the successful accomplishment of the mission. This involves using sound judgment, reasoning logically, and managing resources wisely.
- **Motivating** involves inspiring and guiding others toward mission accomplishment.

OPERATING

2-114. Operating is what you do to accomplish the immediate mission, to get the job done on time and to standard. Operating actions fall into these categories:

- **Planning and preparing** involve developing detailed, executable plans that are feasible, acceptable, and suitable; arranging unit support for the exercise or operation; and conducting rehearsals. During tactical operations, decision making and planning are enhanced by two methodologies: the military decision making process (MDMP) and the troop leading procedures (TLP). Battalion and higher echelons follow the MDMP. Company and lower echelons follow the TLP. (FM 101-5 discusses the MDMP.)
- **Executing** involves meeting mission standards, taking care of people, and efficiently managing resources.
- **Assessing** involves evaluating the efficiency and effectiveness of any system or plan in terms of its purpose and mission.

2-115. Leaders assess, or judge, performance so they can determine what needs to be done to sustain the strong areas and improve weak ones. This kind of forward thinking is linked to the last leader action, improving.

IMPROVING

2-116. Good leaders strive to leave an organization better than they found it. A child struggling to understand why it is better to put money in a piggy bank is learning what leaders know: plan and sacrifice now for the sake of the future. All leaders are tempted to focus on the short-term gain that makes them and their organizations look good today: "Why bother to fix it now? By the time next year rolls around, it will be someone else's problem." But that attitude doesn't serve either your subordinates or the Army well. When an organization sacrifices important training with long-term effects—say, training that leads to true marksmanship skill—and focuses exclusively on short-term appearances—such as qualification scores—the organization's capabilities suffers.

2-117. The results of shortsighted priorities may not appear immediately, but they will appear. Loyalty to your people as well as the Army as an institution demands you consider the long-term effects of your actions. Some of your people will remain in the organization after you've moved on. Some will still be in the Army after you're long gone. Soldiers and DA civilians tomorrow must live with problems leaders don't fix today.

2-118. Army leaders set priorities and balance competing demands. They focus their organizations' efforts on short- and long-term goals while continuing to meet requirements that may or may not contribute directly to achieving those goals. In the case of weapons proficiency, qualification is a requirement but true marksmanship skill is the goal. For battlefield success, soldiers need training that leads to understanding and mastery of technical and tactical skills that hold up under the stress of combat. Throw in all the other things vying for an organization's time and resources and your job becomes even more difficult. Guidance from higher headquarters may help, but you must make the tough calls. Improving actions fall into these categories:

- **Developing** involves investing adequate time and effort to develop individual subordinates as leaders. It includes mentoring.
- **Building** involves spending time and resources to improve teams, groups, and units and to foster an ethical climate.
- **Learning** involves seeking self-improvement and organizational growth. It includes envisioning, adapting, and leading change.

SUMMARY

2-119. As an Army leader, leadership in combat is your primary and most important challenge. It requires you to accept a set of values that contributes to a core of motivation and will. If you fail to accept and live these Army values, your soldiers may die unnecessarily and you may fail to accomplish your mission.

2-120. What must you, as an Army leader, BE, KNOW, and DO? You must have character, that combination of values and attributes that underlie your ability to see what needs to be done, decide to do it, and influence others to follow you. You must be competent, that is, possess the knowledge and skills required to do your job right. And you must lead, take the proper actions to accomplish the mission based on what your character tells you is ethically right and appropriate for the situation.

2-121. Leadership in combat, the greatest challenge, requires a basis for your motivation and will. That foundation is Army values. In them are rooted the basis for the character and self-discipline that generate the will to succeed and the motivation to persevere. From this motivation derives the lifelong work of self-development in the skills that make a successful Army leader, one who walks the talk of BE, KNOW, DO. Chapter 3 examines the environment that surrounds your people and how what you do as a leader affects it. Understanding the human dimension is essential to mastering leader skills and performing leader actions.

Chapter 3

The Human Dimension

All soldiers are entitled to outstanding leadership; I will provide that leadership. I know my soldiers and I will always place their needs above my own. I will communicate consistently with my soldiers and never leave them uninformed.

Creed of the Noncommissioned Officer

3-1. Regardless of the level, keep in mind one important aspect of leadership: you lead people. In the words of former Army Chief of Staff Creighton W. Abrams,

The Army is not made up of people; the Army is people...living, breathing, serving human beings. They have needs and interests and desires. They have spirit and will, strengths and abilities. They have weaknesses and faults, and they have means. They are the heart of our preparedness...and this preparedness—as a nation and as an Army—depends upon the spirit of our soldiers. It is the spirit that gives the Army...life. Without it we cannot succeed.

3-2. GEN Abrams could not have been more clear about what's important. To fully appreciate the human dimension of leadership, you must understand two key elements: *leadership* itself and the *people* you lead. Leadership—what this manual is about—is far from an exact science; every person and organization is different. Not only that, the environment in which you lead is shaped first by who you are and what you know; second, by your people and what they know; and third, by everything that goes on around you.

3-3. This chapter examines this all-important human dimension. Later chapters discuss the levels of Army leadership and the skills and actions required of leaders at each level.

PEOPLE, THE TEAM, AND THE INSTITUTION

3-4. Former Army Chief of Staff John A. Wickham Jr. described the relationship between the people who are the Army and the Army as an institution this way:

The Army is an institution, not an occupation. Members take an oath of service to the nation and the Army, rather than simply accept a job...the Army has moral and ethical obligations to those who serve and their families; they, correspondingly, have responsibilities to the Army.

3-5. The Army has obligations to soldiers, DA civilians, and their families that most organizations don't have; in return, soldiers and DA civilians have responsibilities to the Army that far exceed those of an employee to most employers. This relationship, one of mutual obligation and responsibility, is at the very center of what

makes the Army a team, an institution rather than an occupation.

3-6. Chapter 2 discussed how the Army can't function except as a team. This team identity doesn't come about just because people take an oath or join an organization; you can't force a team to come together any more than you can force a plant to grow. Rather, the team identity comes out of mutual respect among its members and a trust between leaders and subordinates. That bond between leaders and subordinates likewise springs from mutual respect as well as from discipline. The highest form of discipline is the willing obedience of subordinates who trust their leaders, understand and believe in the mission's purpose, value the team and their place in it, and have the will to see the mission through. This form of discipline produces individuals and teams who—in the really tough moments—come up with solutions themselves.

Soldiers Are Our Credentials

In September 1944 on the Cotentin Peninsula in France, the commander of a German stronghold under siege by an American force sent word that he wanted to discuss surrender terms. German MG Hermann Ramcke was in his bunker when his staff escorted the assistant division commander of the US 8th Infantry Division down the concrete stairway to the underground headquarters. MG Ramcke addressed BG Charles D. W. Canham through an interpreter: "I am to surrender to you. Let me see your credentials." Pointing to the dirty, tired, disheveled—but victorious—American infantrymen who had accompanied him and were now crowding the dugout entrance, the American officer replied, "These are my credentials."

DISCIPLINE

I am confident that an army of strong individuals, held together by a sound discipline based on respect for personal initiative and rights and dignity of the individual, will never fail this nation in time of need.

General J. Lawton Collins
Former Army Chief of Staff

3-7. People are our most important resource; soldiers are in fact our "credentials." Part of knowing how to use this most precious resource is understanding the stresses and demands that influence people.

3-8. One sergeant major has described discipline as "a moral, mental, and physical state in which all ranks respond to the will of the [leader], whether he is there or not." Disciplined people take the right action, even if they don't feel like it. True discipline demands habitual and reasoned obedience, an obedience that preserves initiative and works, even when the leader isn't around. Soldiers and DA civilians who understand the purpose of the mission, trust the leader, and share Army values will do the right thing because they're truly committed to the organization.

3-9. Discipline doesn't just mean barking orders and demanding an instant response—it's more complex than that. You build discipline by training to standard, using rewards and punishment judiciously, instilling confidence in and building trust among team members, and creating a knowledgeable collective will. The confidence, trust, and collective will of a disciplined, cohesive unit is crucial in combat.

3-10. You can see the importance of these three characteristics in an example that occurred during the 3 October 1993 American raid in Somalia. One soldier kept fighting despite his wounds. His comrades remembered that he seemed to stop caring about himself, that he had to keep fighting because the other guys—his buddies—were all that mattered. When things go badly, soldiers draw strength from their own and their unit's discipline; they know that other members of the team are depending on them.

3-11. Soldiers—like those of Task Force Ranger in Somalia (which you'll read about later in this chapter) and SGT Alvin York (whose story is in Chapter 5)—persevere in tough situations. They fight through because they have confidence in themselves, their buddies, their leaders, their equipment, and their training—and because they have discipline and will. A young sergeant who participated in Operation Uphold Democracy in Haiti in 1994 asserted this fact when interviewed by the media. The soldier said that operations went well because his unit did things just the way they did them in training and that his training never let him down.

3-12. Even in the most complex operations, the performance of the Army comes down to the training and disciplined performance of individuals and teams on the ground. One example of this fact occurred when a detachment of American soldiers was sent to guard a television tower in Udrigovo, Bosnia-Herzegovina.

3-13. After the soldiers had assumed their posts, a crowd of about 100 people gathered, grew to about 300, and began throwing rocks at the Americans. However, the soldiers didn't overreact. They prevented damage to the tower without creating an international incident. There was no "Boston Massacre" in Udrigovo. The discipline of American soldiers sent into this and other highly volatile situations in Bosnia kept the lid on that operation. The bloody guerrilla war predicted by some didn't materialize. This is a testament to the professionalism of today's American soldiers—your soldiers—and the quality of their leaders—you.

MORALE

NSDQ [Night Stalkers Don't Quit]

Motto of the 160th Special Operations
Aviation Regiment, "The Night Stalkers"
Message sent by Chief Warrant Officer Mike Durant,
held by Somali guerrillas, to his wife, October 1993

3-14. When military historians discuss great armies, they write about weapons and equipment, training and the national cause. They may mention sheer numbers (Voltaire said, "God is always on the side of the heaviest battalions") and all sorts of other things that can be analyzed, measured, and compared. However, some also write about another factor equally important to success in battle, something that can't be measured: the emotional element called morale.

3-15. Morale is the human dimension's most important intangible element. It's a measure of how people feel about themselves, their team, and their leaders. High morale comes from good leadership, shared hardship, and mutual respect. It's an emotional bond that springs from common values like loyalty to fellow soldiers and a belief that the organization will care for families. High morale results in a cohesive team that enthusiastically strives to achieve common goals. Leaders know that morale, the essential human element, holds the team together and keeps it going in the face of the terrifying and dispiriting things that occur in war.

You have a comradeship, a rapport that you'll never have again...There's no competitiveness, no money values. You trust the man on your left and your right with your life.

Captain Audie Murphy
Medal of Honor recipient and most decorated
American soldier of World War II

TAKING CARE OF SOLDIERS

Readiness is the best way of truly taking care of soldiers.

Former Sergeant Major of the Army
Richard A. Kidd

3-16. Sending soldiers in harm's way, into places where they may be killed or wounded, might seem to contradict all the emphasis on taking care of soldiers. Does it? How can you truly care for your comrades and send them on missions that might get them killed? Consider this important and fundamental point as you read the next few paragraphs.

3-17. Whenever the talk turns to what leaders do, you'll almost certainly hear someone say, "Take care of your soldiers." And that's good advice. In fact, if you add one more clause, "Accomplish the mission *and* take care of your soldiers," you have guidance for a career. But "taking care of soldiers" is one of those slippery phrases, like the word "honor," that lots of people talk about but few take the trouble to explain. So what does taking care of soldiers mean?

3-18. Taking care of soldiers means creating a disciplined environment where they can learn and grow. It means holding them to high standards, training them to do their jobs so they can function in peace and win in war. You take care of soldiers when you treat them fairly, refuse to cut corners, share their hardships, and set the example. Taking care of soldiers encompasses everything from making sure a soldier has time for an annual dental exam to visiting off-post housing to make sure it's adequate. It also means providing the family support that assures soldiers their families will be taken care of, whether the soldier is home or deployed. Family support means ensuring there's a support group in place, that even the most junior soldier and most inexperienced family members know where to turn for help when their soldier is deployed.

3-19. Taking care of soldiers also means demanding that soldiers do their duty, even at the risk of their lives. It doesn't mean coddling them or making training easy or comfortable. In fact, that kind of training can get soldiers killed. Training must be rigorous and as much like combat as is possible while being safe. Hard training is one way of preparing soldiers for the rigors of combat. Take care of soldiers by giving them the training, equipment, and support they need to keep them alive in combat.

3-20. In war, soldiers' comfort is important because it affects morale and combat effectiveness, but comfort takes a back seat to the mission. Consider this account of the 1944 landings on the island of Leyte in the Philippines, written more than 50 years later by Richard Gerhardt. Gerhardt, who was an 18-year-old rifleman in the 96th Infantry Division, survived two amphibious landings and months of close combat with the Japanese.

The 96th Division on Leyte

By the time we reached the beach, the smoke and dust created by the preparation fire had largely dissipated and we could see the terrain surrounding the landing area, which was flat and covered with some underbrush and palm trees. We were fortunate in that our sector of the beach was not heavily defended, and in going ashore there were few casualties in our platoon. Our company was engaged by small arms fire and a few mortar rounds, but we were able to move forward and secure the landing area in short order. Inland from the beach, however, the terrain turned into swamps, and as we moved ahead it was necessary to wade through muck and mud that was knee-deep at times....Roads in this part of the island were almost nonexistent, with the area being served by dirt trails around the swamps, connecting the villages....The Japanese had generally backed off the beaches and left them lightly defended, setting up their defense around certain villages which were at the junctions of the road system, as well as dug-in positions at points along the roads and trails. Our strategy was to...not use the roads and trails, but instead to move through the swamps and rice paddies and attack the enemy strong points from directions not as strongly defended. This was slow, dirty, and extremely fatiguing, but by this tactic we reduced our exposure to the enemy defensive plan, and to heavy fire from their strong points. It must be recognized that in combat the comfort of the front-line troops isn't part of the...planning process, but only what they can endure and still be effective. Conditions that seriously [affect] the combat efficiency of the troops then become a factor.

3-21. Gerhardt learned a lifetime's worth of lessons on physical hardship in the Pacific. Mud, tropical heat, monsoon rains, insects, malaria, Japanese snipers, and infiltrators—the details are still clear in his mind half a century later. Yet he knows—and he tells you—that soldiers must endure physical hardship when the best plan calls for it. In the Leyte campaign, the best plan was extremely difficult to execute, but it was tactically sound and it saved lives.

3-22. This concept doesn't mean that leaders sit at some safe, dry headquarters and make plans without seeing what their soldiers are going through, counting on them to tough out any situation. Leaders know that graphics on a map symbolize soldiers going forward to fight. Leaders get out with the soldiers to see and feel what they're experiencing as well as to influence the battle by their presence. (Gerhardt and numerous other front-line writers refer to the rear echelon as "anything behind my foxhole.") Leaders who stay a safe distance from the front jeopardize operations because they don't know what's going on. They risk destroying their soldiers' trust, not to mention their unit.

The K Company Visit

1LT Harold Leinbaugh, commander of K Company, 333d Infantry Regiment, 84th Division, related this experience from the ETO in January, 1945, during the coldest winter in Europe in nearly 50 years:

On a front-line visit, the battalion commander criticized 1LT Leinbaugh and CPT Jay Prophet, the A Company Commander, for their own and their men's appearance. He said it looked like no one had shaved for a week. 1LT Leinbaugh replied that there was no hot water. Sensing a teaching moment, the colonel responded: "Now if you men would save some of your morning coffee it could be used for shaving." Stepping over to a snowbank, 1LT Leinbaugh picked up a five-gallon GI [general issue] coffee can brought up that morning, and shook it in the colonel's face. The frozen coffee produced a thunk. 1LT Leinbaugh shook it again.

"That's enough," said the colonel, "...I can hear."

3-23. This example illustrates three points:

- The importance of a leader going to where the action is to see and feel what's really going on.

- The importance of a first-line leader telling the boss something he doesn't want to hear.

- The importance of a leader accepting information that doesn't fit his preconceived notions.

3-24. Soldiers are extremely sensitive to situations where their leaders are not at risk, and they're not likely to forget a mistake by a leader they haven't seen. Leaders who are out with their soldiers—in the same rain or snow, under the same blazing sun or in the same dark night, under the same threat of enemy artillery or small arms fire—will not fall into the trap of ignorance. Those who lead from the front can better motivate their soldiers to carry on under extreme conditions.

3-25. Taking care of soldiers is every leader's business. A DA civilian engineering team chief volunteered to oversee the installation of six Force Provider troop life support systems in the vicinity of Tuzla, Bosnia-Herzegovina. Using organizational skills, motivational techniques, and careful supervision, the team chief ensured that the sites were properly laid out, integrated, and installed. As a result of thorough planning and the teamwork the DA civilian leader generated, the morale and quality of life of over 5,000 soldiers were significantly improved.

COMBAT STRESS

All men are frightened. The more intelligent they are, the more they are frightened. The courageous man is the man who forces himself, in spite of his fear, to carry on.

General George S. Patton Jr.
War As I Knew It

3-26. Leaders understand the human dimension and anticipate soldiers' reactions to stress, especially to the tremendous stress of combat. The answers may look simple as you sit somewhere safe and read this manual, but be sure easy answers don't come in combat. However, if you think about combat stress and its effects on you and your soldiers ahead of time, you'll be less surprised and better prepared to deal with and reduce its effects. It takes mental discipline to imagine the unthinkable—the plan going wrong, your soldiers wounded or dying, and the enemy coming after YOU. But in combat all of these things can happen, and your soldiers expect you, their leader, to have thought through each of them. Put yourself in the position of the squad leader in the following example.

Task Force Ranger in Somalia, 1993

"Sarge" was a company favorite, a big powerful kid from New Jersey who talked with his hands and played up his "Joy-zee" accent. He loved practical jokes. One of his favorites was to put those tiny charges in guys' cigarettes, the kind that would explode with a loud "POP!" about halfway through a smoke. If anyone else had done it, it would have been annoying; Sarge usually got everyone to laugh—even the guy whose cigarette he destroyed.

During the 3 October 1993 raid in Mogadishu, Sarge was manning his Humvee's .50 cal when he was hit and killed. The driver and some of the guys in back screamed, "He's dead! He's dead!" They panicked and were not responding as their squad leader tried to get someone else up and behind the gun. The squad leader had to yell at them, "Just calm down! We've got to keep fighting or none of us will get back alive."

3-27. Consider carefully what the squad leader did. First he told his squad to calm down. Then he told them why it was important: they had to continue the fight if they wanted to make it back to their base alive. In this way he jerked his soldiers back to a conditioned response, one that had been drilled during training and that took their minds off the loss. The squad leader demonstrated the calm, reasoned leadership under stress that's critical to mission success. In spite of the loss, the unit persevered.

WILL AND WINNING IN BATTLE

3-28. The Army's ultimate responsibility is to win the nation's wars. And what is it that carries soldiers through the terrible challenges of combat? It's the will to win, the ability to gut it out when things get really tough, even when things look hopeless. It's the will not only to persevere but also to find workable solutions to the toughest problems. This drive is part of the warrior ethos, the ability to forge victory out of the chaos of battle—to overcome fear, hunger, deprivation, and fatigue and accomplish the mission. And the will to win serves you just as well in peacetime, when it's easy to become discouraged, feel let down, and spend your energy complaining instead of using your talents to make things better. Discipline holds a team together; the warrior ethos motivates its members—you and your people—to continue the mission.

3-29. All soldiers are warriors: all need to develop and display the will to win—the desire to do their job well—to persevere, no matter what the circumstances. The Army is a team, and all members' contributions are essential to mission accomplishment. As an Army leader, you're responsible for developing this sense of belonging in your subordinates. Not only that; it's your job to inculcate in your people the winning spirit—the commitment to do their part to accomplish the mission, no matter when, no matter where, no matter what.

3-30. Army operations often involve danger and therefore fear. Battling the effects of fear has nothing to do with denying it and everything to do with recognizing fear and handling it. Leaders let their subordinates know, "You can expect to be afraid; here's what we'll do about it." The Army standard is to continue your mission to successful completion, as GEN Patton said, in spite of your fears. But saying this isn't going to make it happen. Army leaders expect fear to take hold when things go poorly, setbacks occur, the unit fails to complete a mission, or there are casualties. The sights and sounds of the modern battlefield are terrifying. So is fear of the unknown. Soldiers who see their buddies killed or wounded suddenly have a greater burden: they become aware of their own mortality. On top of all these obvious sources of fear is the insecurity before battle that many veterans have written about: "Will I perform well or will I let my buddies down?"

3-31. In the October 1993 fight in Somalia, one soldier who made it back to the safety of the American position was told to prepare to go back out; there were other soldiers in trouble. He had just run a gauntlet of fire, had just seen his friends killed and wounded, and was understandably afraid. "I can't go back out there," he told his sergeant. The leader reassured the soldier while reminding him of the mission and his responsibility to the team: "I know you're scared...I'm scared...I've never been in a situation like this, either. But we've got to go. It's our job. The difference between being a coward and a man isn't whether you're scared; it's what you do while you're scared." That frightened soldier probably wasn't any less afraid, but he climbed back on the vehicle and went out to rescue the other American soldiers.

3-32. Will and a winning spirit apply in more situations than those requiring physical courage; sometimes you'll have to carry on for long periods in very difficult situations. The difficulties soldiers face may not be ones of physical danger, but of great physical, emotional, and mental strain. Physical courage allowed the soldier in the situation described above to return to the fight; will allowed his leader to say the right thing, to influence his frightened subordinate to do the right thing. Physical courage causes soldiers to charge a machine gun; will empowers them to fight on when they're hopelessly outnumbered, under appalling conditions, and without basic necessities.

STRESS IN TRAINING

When the bullets started flying...I never thought about half the things I was doing. I simply relied on my training and concentrated on the mission.

Captain Marie Bezubic
Operation Just Cause, Panama

3-33. Leaders must inject stress into training to prepare soldiers for stress in combat. However, creating a problem for subordinates and having them react to it doesn't induce the kind of stress required for combat training. A meaningful and productive mission, given with detailed constraints and limitations plus high standards of performance, does produce stress. Still, leaders must add unanticipated conditions to that stress to create a real learning environment. Sometimes, you don't even have to add stress; it just happens, as in this example.

Mix-up at the Crossroads

A young transportation section chief was leading a convoy of trucks on a night move to link up with several rifle companies. He was to transport the infantry to a new assembly area. When a sudden rainstorm dropped visibility to near zero, the section chief was especially glad that he had carefully briefed his drivers, issued strip maps, and made contingency plans. At a road intersection, his northbound convoy passed through an artillery battery moving east. When his convoy reached the rendezvous and the section chief got out to check his vehicles, he found he was missing two of his own trucks but had picked up three others towing howitzers. The tired and wet infantry commander was concerned that his unit would be late crossing the line of departure and forcefully expressed that concern to the section chief. The section chief now had to accomplish the same mission with fewer resources as well as run down his lost trucks and soldiers. There was certainly enough stress to go around.

After the section chief sent one of his most reliable soldiers with the artillery vehicles to find his missing trucks, he started shuttling the infantrymen to their destination. Later, after the mission was accomplished, the section chief and his drivers talked about what had happened. The leader admitted that he needed to supervise a convoy more closely under difficult conditions, and his soldiers recognized the need to follow the part of the unit SOP concerning reduced visibility operations.

3-34. The section chief fixed the immediate problem by starting to shuttle the infantry soldiers in the available trucks. During the AAR with the drivers, the leader admitted a mistake and figured out how to prevent similar errors in the future. The section chief also let the team know that sometimes, in spite of the best plans, things go wrong. A well-trained organization doesn't buckle under stress but deals with any setbacks and continues the mission.

THE STRESS OF CHANGE

3-35. Since the end of the Cold War, the Army has gone through tremendous change—dramatic decreases in the number of soldiers and DA civilians in all components, changes in assignment policies, base closings, and a host of other shifts that put stress on soldiers, DA civilians, and families. In those same years, the number of deployments to support missions such as peace operations and nation assistance has increased. And these changes have occurred in a peacetime Army. At the same time, Army leaders have had to prepare their soldiers for the stresses of combat, the ultimate crucible.

3-36. The stresses of combat you read about earlier in this chapter are classic: they've been the same for centuries. However, there's an aspect of the human dimension that has assumed an increasing importance: the effect of technological advances on organizations and people. Military leaders have always had to deal with the effect of technological changes. What's different today is the rate at which technology, to include warfighting technology, is changing. Rapid advances in new technologies are forcing the Army to change many aspects of the way it operates and are creating new leadership challenges.

TECHNOLOGY AND LEADERSHIP

3-37. Technology's presence challenges all Army leaders. Technology is here to stay and you, as an Army leader, need to continually learn how to manage it and make it work for you. The challenges come from many directions. Among them—

- You need to learn the strengths and vulnerabilities of the different technologies that support your team and its mission.

- You need to think through how your organization will operate with organizations that are less or more technologically complex. This situation may take the form of heavy and light Army units working together, operating with elements of another service, or cooperating with elements of another nation's armed forces.

- You need to consider the effect of technology on the time you have to analyze problems, make a decision, and act. Events happen faster today, and the stress you encounter as an Army leader is correspondingly greater.

Technological advances have the potential to permit better and more sustainable operations. However, as an Army leader you must remember the limitations of your people. No matter what technology you have or how it affects your mission, it's still your soldiers and DA civilians—their minds, hearts, courage, and talents—that will win the day.

3-38. Advances in electronic data processing let you handle large amounts of information easily. Today's desktop computer can do more, and do it faster, than the room-sized computers of only 20 years ago. Technology is a powerful tool—if you understand its potential uses and limitations. The challenge for all Army leaders is to overcome confusion on a fast-moving battlefield characterized by too much information coming in too fast.

3-39. Army leaders and staffs have always needed to determine mission-critical information, prioritize incoming reports, and process them quickly. The volume of information that current technology makes available makes this skill even more important than in the past. Sometimes something low-tech can divert the flood of technological help into channels the leader and staff can manage. For example, a well-understood commander's intent and thought-through commander's critical information requirements (CCIR) can help free leaders from nonessential information while pushing decisions to lower levels. As an Army leader, you must work hard to overcome the attractiveness and potential pitfalls of centralized decision making that access to information will appear to make practical.

3-40. Technology is also changing the size of the battlefield and the speed of battle. Instant global communications are increasing the pace of military actions. Global positioning systems and night vision capabilities mean the Army can fight at night and during periods of limited visibility—conditions that used to slow things down. Continuous operations increase the mental and physical stress on soldiers and leaders. Nonlinear operations make it more difficult for commanders to determine critical points on the battlefield. Effective leaders develop techniques to identify and manage stress well before actual conflict occurs. They also find ways to overcome the soldier's increased sense of isolation that comes with the greater breadth and depth of the modern battlefield. (FM 100-34 discusses continuous operations. FM 22-51 discusses combat stress control.)

3-41. Modern technology has also increased the number and complexity of skills the Army requires. Army leaders must carefully manage low-density specialties. They need to ensure that critical positions are filled and that their people maintain perishable skills. Army leaders must bring together leadership, personnel management, and training management to ensure their organizations are assigned people with the right specialties and that the entire organization is trained and ready. On top of this, the speed and lethality of modern battle have made mental agility and initiative even more necessary for fighting and winning. As in the past, Army leaders must develop these attributes in their subordinates.

3-42. To some, technology suggests a bloodless battlefield that resembles a computer war game more than the battlefields of the past. That isn't true now and it won't be true in the immediate future. Technology is still directed at answering the same basic questions that Civil War leaders tried to answer when they sent out a line of skirmishers: Where am I? Where are my buddies? Where is the enemy? How do I defeat him? Armed with this information, the soldiers and DA civilians of the Army will continue to accomplish the mission with character, using their technological edge to do the job better, faster, and smarter.

3-43. Modern digital technology can contribute a great deal to the Army leader's understanding of the battlefield; good leaders stay abreast of advances that enhance their tactical abilities. Digital technology has a lot to offer, but don't be fooled. A video image of a place, an action, or an organization can never substitute for the leader's getting down on the ground with the soldiers to find out what's going on. Technology can provide a great deal of information, but it may not present a completely accurate picture. The only way leaders can see the urgency in the faces of their soldiers is to get out and see them. As with any new weapon, the Army leader must know how to use technology without being seduced by it. Technology may be invaluable; however, effective leaders understand its limits.

3-44. Whatever their feeling regarding technology, today's leaders must contend more and more with an increased information flow and operational tempo (OPTEMPO). Pressures to make a decision increase, even as the time to verify and validate information decreases. Regardless of the crunch, Army leaders are responsible for the consequences of their decisions, so they gather, process, analyze, evaluate—and check —information. If they don't, the costs can be disastrous. (FM 100-34 discusses information management and decision making.)

"Superior Technology"

In the late fall of 1950, as United Nations (UN) forces pushed the North Korean People's Army northward, the People's Republic of China prepared to enter the conflict in support of its ally. The UN had air superiority, a marked advantage that had contributed significantly to the UN tactical and operational successes of the summer and early fall. Nonetheless, daily reconnaissance missions over the rugged North Korean interior failed to detect the Chinese People's Liberation Army's movement of nearly a quarter of a million ground troops across the border and into position in the North Korean mountains.

When the first reports of Chinese soldiers in North Korea arrived at Far East Command in Tokyo, intelligence analysts ignored them because they contradicted the information provided by the latest technology—aerial surveillance. Tactical commanders failed to send ground patrols into the mountains. They assumed the photos gave an accurate picture of the enemy situation when, in fact, the Chinese were practicing strict camouflage discipline. When the Chinese attacked in late November, UN forces were surprised, suffered heavy losses, and were driven from the Chinese border back to the 38th parallel.

When GEN Matthew B. Ridgway took over the UN forces in Korea in December, he immediately visited the headquarters of every regiment and many of the battalions on the front line. This gave GEN Ridgway an unfiltered look at the situation, and it sent a message to all his commanders: get out on the ground and find out what's going on.

3-45. The Chinese counterattack undid the results of the previous summer's campaign and denied UN forces the opportunity for a decisive victory that may have ended the war. The UN forces, under US leadership, enjoyed significant technological advantages over the Chinese. However, failure to verify the information provided by aerial photography set this advantage to zero. And this failure was one of leadership, not technology. Questioning good news provided by the latest "gee-whiz" system and ordering reconnaissance patrols to go out in lousy weather both require judgment and moral courage: judgment as to when a doubt is reasonable and courage to order soldiers to risk their lives in cold, miserable weather. But Army leaders must make those judgments and give those orders. Technology has not changed that.

3-46. Technology and making the most of it will become increasingly important. Today's Army leaders require systems understanding and more technical and tactical skills. Technical skill: What does this system do? What does it not do? What are its strengths? What are its weaknesses? What must I check? Tactical skill:

How do this system's capabilities support my organization? How should I employ it to support this mission? What must I do if it fails? There's a fine line between a healthy questioning of new systems' capabilities and an unreasoning hostility that rejects the advantages technology offers. You, as an Army leader, must stay on the right side of that line, the side that allows you to maximize the advantages of technology. You need to remain aware of its capabilities and shortcomings, and you need to make sure your people do as well.

LEADERSHIP AND THE CHANGING THREAT

3-47. Another factor that will have a major impact on Army leadership in the near future is the changing nature of the threat. For the Army, the twenty-first century began in 1989 with the fall of the Berlin Wall and subsequent collapse of the Soviet Union. America no longer defines its security interests in terms of a single, major threat. Instead, it faces numerous, smaller threats and situations, any of which can quickly mushroom into a major security challenge.

3-48. The end of the Cold War has increased the frequency and variety of Army missions. Since 1989, the Army has fought a large-scale land war and been continually involved in many different kinds of stability operations and support operations. There has been a greater demand for special, joint, and multinational operations as well. Initiative at all levels is becoming more and more important. In many instances, Army leaders on the ground have had to invent ways of doing business for situations they could not have anticipated.

3-49. Not only that, the importance of direct leaders—NCOs and junior officers—making the right decisions in stressful situations has increased. Actions by direct-level leaders—sergeants, warrant officers, lieutenants, and captains—can have organizational- and strategic-level implications. Earlier in this chapter, you read about the disciplined soldiers and leaders who accomplished their mission of securing a television tower in Udrigovo, Bosnia-Herzegovina. In that case, the local population's perception of how American soldiers secured the tower was just as important as securing the tower itself. Had the American detachment created an international incident by using what could have been interpreted as excessive force, maintaining order throughout Bosnia Herzegovina would have become more difficult. The Army's organizational and strategic leaders count on direct leaders. It has always been important to accomplish the mission the right way the first time; today it's more important than ever.

3-50. The Army has handled change in the past. It will continue to do so in the future as long as Army leaders emphasize the constants—Army values, teamwork, and discipline—and help their people anticipate change by seeking always to improve. Army leaders explain, to the extent of their knowledge and in clear terms, what may happen and how the organization can effectively react if it does. Change is inevitable; trying to avoid it is futile. The disciplined, cohesive organization rides out the tough times and will emerge even better than it started. Leadership, in a very real sense, includes managing change and making it work for you. To do that, you must know what to change and what not to change.

3-51. FM 100-5 provides a doctrinal framework for coping with these challenges while executing operations. It gives Army leaders clues as to what they will face and what will be required of them, but as COL Chamberlain found on Little Round Top, no manual can cover all possibilities. The essence of leadership remains the same: Army leaders create a vision of what's necessary, communicate it in a way that makes their intent clear, and vigorously execute it to achieve success.

CLIMATE AND CULTURE

3-52. Climate and culture describe the environment in which you lead your people. Culture refers to the environment of the Army as an institution and of major elements or communities within it. Strategic leaders maintain the Army's institutional culture. (Chapter 7 discusses their role.) Climate refers to the environment of units and organizations. All organizational and direct leaders establish their organization's climate, whether purposefully or unwittingly. (Chapters 5 and 6 discuss their responsibilities.)

CLIMATE

3-53. Taking care of people and maximizing their performance also depends on the climate a leader creates in the organization. An organization's climate is the way its members feel about their organization. Climate comes from people's shared perceptions and attitudes, what they believe about the day-to-day functioning of their outfit. These things have a great impact on their motivation and the trust they feel for their team and their leaders. Climate is generally short-term: it depends on a network of the personalities in a small organization. As people come and go, the climate changes. When a soldier says "My last platoon sergeant was pretty good, but this new one is great," the soldier is talking about one of the many elements that affect organizational climate.

3-54. Although such a call seems subjective, some very definite things determine climate. The members' collective sense of the organization—its organizational climat —is directly attributable to the leader's values, skills, and actions. As an Army leader, you establish the climate of your organization, no matter how small it is or how large. Answering the following questions can help you describe an organization's climate:

- Does the leader set clear priorities and goals?

- Is there a system of recognition, rewards and punishments? Does it work?

- Do the leaders know what they're doing? Do they admit when they're wrong?

- Do leaders seek input from subordinates? Do they act on the feedback they're provided?

- In the absence of orders, do junior leaders have authority to make decisions that are consistent with the leader's intent?

- Are there high levels of internal stress and negative competition in the organization? If so, what's the leader doing to change that situation?

- Do the leaders behave the way they talk? Is that behavior consistent with Army values? Are they good role models?

- Do the leaders lead from the front, sharing hardship when things get tough?

- Do leaders talk to their organizations on a regular basis? Do they keep their people informed?

3-55. Army leaders who do the right things for the right reasons—even when it would be easier to do the wrong thing—create a healthy organizational climate. In fact, it's the leader's behavior that has the greatest effect on the organizational climate. That behavior signals to every member of the organization what the leader will and will not tolerate. Consider this example.

Changing a Unit Climate—The New Squad Leader

SSG Withers was having a tough week. He had just been promoted to squad leader in a different company; he had new responsibilities, new leaders, and new soldiers. Then, on his second day, his unit was alerted for a big inspection in two days. A quick check of the records let him know that the squad leader before him had let maintenance slip; the records were sloppy and a lot of the scheduled work had not been done. On top of that, SSG Withers was sure his new platoon sergeant didn't like him. SFC King was professional but gruff, a person of few words. The soldiers in SSG Withers' squad seemed a little afraid of the platoon sergeant.

After receiving the company commander's guidance about the inspection, the squad leaders briefed the platoon sergeant on their plans to get ready. SSG Withers had already determined that he and his soldiers would have to work late. He could have complained about his predecessor, but he thought it would be best just to stick to the facts and talk about what he had found in the squad. For all he knew, the old squad leader might have been a favorite of SFC King.

SFC King scowled as he asked, "You're going to work late?"

SSG Withers had checked his plan twice: "Yes, sergeant. I think it's necessary."

SFC King grunted, but the sound could have meant "okay" or it could have meant "You're being foolish." SSG Withers wasn't sure.

The next day SSG Withers told his soldiers what they would have to accomplish. One of the soldiers said that the old squad leader would have just fudged the paperwork. "No kidding," SSG Withers thought. He wondered if SFC King knew about it. Of course, there was a good chance he would fail the inspection if he didn't fudge the paperwork—and wouldn't *that* be a good introduction to the new company? But he told his squad that they would do it right: "We'll do the best we can. If we don't pass, we'll do better next time."

SSG Withers then asked his squad for their thoughts on how to get ready. He listened to their ideas and offered some of his own. One soldier suggested that they could beat the other squads by sneaking into the motor pool at night and lowering the oil levels in their vehicles. "SFC King gives a half day off to whatever squad does best," the soldier explained. SSG Withers didn't want to badmouth the previous squad leader; on the other hand, the squad was his responsibility now. "It'd be nice to win," SSG Withers said, "but we're not going to cheat."

The squad worked past 2200 hours the night before the inspection. At one point SSG Withers found one of the soldiers sleeping under a vehicle. "Don't you want to finish and go home to sleep?" he asked the soldier.

"I...uh...I didn't think you'd still be here," the soldier answered.

"Where else would I be?" replied the squad leader.

The next day, SFC King asked SSG Withers if he thought his squad's vehicle was going to pass the inspection.

"Not a chance," SSG Withers said.

SFC King gave another mysterious grunt.

Later, when the inspector was going over his vehicle, SSG Withers asked if his soldiers could follow along. "I want them to see how to do a thorough inspection," he told the inspector. As the soldiers followed the inspector around and learned how to look closely at the vehicle, one of them commented that the squad had never been around for any inspection up to that point. "We were always told to stay away," he said.

Later, when the company commander went over the results of the inspection, he looked up at SSG Withers as he read the failing grade. SSG Withers was about to say, "We'll try harder next time, sir," but he decided that sounded lame, so he said nothing. Then SFC King spoke up.

"First time that squad has ever failed an inspection," the platoon sergeant said, "but they're already better off than they were the day before yesterday, failing grade and all."

3-56. SFC King saw immediately that things had changed for the better in SSG Withers' squad. The failing grade was real; previous passing grades had not been. The new squad leader told the truth and expected his soldiers to do the same. He was there when his people were working late. He acted to improve the squad's ethical and performance standards (by clearly stating and enforcing them). He moved to teach his soldiers the skills and standards associated with vehicle maintenance (by asking the inspector to show them how to look at a vehicle). And not once did SSG Withers whine that the failing grade was not his fault; instead, he focused on how to make things better. SSG Withers knew how to motivate soldiers to perform to standard and had the strength of character to do the right thing. In addition, he trusted the chain of command to take the long-term view. Because of his decisive actions, based on his character and competence, SSG Withers was well on his way to creating a much healthier climate in his squad.

3-57. No matter how they complain about it, soldiers and DA civilians expect to be held to standard; in the long run they feel better about themselves when they do hard work successfully. They gain confidence in leaders who help them achieve standards and lose confidence in leaders who don't know the standards or who fail to demand performance.

CULTURE

When you're first sergeant, you're a role model whether you know it or not. You're a role model for the guy that will be in your job. Not next month or next year, but ten years from now. Every day soldiers are watching you and deciding if you are the kind of first sergeant they want to be.

An Army First Sergeant
1988

3-58. Culture is a longer lasting, more complex set of shared expectations than climate. While climate is how people feel about their organization right now, culture consists of the shared attitudes, values, goals, and practices that characterize the larger institution. It's deeply rooted in long-held beliefs, customs, and practices. For instance, the culture of the armed forces is different from that of the business world, and the culture of the Army is different from that of the Navy. Leaders must establish a climate consistent with the culture of the larger institution. They also use the culture to let their people know they're part of something bigger than just themselves, that they have responsibilities not only to the people around them but also to those who have gone before and those who will come after.

3-59. Soldiers draw strength from knowing they're part of a tradition. Most meaningful traditions have their roots in the institution's culture. Many of the Army's everyday customs and traditions are there to remind you that you're just the latest addition to a long line of American soldiers. Think of how much of your daily life connects you to the past and to American soldiers not yet born: the uniforms you wear, the martial music that punctuates your day, the way you salute, your title, your organization's history, and Army values such as selfless service. Reminders of your place in history surround you.

3-60. This sense of belonging is vitally important. Visit the Vietnam Memorial in Washington, DC, some Memorial Day weekend and you'll see dozens of veterans, many of them wearing bush hats or campaign ribbons or fatigue jackets decorated with unit patches. They're paying tribute to their comrades in this division or that company. They're also acknowledging what for many of them was the most intense experience of their lives.

3-61. Young soldiers want to belong to something bigger than themselves. Look at them off duty, wearing tee shirts with names of sports teams and famous athletes. It's not as if an 18-year-old who puts on a jacket with a professional sports team's logo thinks anyone will mistake him for a professional player; rather, that soldier wants to be associated with a winner. Advertising and mass media make heroes of rock stars, athletes, and actors. Unfortunately, it's easier to let some magazine or TV show tell you whom to admire than it is to dig up an organization's history and learn about heroes.

3-62. Soldiers want to have heroes. If they don't know about SGT Alvin York in World War I, about COL Joshua Chamberlain's 20th Maine during the Civil War, about MSG Gary Gordon and SFC Randall Shughart in the 1993 Somalia fight, then it's up to you, their leaders, to teach them. (The bibliography lists works you can use to learn more about your profession, its history, and the people who made it.)

3-63. When soldiers join the Army, they become part of a history: the Big Red One, the King of Battle, Sua Sponte. Teach them the history behind unit crests, behind greetings, behind decorations and badges. The Army's culture isn't something that exists apart from you; it's part of who you are, something you can use to give your soldiers pride in themselves and in what they're doing with their lives.

LEADERSHIP STYLES

3-64. You read in Chapter 2 that all people are shaped by what they've seen, what they've learned, and whom they've met. Who you are determines the way you work with other people. Some people are happy and smiling all the time; others are serious. Some leaders can wade into a room full of strangers and inside of five minutes have everyone there thinking, "How have I lived so long without meeting this person?" Other very competent leaders are uncomfortable in social situations. Most of us are somewhere in between. Although Army leadership doctrine describes at great length how you should interact with your subordinates and how you must strive to learn and improve your leadership skills, the Army recognizes that you must always be yourself; anything else comes across as fake and insincere.

3-65. Having said that, effective leaders are flexible enough to adjust their leadership style and techniques to the people they lead. Some subordinates respond best to coaxing, suggestions, or gentle prodding; others need, and even want at times, the verbal equivalent of a kick in the pants. Treating people fairly doesn't mean treating people as if they were clones of one another. In fact, if you treat everyone the same way, you're probably being unfair, because different people need different things from you.

3-66. Think of it this way: Say you must teach map reading to a large group of soldiers ranging in rank from private to senior NCO. The senior NCOs know a great deal about the subject, while the privates know very little. To meet all their needs, you must teach the privates more than you teach the senior NCOs. If you train the privates only in the advanced skills the NCOs need, the privates will be lost. If you make the NCOs sit through training in the basic tasks the privates need, you'll waste the NCOs' time. You must fit the training to the experience of those being trained. In the same way, you must adjust your leadership style and techniques to the experience of your people and characteristics of your organization.

3-67. Obviously, you don't lead senior NCOs the same way you lead privates. But the easiest distinctions to make are those of rank and experience. You must also take into account personalities, self-confidence, self-esteem—all the elements of the complex mix of character traits that makes dealing with people so difficult and so rewarding. One of the many things that makes your job tough is that, in order to get their best performance, you must figure out what your subordinates need and what they're able to do—even when they don't know themselves.

3-68. When discussing leadership styles, many people focus on the extremes: autocratic and democratic. Autocratic leaders tell people what to do with no explanation; their message is, "I'm the boss; you'll do it because I said so." Democratic leaders use their personalities to persuade subordinates. There are many shades in between; the following paragraphs discuss five of them. However, bear in mind that competent leaders mix elements of all these styles to match to the place, task, and people involved. Using different leadership styles in different situations or elements of different styles in the same situation isn't inconsistent. The opposite is true: if you can use only one leadership style, you're inflexible and will have difficulty operating in situations where that style doesn't fit.

DIRECTING LEADERSHIP STYLE

3-69. The directing style is leader-centered. Leaders using this style don't solicit input from subordinates and give detailed instructions on how, when, and where they want a task performed. They then supervise its execution very closely.

3-70. The directing style may be appropriate when time is short and leaders don't have a chance to explain things. They may simply give orders: Do this. Go there. Move. In fast-paced operations or in combat, leaders may revert to the directing style, even with experienced subordinates. This is what the motor sergeant you read about in Chapter 1 did. If the leader has created a climate of trust, subordinates will assume the leader has switched to the directing style because of the circumstances.

3-71. The directing style is also appropriate when leading inexperienced teams or individuals who are not yet trained to operate on their own. In this kind of situation, the leader will probably remain close to the action to make sure things go smoothly.

3-72. Some people mistakenly believe the directing style means using abusive or demeaning language or includes threats and intimidation. This is wrong. If you're ever tempted to be abusive, whether because of pressure or stress or what seems like improper behavior by a subordinate, ask yourself these questions: Would I want to work for someone like me? Would I want my boss to see and hear me treat subordinates this way? Would I want to be treated this way?

PARTICIPATING LEADERSHIP STYLE

3-73. The participating style centers on both the leader and the team. Given a mission, leaders ask subordinates for input, information, and recommendations but make the final decision on what to do themselves. This style is especially appropriate for leaders who have time for such consultations or who are dealing with experienced subordinates.

3-74. The team-building approach lies behind the participating leadership style. When subordinates help create a plan, it becomes—at least in part—their plan. This ownership creates a strong incentive to invest the effort necessary to make the plan work. Asking for this kind of input is a sign of a leader's strength and self-confidence. But asking for advice doesn't mean the leader is obligated to follow it; the leader alone is always responsible for the quality of decisions and plans.

DELEGATING LEADERSHIP STYLE

3-75. The delegating style involves giving subordinates the authority to solve problems and make decisions without clearing them through the leader. Leaders with mature and experienced subordinates or who want to create a learning experience for subordinates often need only to give them authority to make decisions, the necessary resources, and a clear understanding of the mission's purpose. As always, the leader is ultimately responsible for what does or does not happen, but in the delegating leadership style, the leader holds subordinate leaders accountable for their actions. This is the style most often used by officers dealing with senior NCOs and by organizational and strategic leaders.

TRANSFORMATIONAL AND TRANSACTIONAL LEADERSHIP STYLES

A man does not have himself killed for a few halfpence a day or for a petty distinction. You must speak to the soul in order to electrify the man.

Napoleon Bonaparte

3-76. These words of a distinguished military leader capture the distinction between the transformational leadership style, which focuses on inspiration and change, and the transactional leadership style, which focuses on rewards and punishments. Of course Napoleon understood the importance of rewards and punishments. Nonetheless, he also understood that carrots and sticks alone don't inspire individuals to excellence.

Transformational Leadership Style

3-77. As the name suggests, the transformational style "transforms" subordinates by challenging them to rise above their immediate needs and self-interests. The transformational style is developmental: it emphasizes individual growth (both professional and personal) and organizational enhancement. Key features of the transformational style include empowering and mentally stimulating subordinates: you consider and motivate them first as individuals and then as a group. To use the transformational style, you must have the courage to communicate your intent and then step back and let your subordinates work. You must also be aware that immediate benefits are often delayed until the mission is accomplished.

3-78. The transformational style allows you to take advantage of the skills and knowledge of experienced subordinates who may have better ideas on how to accomplish a mission. Leaders who use this style communicate reasons for their decisions or actions and, in the process, build in subordinates a broader understanding and ability to exercise initiative and operate effectively. However, not all situations lend themselves to the transformational leadership style. The transformational style is most effective during periods that call for change or present new opportunities. It also works well when organizations face a crisis, instability, mediocrity, or disenchantment. It may not be effective when subordinates are inexperienced, when the mission allows little deviation from accepted procedures, or when subordinates are not motivated. Leaders who use only the transformational leadership style limit their ability to influence individuals in these and similar situations.

Transactional Leadership Style

3-79. In contrast, some leaders employ only the transactional leadership style. This style includes such techniques as—

- Motivating subordinates to work by offering rewards or threatening punishment.

- Prescribing task assignments in writing.

- Outlining all the conditions of task completion, the applicable rules and regulations, the benefits of success, and the consequences—to include possible disciplinary actions—of failure.

- "Management-by-exception," where leaders focus on their subordinates' failures, showing up only when something goes wrong.

The leader who relies exclusively on the transactional style, rather than combining it with the transformational style, evokes only short-term commitment from his subordinates and discourages risk-taking and innovation.

3-80. There are situations where the transactional style is acceptable, if not preferred. For example, a leader who wants to emphasize safety could reward the organization with a three-day pass if the organization prevents any serious safety-related incidents over a two-month deployment. In this case, the leader's intent appears clear: unsafe acts are not tolerated and safe habits are rewarded.

3-81. However, using only the transactional style can make the leader's efforts appear self-serving. In this example, soldiers might interpret the leader's attempt to reward safe practices as an effort to look good by focusing on something that's unimportant but that has the boss's attention. Such perceptions can destroy the trust subordinates have in the leader. Using the transactional style alone can also deprive subordinates of opportunities to grow, because it leaves no room for honest mistakes.

3-82. The most effective leaders combine techniques from the transformational and transactional leadership styles to fit the situation. A strong base of transactional understanding supplemented by charisma, inspiration and individualized concern for each subordinate, produces the most enthusiastic and genuine response. Subordinates will be more committed, creative, and innovative. They will also be more likely to take calculated risks to accomplish their mission. Again referring to the safety example, leaders can avoid any misunderstanding of their intent by combining transformational techniques with transactional techniques. They can explain why safety is important (intellectual stimulation) and encourage their subordinates to take care of each other (individualized concern).

INTENDED AND UNINTENDED CONSEQUENCES

3-83. The actions you take as a leader will most likely have unintended as well as intended consequences. Like a chess player trying to anticipate an opponent's moves three or four turns in advance—if I do this, what will my opponent do; then what will I do next?—leaders think through what they can expect to happen as a result of a decision. Some decisions set off a chain of events; as far as possible, leaders must anticipate the second- and third-order effects of their actions. Even lower-level leaders' actions may have effects well beyond what they expect.

3-84. Consider the case of a sergeant whose team is manning a roadblock as part of a peace operation. The mission has received lots of media attention (Haiti and Bosnia come to mind), and millions of people back home are watching. Early one morning, a truckload of civilians appears, racing toward the roadblock. In the half-light, the sergeant can't tell if the things in the passengers' hands are weapons or farm tools, and the driver seems intent on smashing through the barricade. In the space of a few seconds, the sergeant must decide whether or not to order his team to fire on the truck.

3-85. If the sergeant orders his team to fire because he feels he and his soldiers are threatened, that decision will have international consequences. If he kills any civilians, chances are good that his chain of command from the president on down—not to mention the entire television audience of the developed world—will know about the incident in a few short hours. But the decision is tough for another reason: if the sergeant doesn't order his team to fire and the civilians turn out to be an armed gang, the team may take casualties that could have been avoided. If the only factor involved was avoiding civilian casualties, the choice is simple: don't shoot. But the sergeant must also consider the requirement to protect his force and accomplish the mission of preventing unauthorized traffic from passing the roadblock. So the sergeant must act; he's the leader, and he's in charge. Leaders who have thought through the consequences of possible actions, talked with their own leaders about the commander's intent and mission priorities, and trust their chain of command to support them are less likely to be paralyzed by this kind of pressure.

INTENDED CONSEQUENCES

3-86. Intended consequences are the anticipated results of a leader's decisions and actions. When a squad leader shows a team leader a better way to lead PT, that action will have intended consequences: the team leader will be better equipped to do the job. When leaders streamline procedures, help people work smarter, and get the resources to the right place at the right time, the intended consequences are good.

UNINTENDED CONSEQUENCES

3-87. Unintended consequences are the results of things a leader does that have an unplanned impact on the organization or accomplishment of the mission. Unintended consequences are often more lasting and harder to anticipate than intended consequences. Organizational and strategic leaders spend a good deal of energy considering possible unintended consequences of their actions. Their organizations are complex, so figuring out the effects today's decisions will have a few years in the future is difficult.

3-88. Unintended consequences are best described with an example, such as setting the morning PT formation time: Setting the formation time at 0600 hours results in soldiers standing in formation at 0600 hours, an intended consequence. To not be late, soldiers living off post may have to depart their homes at 0500 hours, a consequence that's probably also anticipated. However, since most junior enlisted soldiers with families probably own only one car, there will most likely be another consequence: entire families rising at 0430 hours. Spouses must drive their soldiers to post and children, who can't be left at home unattended, must accompany them. This is an unintended consequence.

SUMMARY

3-89. The human dimension of leadership, how the environment affects you and your people, affects how you lead. Stress is a major part of the environment, both in peace and war. Major sources of stress include the rapid pace of change and the increasing complexity of technology. As an Army leader, you must stay on top of both. Your character and skills—how you handle stress—and the morale and discipline you develop and your team are more important in establishing the climate in your organization than any external circumstances.

3-90. The organizational climate and the institutional culture define the environment in which you and your people work. Direct, organizational, and strategic leaders all have different responsibilities regarding climate and culture; what's important now is to realize that you, the leader, establish the climate of your organization. By action or inaction, you determine the environment in which your people work.

3-91. Leadership styles are different ways of approaching the DO of BE, KNOW, DO—the actual work of leading people. You've read about five leadership styles: directing, participating, delegating, transformational, and transactional. But remember that you must be able to adjust the leadership style you use to the situation and the people you're leading. Remember also that you're not limited to any one style in a given situation: you should use techniques from different styles if that will help you motivate your people to accomplish the mission. Your leader attributes of judgment, intelligence, cultural awareness, and self-control all play major roles in helping you choose the proper style and the appropriate techniques for the task at hand. That said, you must always be yourself.

3-92. All leader actions result in intended and unintended consequences. Two points to remember: think through your decisions and do your duty. It might not seem that the actions of one leader of one small unit matter in the big picture. But they do. In the words of Confederate COL William C. Oats, who faced COL Joshua Chamberlain at Little Round Top: "Great events sometimes turn on comparatively small affairs."

3-93. In spite of stress and changes, whether social or technological, leadership always involves shaping human emotions and behaviors. As they serve in more complex environments with wider-ranging consequences, Army leaders refine what they've known and done as well as develop new styles, skills, and actions. Parts Two and Three discuss the skills and actions required of leaders from team to Department of the Army level.

PART TWO

Direct Leadership

The first three chapters of this manual cover the constants of leadership. They focus primarily on what a leader must BE. Part Two examines what a direct leader must KNOW and DO. Note the distinction between a skill, *knowing* something, and an action, *doing* something. The reason for this distinction bears repeating: knowledge isn't enough. You can't be a leader until you apply what you know, until you act and DO what you must.

Army leaders are grounded in the heritage, values, and tradition of the Army. They embody the warrior ethos, value continuous learning, and demonstrate the ability to lead and train their subordinates. Army leaders lead by example, train from experience, and maintain and enforce standards. They do these things while taking care of their people and adapting to a changing world. Chapters 4 and 5 discuss these subjects in detail.

The *warrior ethos* is the will to win with honor. Despite a thinking enemy, despite adverse conditions, you accomplish your mission. You express your character— the BE of BE, KNOW, DO—when you and your people confront a difficult mission and persevere. The warrior ethos applies to all soldiers and DA civilians, not just those who close with and destroy the enemy. It's the will to meet mission demands no matter what, the drive to get the job done whatever the cost.

Continuous learning requires dedication to improving your technical and tactical skills through study and practice. It also includes learning about the world around you—mastering new technology, studying other cultures, staying aware of current events at home and abroad. All these things affect your job as a leader.

Continuous learning also means consciously developing your *character* through study and reflection. It means reflecting on Army values and developing leader attributes. Broad knowledge and strong character underlie the right decisions in hard times. Seek to learn as much as you can about your job, your people, and yourself. That way you'll be prepared when the time comes for tough decisions. You'll BE a leader of character, KNOW the necessary skills, and DO the right thing.

Army leaders train and lead people. Part of this responsibility is maintaining and enforcing standards. Your subordinates expect you to show them what the standard is and train them to it: they expect you to lead by example. In addition, as an Army leader you're required to take care of your people. You may have to call on them to do things that seem impossible. You may have to ask them to make extraordinary sacrifices to accomplish the mission. If you train your people to standard, inspire the warrior ethos in them, and consistently look after their interests, they'll be prepared to accomplish the mission—anytime, anywhere.

Chapter 4

Direct Leadership Skills

Never get so caught up in cutting wood that you forget to sharpen your ax.

First Sergeant James J. Karolchyk, 1986

4-1. The Army's direct leaders perform a huge array of functions in all kinds of places and under all kinds of conditions. Even as you read these pages, someone is in the field in a cold place, someone else in a hot place. There are people headed to a training exercise and others headed home. Somewhere a motor pool is buzzing, a medical ward operating, supplies moving. Somewhere a duty NCO is conducting inspections and a sergeant of the guard is making the rounds. In all these places, no matter what the conditions or the mission, direct leaders are guided by the same principles, using the same skills, and performing the same actions.

4-2. This chapter discusses the skills a direct leader must master and develop. It addresses the KNOW of BE, KNOW, and DO for direct leaders. The skills are organized under the four skill groups Chapter 1 introduced: interpersonal, conceptual, technical, and tactical. (Appendix B lists performance indicators for leader skills.)

INTERPERSONAL SKILLS

4-3. A DA civilian supervisor was in a frenzy because all the material needed for a project wasn't available. The branch chief took the supervisor aside and said, "You're worrying about *things*. Things are not important; things will or won't be there. Worry about working with the people who will get the job done."

4-4. Since leadership is about people, it's not surprising to find interpersonal skills, what some call "people skills," at the top of the list of what an Army leader must KNOW. Figure 4-1 (on page 4-3) identifies the direct leader interpersonal skills. All these skills—communicating, team building, supervising, and counseling—require communication. They're all closely related; you can hardly use one without using the others.

COMMUNICATING

4-5. Since leadership is about getting other people to do what you want them to do, it follows that communicating—transmitting information so that it's clearly understood—is an important skill. After all, if people can't understand you, how will you ever let them know what you want? The other interpersonal skills—supervising, team building, and counseling—also depend on your ability to communicate.

4-6. If you take a moment to think about all the training you've received under the heading "communication," you'll see that it probably falls into four broad categories: speaking, reading, writing, and listening. You begin practicing speech early; many children are using words by the age of one. The heavy emphasis on reading and writing begins in school, if not before. Yet how many times have you been taught how to listen? Of the four forms of communication,

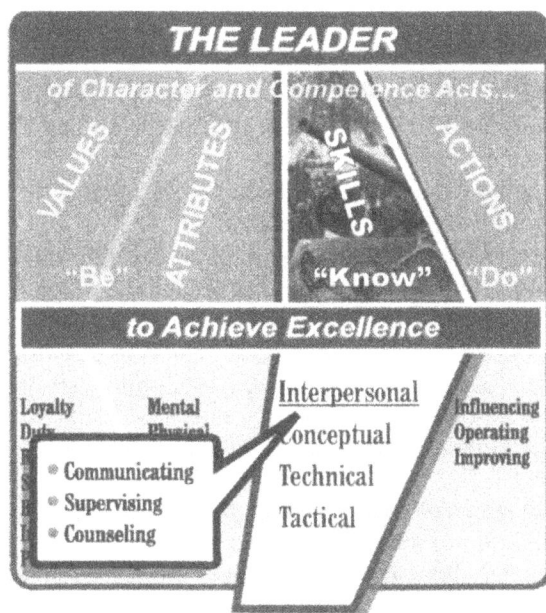

**Figure 4-1. Direct Leader
Skills—Interpersonal**

listening is the one in which most people receive the least amount of formal training. Yet for an Army leader, it's every bit as important as the others. It often comes first because you must listen and understand before you can decide what to say.

One-Way and Two-Way Communication

4-7. There are two common forms of one-way communication that are not necessarily the best way to exchange information: seeing and hearing. The key difference between one-way and two-way communication is that one-way communication—hearing or seeing something on television, reading a copy of a slide presentation, or even watching a training event unfold—may not give you a complete picture. You

may have unanswered questions or even walk away with the wrong concept of what has occurred. That's why two-way communication is preferred when time and resources permit.

Active Listening

4-8. An important form of two-way communication is active listening. When you practice active listening, you send signals to the speaker that say, "I'm paying attention." Nod your head every once in a while, as if to say, "Yes, I understand." When you agree with the speaker, you might use an occasional "uh-huh." Look the speaker in the eye. Give the speaker your full attention. Don't allow yourself to be distracted by looking out the window, checking your watch, playing with something on your desk, or trying to do more than one thing at a time. Avoid interrupting the speaker; that's the cardinal sin of active listening.

4-9. Be aware of barriers to listening. Don't form your response while the other person is still talking. Don't allow yourself to become distracted by the fact that you're angry, or that you have a problem with the speaker, or that you have lots of other things you need to be thinking about. If you give in to these temptations, you'll miss most of what's being said.

Nonverbal Communication

4-10. In face-to-face communication, even in the simplest conversation, there's a great deal going on that has almost nothing to do with the words being used. Nonverbal communication involves all the signals you send with your facial expressions, tone of voice, and body language. Effective leaders know that communication includes both verbal and nonverbal cues. Look for them in this example.

The Checking Account

A young soldier named PVT Bell, new to the unit, approaches his team leader, SGT Adams, and says, "I have a problem I'd like to talk to you about."

The team leader makes time—right then if possible—to listen. Stopping, looking the soldier in the eye, and asking, "What's up?" sends many signals: _I am concerned about your problem. You're part of the team, and we help each other. What can I do to help?_ All these signals, by the way, reinforce Army values.

The Checking Account (continued)

PVT Bell sees the leader is paying attention and continues, "Well, I have this checking account, see, and it's the first time I've had one. I have lots of checks left, but for some reason the PX [post exchange] is saying they're no good."

SGT Adams has seen this problem before: PVT Bell thinks that checks are like cash and has no idea that there must be money in the bank to cover checks written against the account. SGT Adams, no matter how tempted, doesn't say anything that would make PVT Bell think that his difficulty was anything other than the most important problem in the world. He is careful to make sure that PVT Bell doesn't think that he's anyone other than the most important soldier in the world. Instead, SGT Adams remembers life as a young soldier and how many things were new and strange. What may seem like an obvious problem to an experienced person isn't so obvious to an inexperienced one. Although the soldier's problem may seem funny, SGT Adams doesn't laugh at the subordinate. And because nonverbal cues are important, SGT Adams is careful that his tone of voice and facial expressions don't convey contempt or disregard for the subordinate.

Instead, the leader listens patiently as PVT Bell explains the problem; then SGT Adams reassures PVT Bell that it can be fixed and carefully explains the solution. What's more, SGT Adams follows up later to make sure the soldier has straightened things out with the bank.

A few months later, a newly promoted PFC Bell realizes that this problem must have looked pretty silly to someone with SGT Adams' experience. But PFC Bell will always remember the example SGT Adams set. Future leaders are groomed every day and reflect their past leaders. By the simple act of listening and communicating, SGT Adams won the loyalty of PFC Bell. And when the next batch of new soldiers arrives, PFC Bell, now the old-timer, will say to them, "Yeah, in all my experience, I've got to say this is one of the best units in the Army. And SGT Adams is the best team leader around. Why, I remember a time…"

4-11. SGT Adams performed crisis counseling, a leader action Appendix C discusses. Look for the communicating skills in this example. SGT Adams listened actively and controlled his nonverbal communication. He gave PVT Bell his full attention and was careful not to signal indifference or a lack of concern. SGT Adams' ability to do this shows the mental attribute of self-discipline and the emotional attribute of self-control, which you read about in Chapter 2. The leader also displayed empathy, that is, sensitivity to the feelings, thoughts, and experiences of another person. It's an important quality for a counselor.

SUPERVISING

If a squad leader doesn't check, and the guy on point has no batteries for his night vision goggles, he has just degraded the effectiveness of the entire unit.

A Company Commander, Desert Storm

4-12. Direct leaders check and recheck things. Leaders strike a balance between checking too much and not checking enough. Training subordinates to act independently is important; that's why direct leaders give instructions or their intent and then allow subordinates to work without constantly looking over their shoulders. Accomplishing the mission is equally important; that's why leaders check things—especially conditions critical to the mission (fuel levels), details a soldier might forget (spare batteries for night vision goggles), or tasks at the limit of what a soldier has accomplished before (preparing a new version of a report).

4-13. Checking minimizes the chance of oversights, mistakes, or other circumstances that might derail a mission. Checking also gives leaders a chance to see and recognize subordinates who are doing things right or make on-the-spot corrections when necessary. Consider this example: A platoon sergeant delegates to the platoon's squad leaders the authority to get their squads ready for a tactical road march. The platoon sergeant oversees the activity but doesn't intervene unless errors, sloppy work, or lapses occur. The leader is there to answer questions or resolve problems that the squad leaders can't handle. This supervision ensures that the squads are prepared to standard and demonstrates to the squad leaders that the platoon sergeant cares about them and their people.

The Rusty Rifles Incident

While serving in the Republic of Vietnam, SFC Jackson was transferred from platoon sergeant of one platoon to platoon leader of another platoon in the same company. SFC Jackson quickly sized up the existing standards in the platoon. He wasn't pleased. One problem was that his soldiers were not keeping their weapons cleaned properly: rifles were dirty and rusty. He put out the word: weapons would be cleaned to standard each day, each squad leader would inspect each day, and he would inspect a sample of the weapons each day. He gave this order three days before the platoon was to go to the division rest and recuperation (R&R) area on the South China Sea.

The next day SFC Jackson checked several weapons in each squad. Most weapons were still unacceptable. He called the squad leaders together and explained the policy and his reasons for implementing it. SFC Jackson checked again the following day and still found dirty and rusty weapons. He decided there were two causes for the problem. First, the squad leaders were not doing their jobs. Second, the squad leaders and troops were bucking him—testing him to see who would really make the rules in the platoon. He sensed that, because he was new, they resisted his leadership. He knew he had a serious discipline problem he had to handle correctly. He called the squad leaders together again. Once again, he explained his standards clearly. He then said, "Tomorrow we are due to go on R&R for three days and I'll be inspecting rifles. We won't go on R&R until each weapon in this platoon meets the standard."

The next morning SFC Jackson inspected and found that most weapons in each squad were still below standard. He called the squad leaders together. With a determined look and a firm voice, he told them he would hold a formal in-ranks inspection at 1300 hours, even though the platoon was scheduled to board helicopters for R&R then. If every weapon didn't meet the standard, he would conduct another in-ranks inspection for squad leaders and troops with substandard weapons. He would continue inspections until all weapons met the standard.

At 1300 hours the platoon formed up, surly and angry with the new platoon leader, who was taking their hard-earned R&R time. The soldiers could hardly believe it, but his message was starting to sink in. This leader meant what he said. This time all weapons met the standard.

COUNSELING

Nothing will ever replace one person looking another in the eyes and telling the soldier his strengths and weaknesses. [Counseling] charts a path to success and diverts soldiers from heading down the wrong road.

Sergeant Major Randolph S. Hollingsworth

4-14. Counseling is subordinate-centered communication that produces a plan outlining actions necessary for subordinates to achieve individual or organizational goals. Effective counseling takes time, patience, and practice. As with everything else you do, you must develop your skills as a counselor. Seek feedback on how effective you are at counseling, study various counseling techniques, and make efforts to improve. (Appendix C discusses developmental counseling techniques.)

4-15. Proper counseling leads to a specific plan of action that the subordinate can use as a road map for improvement. Both parties, counselor and counseled, prepare this plan of action. The leader makes certain the subordinate understands and takes ownership of it. The best plan of action in the world does no good if the subordinate doesn't understand it, follow it, and believe in it. And once the plan of action is agreed upon, the leader must follow up with one-on-one sessions to ensure the subordinate stays on track.

4-16. Remember the Army values of loyalty, duty, and selfless service require you to counsel your subordinates. The values of honor, integrity, and personal courage require you to give them straightforward feedback. And the Army value of respect requires you to find the best way to communicate that feedback so that your subordinates understand it. These Army values all point to the requirement for you to become a proficient counselor. Effective counseling helps your subordinates develop personally and professionally.

4-17. One of the most important duties of all direct, organizational, and strategic leaders is to develop subordinates. Mentoring, which links the operating and improving leader actions, plays a major part in developing competent and confident future leaders. Counseling is an interpersonal skill essential to effective mentoring. (Chapters 5, 6, and 7 discuss the direct, organizational, and strategic leader mentoring actions.)

CONCEPTUAL SKILLS

4-18. Conceptual skills include competence in handling ideas, thoughts, and concepts. Figure 4-2 (on page 4-7) lists the direct leader conceptual skills.

CRITICAL REASONING

4-19. Critical reasoning helps you think through problems. It's the key to understanding situations, finding causes, arriving at justifiable conclusions, making good judgments, and learning from the experience—in short, solving problems. Critical reasoning is an essential part of effective counseling and underlies ethical reasoning, another conceptual skill. It's also a central aspect of decision making, which Chapter 5 discusses.

4-20. The word "critical" here doesn't mean finding fault; it doesn't have a negative meaning at all. It means getting past the surface of the problem and thinking about it in depth. It means looking at a problem from several points of view instead of just being satisfied with the first answer that comes to mind. Army leaders need this ability because many of the choices they face are complex and offer no easy solution.

4-21. Sometime during your schooling you probably ran across a multiple choice test, one that required you to "choose answer a, b, c, or d" or "choose one response from column a and two from column b." Your job as an Army leader would be a lot easier if the problems you faced were presented that way, but leadership is a lot more complex than that. Sometimes just figuring out the real problem presents a huge hurdle; at other times you have to sort through distracting multiple problems to get to the real difficulty. On some occasions you know what the problem is but have no clue as to what an answer might be. On others you can come up with two or three answers that all look pretty good.

Finding the Real Problem

A platoon sergeant directs the platoon's squad leaders to counsel their soldiers every month and keep written records. Three months later, the leader finds the records are sloppy or incomplete; in many cases, there's no record at all. The platoon sergeant's first instinct is to chew out the squad leaders for ignoring his instructions. It even occurs to him to write a counseling annex to the platoon SOP so he can point to it the next time the squad leaders fail to follow instructions.

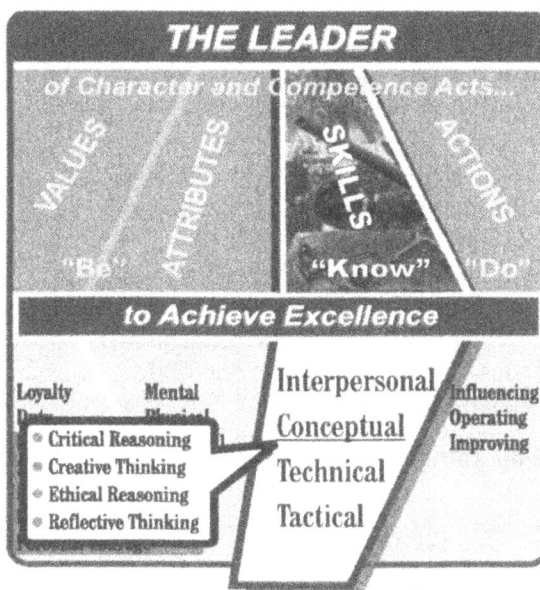

Figure 4-2. Direct Leader Skills—Conceptual

CREATIVE THINKING

4-22. Sometimes you run into a problem that you haven't seen before or an old problem that requires a new solution. Here you must apply imagination; a radical departure from the old way of doing things may be refreshing. Army leaders prevent complacency by finding ways to challenge subordinates with new approaches and ideas. In these cases, rely on your intuition, experience, and knowledge. Ask for input from your subordinates. Reinforce team building by making everybody responsible for, and shareholders in, the accomplishment of difficult tasks.

4-23. Creative thinking isn't some mysterious gift, nor does it have to be outlandish. It's not reserved for senior officers; all leaders think creatively. You employ it every day to solve small problems. A unit that deploys from a stateside post on a peace operation, for instance, may find itself in a small compound with limited athletic facilities and no room to run. Its leaders must devise new ways for their soldiers to maintain physical fitness. These may include sports and games, even games the local nationals play.

ETHICAL REASONING

4-24. Ethical leaders do the right things for the right reasons all the time, even when no one is watching. But figuring out what's the "right" thing is often, to put it mildly, a most difficult task. To fulfill your duty, maintain your integrity, and serve honorably, you must be able to reason ethically.

4-25. Occasionally, when there's little or no time, you'll have to make a snap decision based on your experience and intuition about what feels right. For Army leaders, such decisions are guided by Army values (discussed in Chapter 2), the institutional culture, and the organizational climate (discussed in Chapter 3). These shared values then serve as a basis for the whole team's buying into the leader's decision. But comfortable as this might be, you should not make all decisions on intuition.

4-26. When there's time to consider alternatives, ask for advice, and think things through, you can make a deliberate decision. First determine what's legally right by law and regulation. In gray areas requiring interpretation, apply Army values to the situation. Inside those boundaries, determine the best possible answer from among competing solutions, make your decision, and act on it.

4-27. The distinction between snap and deliberate decisions is important. In many decisions, you must think critically because your intuition—what feels right—may lead to the wrong answer. In combat especially, the intuitive response won't always work.

4-28. The moral application of force goes to the heart of military ethics. S. L. A. Marshall, a military historian as well as a brigadier general, has written that the typical soldier is often at a disadvantage in combat because he "comes from a civilization in which aggression, connected with the taking of a human life, is prohibited and unacceptable." Artist Jon Wolfe, an infantryman in Vietnam, once said that the first time he aimed his weapon at another human being, a "little voice" in the back of his mind asked, "Who gave you permission to do this?" That "little voice" comes, of course, from a lifetime of living within the law. You can determine the right thing to do in these very unusual circumstances only when you apply ethical as well as critical reasoning.

4-29. The right action in the situation you face may not be in regulations or field manuals. Even the most exhaustive regulations can't predict every situation. They're designed for the routine, not the exceptional. One of the most difficult tasks facing you as an Army leader is determining when a rule or regulation simply doesn't apply because the situation you're facing falls outside the set of conditions envisioned by those who wrote the regulation. Remember COL Chamberlain on Little Round Top. The drill manuals he had studied didn't contain the solution to the tactical problem he faced; neither this nor any other manual contain "cookbook" solutions to ethical questions you will confront. COL Chamberlain *applied* the doctrine he learned from the drill manuals. So you should apply Army values, your knowledge, and your experience to any decision you make and be prepared to accept the consequences of your actions. Study, reflection, and ethical reasoning can help you do this.

4-30. Ethical reasoning takes you through these steps:

- Define the problem.
- Know the relevant rules.
- Develop and evaluate courses of action.
- Choose the course of action that best represents Army values.

4-31. These steps correspond to some of the steps of the decision making leadership action in Chapter 5. Thus, ethical reasoning isn't a separate process you trot out only when you think you're facing an ethical question. It should be part of the thought process you use to make any decision. Your subordinates count on you to do more than make tactically sound decisions. They rely on you to make decisions that are ethically sound as well. You should always consider ethical factors and, when necessary, use Army values to gauge what's right.

4-32. That said, not every decision is an ethical problem. In fact, most decisions are ethically neutral. But that doesn't mean you don't have to think about the ethical consequences of your actions. Only if you reflect on whether what you're asked to do or what you ask your people to do accords with Army values will you develop that sense of right and wrong that marks ethical people and great leaders. That sense of right and wrong alerts you to the presence of ethical aspects when you face a decision.

4-33. Ethical reasoning is an art, not a science, and sometimes the best answer is going to be hard to determine. Often, the hardest decisions are not between right and wrong, but between shades of right. Regulations may allow more than one choice. There may even be more than one good answer, or there may not be enough time to conduct a long review. In those cases, you must rely on your judgment.

Define the Problem

4-34. Defining the problem is the first step in making any decision. When you think a decision may have ethical aspects or effects, it's especially important to define it precisely. Know who said what—and what specifically was said, ordered, or demanded. Don't settle for secondhand information; get the details. Problems can be described in more than one way. This is the hardest step in solving any problem. It's especially difficult for decisions in the face of potential ethical conflicts. Too often some people come to rapid conclusions about the nature of a problem and end up applying solutions to what turn out to be only symptoms.

Know the Relevant Rules

4-35. This step is part of fact gathering, the second step in problem solving. Do your homework. Sometimes what looks like an ethical problem may stem from a misunderstanding of a regulation or policy, frustration, or overenthusiasm. Sometimes the person who gave an order or made a demand didn't check the regulation and a thorough reading may make the problem go away. Other times, a difficult situation results from trying to do something right in the wrong way. Also, some regulations leave room for interpretation; the problem then becomes a policy matter rather than an ethical one. If you do perceive an ethical problem, explain it to the person you think is causing it and try to come up with a better way to do the job.

Develop and Evaluate Courses of Action

4-36. Once you know the rules, lay out possible courses of action. As with the previous steps, you do this whenever you must make a decision. Next, consider these courses of action in view of Army values. Consider the consequences of your courses of action by asking yourself a few practical questions: Which course of action best upholds Army values? Do any of the courses of action compromise Army values? Does any course of action violate a principle, rule, or regulation identified in Step 2? Which course of action is in the best interest of the Army and of the nation? This part will feel like a juggling act; but with careful ethical reflection, you can reduce the chaos, determine the essentials, and choose the best course—even when that choice is the least bad of a set of undesirable options.

Choose the Course of Action That Best Represents Army Values

4-37. The last step in solving any problem is making a decision and acting on it. Leaders are paid to make decisions. As an Army leader, you're expected—by your bosses and your people—to make decisions that solve problems without violating Army values.

4-38. As a values-based organization, the Army uses expressed values—Army values—to provide its fundamental ethical framework. Army values lay out the ethical standards expected of soldiers and DA civilians. Taken together, Army values and ethical decision making provide a moral touchstone and a workable process that enable you to make sound ethical decisions and take right actions confidently.

4-39. The ethical aspects of some decisions are more obvious that those of others. This example contains an obvious ethical problem. The issues will seldom be so clear-cut; however, as you read the example, focus on the steps SGT Kirk follows as he moves toward an ethical decision. Follow the same steps when you seek to do the right thing.

The EFMB Test

SGT Kirk, who has already earned the Expert Field Medical Badge (EFMB), is assigned as a grader on the division's EFMB course. Sergeant Kirk's squad leader, SSG Michaels, passes through SGT Kirk's station and fails the task. Just before SGT Kirk records the score, SSG Michaels pulls him aside.

"I need my EFMB to get promoted," SSG Michaels says. "You can really help me out here; it's only a couple of points anyway. No big deal. Show a little loyalty."

SGT Kirk wants to help SSG Michaels, who's been an excellent squad leader and who's loyal to his subordinates. SSG Michaels even spent two Saturdays helping SGT Kirk prepare for his promotion board. If SGT Kirk wanted to make this easy on himself, he would say the choice is between honesty and loyalty. Then he could choose loyalty, falsify the score, and send everyone home happy. His life under SSG Michaels would probably be much easier too.

However, SGT Kirk would not have defined the problem correctly. (Remember, defining the problem is often the hardest step in ethical reasoning.) SGT Kirk knows the choice isn't between loyalty and honesty. Loyalty doesn't require that he lie. In fact, lying would be disloyal to the Army, himself, and the soldiers who met the standard. To falsify the score would also be a violation of the trust and confidence the Army placed in him when he was made an NCO and a grader. SGT Kirk knows that loyalty to the Army and the NCO corps comes first and that giving SSG Michaels a passing score would be granting the squad leader an unfair advantage. SGT Kirk knows it would be wrong to be a coward in the face of this ethical choice, just as it would be wrong to be a coward in battle. And if all that were not enough, when SGT Kirk imagines seeing the incident in the newspaper the next morning—Trusted NCO Lies to Help Boss—he knows what he must do.

4-40. When SGT Kirk stands his ground and does the right thing, it may cost him some pain in the short run, but the entire Army benefits. If he makes the wrong choice, he weakens the Army. Whether or not the Army lives by its values isn't just up to generals and colonels; it's up to each of the thousands of SGT Kirks, the Army leaders who must make tough calls when no one is watching, when the easy thing to do is the wrong thing to do.

REFLECTIVE THINKING

4-41. Leader development doesn't occur in a vacuum. All leaders must be open to feedback on their performance from multiple perspectives—seniors, peers, and subordinates. But being open to feedback is only one part of the equation. As a leader, you must also listen to and use the feedback: you must be able to reflect. Reflecting is the ability to take information, assess it, and apply it to behavior to explain why things did or did not go well. You can then use the resulting explanations to improve future behavior. Good leaders are always striving to become better leaders. This means you need consistently to assess your strengths and weaknesses and reflect on what you can do to sustain your strengths and correct your weaknesses. To become a better leader, you must be willing to change.

4-42. For reasons discussed fully in Chapter 5, the Army often places a premium on doing—on the third element of BE, KNOW, DO. All Army leaders are busy dealing with what's on their plates and investing a lot of energy in accomplishing tasks. But how often do they take the time to STOP and really THINK about what they are doing? How often have you seen this sign on a leader's door: Do Not Disturb—Busy Reflecting? Not often. Well, good leaders need to take the time to think and reflect. Schedule it; start really exercising your capacity to get feedback. Then reflect on it and use it to improve. There's nothing wrong with making mistakes, but there's plenty wrong with not learning from those mistakes. Reflection is the means to that end.

TECHNICAL SKILLS

The first thing the senior NCOs had to do was to determine who wasn't qualified with his weapon, who didn't have his protective mask properly tested and sealed—just all the basic little things. Those things had to be determined real fast.

A Command Sergeant Major, Desert Storm

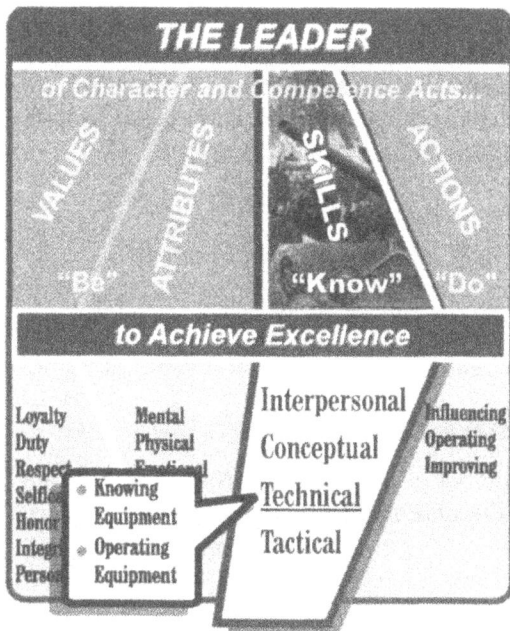

Figure 4-3. Direct Leader Skills—Technical

KNOWING EQUIPMENT

4-43. Technical skill is skill with things—equipment, weapons, systems—everything from the towing winch on the front of a vehicle to the computer that keeps track of corps personnel actions. Direct leaders must know their equipment and how to operate it. Figure 4-3 highlights direct leader technical skills. Technical manuals, training circulars, SOPs, and all the other publications necessary for efficient, effective performance explain specific skills more completely.

4-44. Direct leaders are closer to their equipment than organizational and strategic leaders. Thus, they have a greater need to know how it works and how to use it. In addition, direct leaders are the experts who are called upon to solve problems with the equipment, the ones who figure out how to make it work better, how to apply it, how to fix it—even how to modify it. Sergeants, junior officers, warrant officers, wage grade employees, and journeymen are the Army's technical experts and best teachers. Subordinates expect their first-line leaders to know their equipment and be experts in all the applicable technical skills.

OPERATING EQUIPMENT

4-45. Direct leaders know how to operate their equipment and make sure their people do as well. They set the example with a hands-on approach. When new equipment arrives, direct leaders find out how it works, learn how to use it themselves, and train their subordinates to do the same.

Technical Skill into Combat Power

Technical skill gave the Army a decided advantage in the 1944 battle for France. For example, the German Army had nothing like the US Army's maintenance battalions. Such an organization was a new idea, and a good one. These machine-age units were able to return almost half the battle-damaged tanks to action within two days. The job was done by young men who had been working at gas stations and body shops two years earlier and had brought their skill into the service of their country. Instead of fixing cars, they replaced damaged tank tracks, welded patches on the armor, and repaired engines. These combat supporters dragged tanks that were beyond repair to the rear and stripped them for parts. The Germans just left theirs in place.

I felt we had to get back to the basic soldier skills. The basics of setting up a training schedule for every soldier every day. We had to execute the standard field disciplines, such as NCOs checking weapons cleanliness and ensuring soldiers practiced personal hygiene daily. Our job is to go out there and kill the enemy. In order to do that, as Fehrenbach writes in [his study of the Korean Conflict entitled] This Kind of War, *we have to have disciplined teams; discipline brings pride to the unit. Discipline coupled with tough, realistic training is the key to high morale in units. Soldiers want to belong to good outfits, and our job as leaders is to give them the best outfit we can.*

A Company Commander, Desert Storm

4-46. This company commander is talking about two levels of skill. First is the individual level: soldiers are trained with their equipment and know how to do their jobs. Next is the collective level: leaders take these trained individuals and form them into teams. The result: a whole greater than the sum of its parts, a team that's more than just a collection of trained individuals, an organization that's capable of much more than any one of its elements. (FM 25-101 discusses how to integrate individual, collective, and leader training).

TACTICAL SKILLS

Man is and always will be the supreme element in combat, and upon the skill, the courage and endurance, and the fighting heart of the individual soldier the issue will ultimately depend.

General Matthew B. Ridgway
Former Army Chief of Staff

DOCTRINE

4-47. Tactics is the art and science of employing available means to win battles and engagements. The science of tactics encompasses capabilities, techniques, and procedures that can be codified. The art of tactics includes the creative and flexible array of means to accomplish assigned missions, decision making when faced with an intelligent enemy, and the effects of combat on soldiers. Together, FM 100-34, FM 100-40, and branch-specific doctrinal manuals capture the tactical skills that are essential to mastering both the science and the art of tactics. Figure 4-4 highlights direct leader tactical skills.

FIELDCRAFT

4-48. Fieldcraft consists of the skills soldiers need to sustain themselves in the field. Proficiency in fieldcraft reduces the likelihood soldiers will become casualties. The requirement to be able to do one's job in a field environment distinguishes the soldier's profession from most civilian occupations. Likewise, the

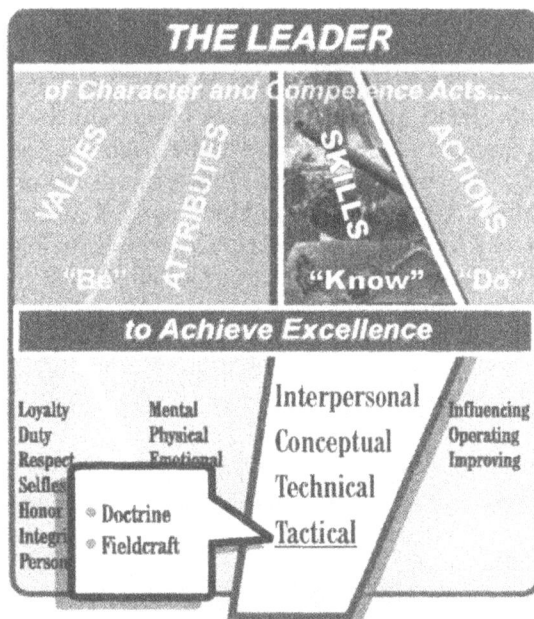

Figure 4-4. Direct Leader Skills—Tactical

requirement that Army leaders make sure their soldiers take care of themselves and provide them with the means to do so is unique.

4-49. The Soldier's Manual of Common Tasks lists the individual skills all soldiers must master to operate effectively in the field. Those skills include everything from how to stay healthy, to how to pitch a tent, to how to run a heater. Some military occupational specialties (MOS) require proficiency in additional fieldcraft skills. Soldier's Manuals for these MOS list them.

4-50. Army leaders gain proficiency in fieldcraft through schooling, study, and practice. Once learned, few fieldcraft skills are difficult. However, they are easy to neglect during exercises, when everyone knows that the exercise will end at a specific time, sick and injured soldiers are always evacuated, and the adversary isn't using real ammunition. During peacetime, it's up to Army leaders to enforce tactical discipline, to make sure their soldiers practice the fieldcraft skills that will keep them from becoming casualties—battle or nonbattle—during operations.

TACTICAL SKILLS AND TRAINING

4-51. Direct leaders are the Army's primary tactical trainers, both for individuals and for teams. Practicing tactical skills is often challenging. The best way to improve individual and collective skills is to replicate operational conditions. Unfortunately, Army leaders can't always get the whole unit out in the field to practice maneuvers, so they make do with training parts of it separately. Sometimes they can't get the people, the time, and the money all together at the right time and the right place to train the entire team. There are always training distracters. There will always be a hundred excuses not to train together and one reason why such training must occur: units fight as they train. (FM 25-100 and FM 25-101 discuss training principles and techniques.)

4-52. Unfortunately, the Army has been caught unprepared for war more than once. In July 1950, American troops who had been on occupation duty in Japan were thrown into combat when North Korean forces invaded South Korea. Ill-trained, ill-equipped, and out of shape, they went into action and were overrun. However, that same conflict provides another example of how well things can go when a direct leader has tactical skill, the ability to pull people and things together into a team. Near the end of November 1950, American forces were chasing the remnants of the broken North Korean People's Army into the remote northern corners of the Korean Peninsula. Two American units pushed all the way to the Yalu River, which forms the boundary between North Korea and the People's Republic of China. One was the 17th Infantry Regiment. The other was a task force commanded by a 24-year-old first lieutenant named Joseph Kingston.

Task Force Kingston

1LT Joseph Kingston, a boyish-looking platoon leader in K Company, 3d Battalion, 32d Infantry, was the lead element for his battalion's move northward. The terrain was mountainous, the weather bitterly cold—the temperature often below zero—and the cornered enemy still dangerous. 1LT Kingston inched his way forward, with the battalion adding elements to his force. He had antiaircraft jeeps mounted with quad .50 caliber machine guns, a tank, a squad (later a platoon) of engineers, and an artillery forward observer. Some of these attachments were commanded by lieutenants who outranked him, as did a captain who was the tactical air controller. But 1LT Kingston remained in command, and battalion headquarters began referring to Task Force Kingston.

Bogged down in Yongsong-ni with casualties mounting, Task Force Kingston received reinforcements that brought the number of men to nearly 300. Despite tough fighting, the force continued to move northward. 1LT Kingston's battalion commander wanted him to remain in command, even though they sent several more officers who outranked 1LT Kingston. One of the

---

> ### Task Force Kingston (continued)
>
> attached units was a rifle company, commanded by a captain. But the arrangement worked, mostly because 1LT Kingston himself was an able leader. Hit while leading an assault on one enemy stronghold, he managed to toss a grenade just as a North Korean soldier shot him in the head. His helmet, badly grazed, saved his life. His personal courage inspired his men and the soldiers from the widely varied units who were under his control. Task Force Kingston was commanded by the soldier who showed, by courage and personal example, that he could handle the job.

4-53. 1LT Kingston made the task force work by applying skills at a level of responsibility far above what was normal for a soldier of his rank and experience. He knew how to shoot, move, and communicate. He knew the fundamentals of his profession. He employed the weapons under his command and controlled a rather unwieldy collection of combat assets. He understood small-unit tactics and applied his reasoning skills to make decisions. He fostered a sense of teamwork, even in this collection of units that had never trained together. Finally, he set the example with personal courage.

SUMMARY

4-54. Direct leadership is face-to-face, first-line leadership. It takes place in organizations where subordinates are used to seeing their leaders all the time: teams, squads, sections, platoons, companies, and battalions. To be effective, direct leaders must master many interpersonal, conceptual, technical, and tactical skills.

4-55. Direct leaders are first-line leaders. They apply the conceptual skills of critical reasoning and creative thinking to determine the best way to accomplish the mission. They use ethical reasoning to make sure their choice is the right thing to do, and they use reflective thinking to assess and improve team performance, their subordinates, and themselves. They employ the interpersonal skills of communicating and supervising to get the job done. They develop their people by mentoring and counseling and mold them into cohesive teams by training them to standard.

4-56. Direct leaders are the Army's technical experts and best teachers. Both their bosses and their people expect them to know their equipment and be experts in all the applicable technical skills. On top of that, direct leaders combine those skills with the tactical skills of doctrine, fieldcraft, and training to accomplish tactical missions.

4-57. Direct leaders use their competence to foster discipline in their units and to develop soldiers and DA civilians of character. They use their mastery of equipment and doctrine to train their subordinates to standard. They create and sustain teams with the skill, trust, and confidence to succeed—in peace and war.

Chapter 5

Direct Leadership Actions

5-1. Preparing to be a leader doesn't get the job done; the test of your character and competence comes when you act, when you DO those things required of a leader.

5-2. The three broad leader actions that Chapters 1 and 2 introduced—influencing, operating, and improving—contain other activities. As with the skills and attributes discussed previously, none of these exist alone. Most of what you do as a leader is a mix of these actions. This manual talks about them individually to explain them more clearly; in practice they're often too closely connected to sort out.

5-3. Remember that your actions say more about what kind of leader you are than anything else. Your people watch you all the time; you're always on duty. And if there's a disconnect between what you say and how you act, they'll make up their minds about you—and act accordingly—based on how you act. It's not good enough to talk the talk; you have to walk the walk.

> The most important influence you have on your people is the example you set.

INFLUENCING ACTIONS

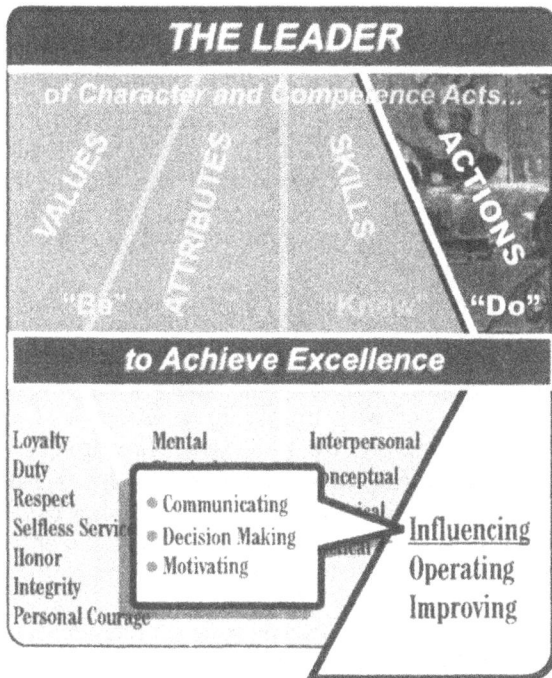

Figure 5-1. Direct Leader Actions—Influencing

5-4. Leadership is both art and science. It requires constant study, hard work, and frequent practice. Since you're dealing with people and their emotions, dreams, and fears, it also calls for imagination and a positive, upbeat approach.

5-5. Effective leaders act competently and confidently. Your attitude sets the tone for the entire unit, and you choose your attitude—day to day, task to task, even minute to minute. Remember that optimism, a positive outlook, and a sense of humor are infectious. This is especially true when you must make unpopular decisions and face the challenge of bringing the team on board.

5-6. Figure 5-1 shows that influencing consists of communicating, decision making and motivating. As a leader, you should be asking several questions: What's happening? What should be happening but isn't? Why are these things happening? Then ask yourself: How can I get this team moving toward the goal? (Appendix B lists leader performance indicators).

COMMUNICATING

You must talk to your soldiers...I don't just mean in formation or groups, but one-on-one. Take time (at least 15 to 30 minutes a day) to really talk to a soldier, one soldier a day.

Command Sergeant Major Daniel E. Wright

5-7. Leaders keep their subordinates informed because doing so shows trust, because sharing information can relieve stress, and because information allows subordinates to determine what they need to do to accomplish the mission when circumstances change. By informing them of a decision—and, as much as possible, the reasons for it—you show your subordinates they're important members of the team. Accurate information also relieves unnecessary stress and helps keep rumors under control. (Without an explanation for what's happening, your people will manufacture one—or several—of their own.) Finally, if something should happen to you, the next leader in the chain will be better prepared to take over and accomplish the mission if everyone knows what's going on. Subordinates must understand your intent. In a tactical setting, leaders must understand the intent of their commanders two levels up.

5-8. In other situations, leaders use a variety of means to keep people informed, from face-to-face talks to published memos and family newsletters. No matter what the method, keep two things in mind:

- As a leader, you are responsible for making sure your subordinates understand you.
- Communication isn't limited to your immediate superiors and subordinates.

5-9. The success or failure of any communication is the responsibility of the leader. If it appears your subordinates don't understand, check to make sure you've made yourself clear. In fact, even if you think your people understand, check anyway; ask for a back-brief.

5-10. Don't assume that communication begins or ends at the next level up or the next level down. If you're a team leader, listen carefully to what your supervisors, platoon sergeants, platoon leaders, and company commanders say. If you're a platoon sergeant, pass the word through your squad leaders or section chiefs, but also watch and *listen to the troops* to see if the information has made it all the way to where it needs to go. Listen carefully at least two levels up and two levels down.

5-11. In combat, subordinates may be out of contact with their leaders. Sometimes the plan falls apart because of something unexpected—weather, terrain, enemy action. Sometimes the leader may be killed or wounded. In those situations, subordinates who know the overall purpose of the mission and the commander's intent have the basic information they need to carry on. And if the leader has established a climate of trust, if the leader has trained the subordinate leaders in how and why decisions are made, one of these subordinates is more likely to step up and take charge.

5-12. To prepare your subordinates for such circumstances, create training situations where they must act on their own with a minimum of guidance—or with no guidance except a clear understanding of the purpose. Follow up these training situations with AARs so that subordinates learn what they did well, what they could have done better, and they should do differently next time.

5-13. Communicating also goes on from bottom to top. Leaders find out what their people are thinking, saying, and doing by using that most important communication tool: listening. By listening carefully, you can even hear those messages behind what a person is actually saying, the equivalent of reading between the lines. Practice "leadership by walking around." Get out and coach, listen, teach, and clarify; pass on what you learn to your superiors. They need to know what's going on to make good plans.

DECISION MAKING

A good leader must sometimes be stubborn. Armed with the courage of his convictions, he must often fight to defend them. When he has come to a decision after thorough analysis—and when he is sure he is right—he must stick to it even to the point of stubbornness.

General of the Army, Omar N. Bradley
Address to the US Army Command and
General Staff College, May 1967

5-14. A problem is an existing condition or situation in which what you want to happen is different from what actually is happening. Decision making is the process that begins to change that situation. Thus, decision making is knowing *whether* to decide, then *when* and *what* to decide. It includes understanding the consequences of your decisions.

5-15. Army leaders usually follow one of two decision-making processes. Leaders at company level and below follow the troop leading procedures (TLP). The TLP are designed to support solving tactical problems. Leaders at battalion level and above follow the military decision making process (MDMP). The MDMP, which FM 101-5 discusses, is designed for organizations with staffs. These established and proven methodologies combine elements of the planning operating action to save time and achieve parallel decision making and planning. Both follow the problem solving steps discussed below.

5-16. Every once in a while, you may come across a decision that's easy to make: yes or no, right or left, on or off. As you gain experience as a leader, some of the decisions you find difficult now will become easier. But there will always be difficult decisions that require imagination, that require rigorous thinking and analysis, or that require you to factor in your gut reaction. Those are the tough decisions, the ones you're getting paid to make. As an experienced first sergeant once said to a brand new company commander, "We get paid the big bucks to

make the hard calls." The next several paragraphs explain the steps you should use to solve a problem; then you'll read about other factors that affect how you make those hard calls and the importance of setting priorities.

Problem Solving Steps

5-17. **Identify the problem**. Don't be distracted by the symptoms of the problem; get at its root cause. There may be more than one thing contributing to a problem, and you may run into a case where there are lots of contributing factors but no real "smoking gun." The issue you choose to address as the root cause becomes the mission (or restated mission for tactical problems). The mission must include a simple statement of who, what, when, where, and why. In addition, it should include your end state, how you want things to look when the mission is complete.

5-18. **Identify facts and assumptions**. Get whatever facts you can in the time you have. Facts are statements of what you know about the situation. Assumptions are statements of what you believe about the situation but don't have facts to support. Make only assumptions that are likely to be true and essential to generate alternatives. Some of the many sources of facts include regulations, policies, and doctrinal publications. Your organization's mission, goals, and objectives may also be a source. Sources of assumptions can be personal experiences, members of the organization, subject matter experts, or written observations. Analyze the facts and assumptions you identify to determine the scope of the problem. (FM 101-5 contains more information on facts and assumptions.)

5-19. **Generate alternatives**. Alternatives are ways to solve the problem. Develop more than one possible alternative. Don't be satisfied with the first thing that comes into your mind. That's lazy thinking; the third or fourth or twentieth alternative you come up with might be the best one. If you have time and experienced subordinates, include them in this step.

5-20. **Analyze the alternatives**. Identify intended and unintended consequences, resource or other constraints, and the advantages and disadvantages of each alternative. Be sure to consider all your alternatives. Don't prejudge the situation by favoring any one alternative over the others.

5-21. **Compare the alternatives**. Evaluate each alternative for its probability of success and its cost. Think past the immediate future. How will this decision change things tomorrow? Next week? Next year?

5-22. **Make and execute your decision**. Prepare a leader's plan of action, if necessary, and put it in motion. (Planning, an operating action, is covered later in this chapter. Appendix C discusses plans of action as part of developmental counseling. Appendix D contains an example of a leader's plan of action.)

5-23. **Assess the results**. Check constantly to see how the execution of your plan of action is going. Keep track of what happens. Adjust your plan, if necessary. Learn from the experience so you'll be better equipped next time. Follow up on results and make further adjustments as required.

Factors to Consider

5-24. All of this looks great on paper; and it's easy to talk about when things are calm, when there's plenty of time. But even when there isn't a great deal of time, you'll come up with the best solution if you follow this process to the extent that time allows.

5-25. Even following these steps, you may find that with some decisions you need to take into account your knowledge, your intuition, and your best judgment. Intuition tells you what feels right; it comes from accumulated experience, often referred to as "gut feeling." However, don't be fooled into relying only on intuition, even if it has worked in the past. A leader who says "Hey, I just do what feels right"

may be hiding a lack of competence or may just be too lazy to do the homework needed to make a reasoned, thought-out decision. Don't let that be you. Use your experience, listen to your instincts, but do your research as well. Get the facts and generate alternatives. Analyze and compare as many as time allows. Then make your decision and act.

5-26. Remember also that any decision you make must reflect Army values. Chapter 4 discusses ethical reasoning. Its steps match the problem solving steps outlined here. Most problems are not ethical problems, but many have ethical aspects. Taking leave for example, is a right soldiers and DA civilians enjoy, but leaders must balance mission requirements with their people's desires and their own. Reconciling such issues may require ethical reasoning. As a leader, your superiors and your people expect you to take ethical aspects into account and make decisions that are right as well as good.

Setting Priorities

5-27. Decisions are not often narrowly defined, as in "Do I choose A or B?" Leaders make decisions when they establish priorities and determine what's important, when they supervise, when they choose someone for a job, when they train.

5-28. As a leader, you must also set priorities. If you give your subordinates a list of things to do and say "They're all important," you may be trying to say something about urgency. But the message you actually send is "I can't decide which of these are most important, so I'll just lean on you and see what happens."

5-29. Sometimes all courses of action may appear equally good (or equally bad) and that any decision will be equally right (or equally wrong). Situations like that may tempt you to sit on the fence, to make no decision and let things work themselves out. Occasionally that may be appropriate; remember that decision making involves judgment, knowing *whether* to decide. More often, things left to themselves go from bad to worse. In such situations, the decision you make may be less important than simply deciding to do something. Leaders must have the personal courage to say which tasks are more important than others. In the absence of a clear priority, you must set one; not everything can be a top priority, and you can't make progress without making decisions.

Solving a Training Problem

A rifle platoon gets a new platoon leader and a new platoon sergeant within days of a poor showing in the division's military operations on urbanized terrain (MOUT) exercise. The new leaders assume the platoon's poor showing is a problem. Feedback from the evaluators is general and vague. The platoon's squad and fire team leaders are angry and not much help in assessing what went wrong, so the new leaders begin investigating. In their fact-finding step they identify the following facts: (1) The soldiers are out of shape and unable to complete some of the physical tasks. (2) The fire team leaders don't know MOUT tactics, and some of the squad leaders are also weak. (3) Third Squad performed well, but didn't help the other squads. (4) The soldiers didn't have the right equipment at the training site.

Pushing a bit further to get at the root causes of these problems, the new leaders uncover the following: (1) Platoon PT emphasizes preparation for the APFT only. (2) Third Squad's leaders know MOUT techniques, and had even developed simple drills to help their soldiers learn, but because of unhealthy competition encouraged by the previous leaders, Third Squad didn't share the knowledge. (3) The company supply sergeant has the equipment the soldiers needed, but because the platoon had lost some equipment on the last field exercise, the supply sergeant didn't let the platoon sign out the equipment.

The new platoon leader and platoon sergeant set a goal of successfully meeting the exercise standard in two months. To generate alternatives, they meet with the squad leaders and ask for suggestions to improve training. They use all their available resources to develop solutions. Among the things suggested was to shuffle some of the team leaders to break up Third Squad's clique and spread some of the tactical knowledge around. When squad leaders complained, the platoon sergeant emphasized that they must think as a platoon, not just a collection of squads.

The platoon sergeant talks to the supply sergeant, who tells him the platoon's previous leadership had been lax about property accountability. Furthermore, the previous leaders didn't want to bother keeping track of equipment, so they often left it in garrison. The platoon sergeant teaches his squad leaders how to keep track of equipment and says that, in the future, soldiers who lose equipment will pay for it: "We wouldn't leave our stuff behind in war, so we're not going to do it in training."

Building on Third Squad's experience, the platoon leader works with the squad and fire team leaders to come up with some simple drills for the platoon's missions. He takes the leaders to the field and practices the drills with them so they'll be able to train their soldiers to the new standard.

The platoon sergeant also goes to the brigade's fitness trainers and, with their help, develops a PT program that emphasizes skills the soldiers need for their combat tasks. The new program includes rope climbing, running with weapons and equipment, and road marches. Finally, the leaders monitor how their plan is working. A few weeks before going through the course again, they decide to eliminate one of the battle drills because the squad leaders suggested that it wasn't necessary after all.

5-30. The platoon leader and platoon sergeant followed the problem solving steps you just read about. Given a problem (poor performance), they identified the facts surrounding it (poor PT practices, poor property accountability, and unhealthy competition), developed a plan of action, and executed it. Where appropriate, they analyzed and compared different alternatives (Third Squad's drills). They included their subordinates in the process, but had the moral courage to make unpopular decisions (breaking up the Third Squad clique). Will the platoon do better the next time out? Probably, but before then the new leaders will have to assess the results of their actions to make sure they're accomplishing what the leaders want. There may be other aspects of this problem that were not apparent at first. And following this or any process doesn't guarantee success. The process is only a framework that helps you make a plan and act. Success depends on your ability to apply your attributes and skills to influencing and operating actions.

5-31. Army leaders also make decisions when they evaluate subordinates, whether it's with a counseling statement, an evaluation report, or even on-the-spot encouragement. At an in-ranks inspection, a new squad leader takes a second look at a soldier's haircut—or lack of one. The squad leader's first reaction may be to ask, "Did you get your haircut lately?" But that avoids the problem. The soldier's haircut is either to standard or not—the NCO must decide. The squad leader either says—without apologizing or dancing around the subject—"You need a haircut" or else says nothing. Either way, the decision communicates the leader's standard. Looking a subordinate in the eye and making a necessary correction is a direct leader hallmark.

MOTIVATING

A unit with a high esprit de corps can accomplish its mission in spite of seemingly insurmountable odds.

FM 22-10, 1951

5-32. Recall from Chapter 1 that motivation involves using word and example to give your subordinates the will to accomplish the mission. Motivation grows out of people's confidence in themselves, their unit, and their leaders. This confidence is born in hard, realistic training; it's nurtured by constant reinforcement and through the kind of leadership—consistent, hard, and fair—that promotes trust. Remember that trust, like loyalty, is a gift your soldiers give you only when you demonstrate that you deserve it. Motivation also springs from the person's faith in the larger mission of the organization—a sense of being a part of the big picture.

Empowering People

5-33. People want to be recognized for the work they do and want to be empowered. You empower subordinates when you train them to do a job, give them the necessary resources and authority, get out of their way, and let them work. Not only is this a tremendous statement of the trust you have in your subordinates; it's one of the best ways to develop them as leaders. Coach and counsel them, both when they succeed and when they fail.

Positive Reinforcement

5-34. Part of empowering subordinates is finding out their needs. Talk to your people: find out what's important to them, what they want to accomplish, what their personal goals are. Give them feedback that lets them know how they're doing. Listen carefully so that you know what they mean, not just what they say. Use their feedback when it makes sense, and if you change something in the organization because of a subordinate's suggestion, let everyone know where the good idea came from. Remember, there's no limit to the amount of good you can do as long as you don't worry about who gets the credit. Give the credit to those who deserve it and you'll be amazed at the results.

5-35. You recognize subordinates when you give them credit for the work they do, from a pat on the back to a formal award or decoration. Don't underestimate the power of a few choice words of praise when a person has done a good job. Don't hesitate to give out awards—commendations, letters, certificates—when appropriate. (Use good judgment, however. If you give out a medal for every little thing, pretty soon the award becomes meaningless. Give an award for the wrong thing and you show you're out of touch.) Napoleon marveled at the motivational power of properly awarded ribbons and medals. He once said that if he had enough ribbon, he could rule the world.

5-36. When using rewards, you have many options. Here are some things to consider:

- Consult the leadership chain for recommendations.

- Choose a reward valued by the person receiving it, one that appeals to the individual's personal pride. This may be a locally approved award that's more respected than traditional DA awards.

- Use the established system of awards (certificates, medals, letters of commendation, driver and mechanic badges) when appropriate. These are recognized throughout the Army; when a soldier goes to a new unit, the reward will still be valuable.

- Present the award at an appropriate ceremony. Emphasize its importance. Let others see how hard work is rewarded.

- Give rewards promptly.

- Praise only good work or honest effort. Giving praise too freely cheapens its effect.

- Promote people who get the job done and who influence others to do better work.

- Recognize those who meet the standard and improve their performance. A soldier who works hard and raises his score on the APFT deserves some recognition, even if the soldier doesn't achieve the maximum score. Not everyone can be soldier of the quarter.

Negative Reinforcement

5-37. Of course, not everyone is going to perform to standard. In fact, some will require punishment. Using punishment to motivate a person away from an undesirable behavior is effective, but can be tricky. Sound judgment must guide you when administering punishment. Consider these guidelines:

- Before you punish a subordinate, make sure the subordinate understands the reason for the punishment. In most—although not all—cases, you'll want to try to change the subordinate's behavior by counseling or retraining before resulting to punishment.

- Consult your leader or supervisor before you punish a subordinate. They'll be aware of policies you need to consider and may be able to assist you in changing the subordinate's behavior.

- Avoid threatening a subordinate with punishment. Making a threat puts you in the position of having to deliver on that threat. In such a situation you may end up punishing because you said you would rather than because the behavior merits punishment. This undermines your standing as a leader.

- Avoid mass punishment. Correctly identify the problem, determine if an individual or individuals are responsible, and use an appropriate form of correction.

- With an open mind and without prejudging, listen to the subordinate's side of the story.

- Let the subordinate know that it's the behavior—not the individual—that is the problem. "You let the team down" works; "You're a loser" sends the wrong message.

- Since people tend to live up to their leader's expectations, tell them, "I know you can do better than that. I expect you to do better than that."

- Punish those who are able but *unwilling* to perform. Retrain a person who's *unable* to complete a task.

- Respond immediately to undesirable behavior. Investigate fully. Take prompt and prudent corrective action in accordance with established legal or regulatory procedures.

- Never humiliate a subordinate; avoid public reprimand.

- Ensure the person knows exactly what behavior got the person in trouble.

- Make sure the punishment isn't excessive or unreasonable. It's not only the severity of punishment that keeps subordinates in line; it's the certainty that they can't get away with undesirable behavior.

- Control your temper and hold no grudges. Don't let your personal feelings interfere; whether you like or dislike someone has nothing to do with good order and discipline.

5-38. If you were surprised to find a discussion of punishment under the section on motivation, consider this: good leaders are always on the lookout for opportunities to develop subordinates, even the ones who are being punished. Your people—even the ones who cause you problems—are still the most important resource you have. When a vehicle is broken, you don't throw it out; you fix it. If one of your people is performing poorly, don't just get rid of the person; try to help fix the problem.

OPERATING ACTIONS

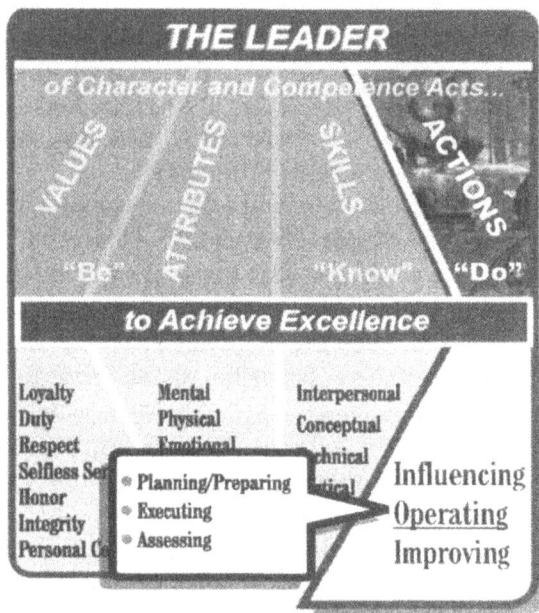

Figure 5-2. Direct Leader Actions—Operating

5-39. You're operating when you act to achieve an immediate objective, when you're working to get today's job done. Although FM 25-100 is predominantly a training tool, its methodology applies to a unit's overall operational effectiveness. Because operating includes planning, preparing, executing, and assessing (see Figure 5-2), you can use the FM 25-100 principles as a model for operations other than training. Sometimes these elements are part of a cycle; other times they happen simultaneously.

5-40. You'll often find yourself influencing after you've moved on to operating. In practice, the nice, neat divisions in this manual are not clear-cut; you often must handle multiple tasks requiring different skills at the same time. (Appendix B lists operating actions and some indicators of effectiveness.)

PLANNING AND PREPARING

5-41. In peacetime training, in actual operations, and especially in combat, your job is to help your organization function effectively—accomplish the mission—in an environment that can be chaotic. That begins with a well thought-out plan and thorough preparation. A well-trained organization with a sound plan is much better prepared than one without a plan. Planning ahead reduces confusion, builds subordinates' confidence in themselves and the organization, and helps ensure success with a minimum of wasted effort—or in combat, the minimum number of casualties. (FM 101-5 discusses the different types of plans.)

5-42. A plan is a proposal for executing a command decision or project. Planning begins with a mission, specified or implied. A specified mission comes from your boss or from higher headquarters. An implied mission results when the leader, who may be you, sees something within his area of responsibility that needs to be done and, on his own initiative, develops a leader plan of action. (Remember that a problem exists when you're not satisfied with the way things are or the direction they're heading.) Either type of mission contains implied and specified tasks, actions that must be completed to accomplish the mission. (FM 101-5 discusses how the MDMP supports planning.)

Reverse Planning

5-43. When you begin with the goal in mind, you often will use the reverse planning method. Start with the question "Where do I want to end up?" and work backward from there until you reach "We are here right now."

5-44. Along the way, determine the basics of what's required: who, what, when, where, and why. You may also want to consider how to accomplish the task, although the "how" is usually not included in a mission to a subordinate. As you plan, consider the amount of time needed to coordinate and conduct each step. For instance, a tank platoon sergeant whose platoon has to spend part of a field exercise on the firing range might have to arrange, among other things, refueling at the range. No one explicitly said to refuel at the range, but the platoon sergeant knows what needs to happen. The platoon sergeant must think through the steps from the last to the first: (1) when the refueling must be complete, (2) how long the refueling will take, (3) how long it takes the refueling unit to get set up, and finally (4) when the refueling vehicles should report to the range.

5-45. After you have figured out what must happen on the way to the goal, put the tasks in sequence, set priorities, and determine a schedule. Look at the steps in the order they will occur. Make sure events are in logical order and you have allotted enough time for each one. As always, a good leader asks for input from subordinates when time allows. Getting input not only acts as a check on the your plan (you may have overlooked something), but also gets your people involved; involvement builds trust, self-confidence, and the will to succeed.

Preparing

5-46. While leaders plan, subordinates prepare. Leaders can develop a plan while their organization is preparing if they provide advance notice of the task or mission and initial guidance for preparation in a warning order. (Warning orders are part of the TLP and MDMP; however, any leader—uniformed or DA civilian—can apply the principle of the warning order by giving subordinates advance notice of an impending requirement and how they'll be expected to contribute to it. FM 101-5 discusses warning orders.) Based on this guidance, subordinates can draw ammunition, rehearse key actions, inspect equipment, conduct security patrols, or begin movement while the leader completes the plan. In the case of a nontactical requirement, preparation may include making sure the necessary facilities and other resources are available to support it. In all cases, preparation includes coordinating with people and organizations that are involved or might be affected by the operation or project. (TC 25-30 discusses preparing for company- and platoon-level training).

5-47. Rehearsal is an important element of preparation. Rehearsing key combat actions lets subordinates see how things are supposed to work and builds confidence in the plan for both soldiers and leaders. Even a simple walk-through helps them visualize who's supposed to be where and do what when. Mobilization exercises provide a similar function for DA civilians and reserve component soldiers: they provide a chance to understand and rehearse mobilization and deployment support functions. Execution goes more smoothly because everyone has a mental picture of what's supposed to happen. Rehearsals help people remember their responsibilities. They also help leaders see how things might happen, what might go wrong, how the plan needs to be changed, and what things the leader didn't think of. (FM 101-5 contains an in-depth discussion of rehearsals.)

An Implied Mission and Leader Plan of Action

Not all missions originate with higher headquarters; sometimes the leader sees what's required and, exercising initiative, develops a leader plan of action.

Suppose a platoon sergeant's soldiers had trouble meeting minimum weapons qualification requirements. Since everyone qualified, no one has said to work on marksmanship. But the platoon sergeant knows the platoon made it through range week on sheer luck. The leader develops a training

---

An Implied Mission and Leader Plan of Action (continued)

plan to work on basic marksmanship, then goes to the platoon leader and presents it. Together the two leaders figure out a way to make sure their soldiers get a chance to train, even with all the other mission requirements. After they've talked it over, they bring in their subordinate leaders and involve them in the planning process. The platoon sergeant keeps track of the progress and effectiveness of the leader plan of action, making sure it accomplishes the intent and changing it when necessary. Later, the platoon leader and platoon sergeant meet to assess the plan's results and to decide if further action is required.

5-48. Leader plans of action can be used to reinforce positive behavior, improve performance, or even change an aspect of the organizational climate. A leader plan of action may also be personal—as when the leader decides "I need to improve my skills in this area."

Brief Solutions, Not Problems

Leaders develop their subordinates by requiring those subordinates to plan. A lieutenant, new to the battalion staff, ran into a problem getting all the resources the unit was going to need for an upcoming deployment. The officer studied the problem, talked to the people involved, checked his facts, and generally did a thorough analysis—of which he was very proud. Then he marched into the battalion executive officer's (XO's) office and laid it all out in a masterly fashion. The XO looked up from his desk and said, "Great. What are you going to do about it?"

The lieutenant was back in a half-hour with three possible solutions he had worked out with his NCOs. From that day on, the officer never presented a problem to any boss without offering some solutions as well. The lieutenant learned a useful technique from the XO. He learned it so well he began using it with his soldiers and became a better coach and mentor because of it.

5-49. No matter what your position is, part of your duty is making your boss's job easier. Just as you loyally provide resources and authority for your subordinates to do their jobs, you leave the boss free to do his. Ask only for decisions that fall outside your scope of authority—not those you want to avoid. Forward only problems you can't fix—not those whose solutions are just difficult. Ask for advice from others with more experience or seek clarification when you don't understand what's required. Do all that and exercise disciplined initiative within your boss's intent. (Appendix A discusses delegation of authority.)

EXECUTING

Soldiers do what they are told to do. It's leadership that's the key. Young men and women join the Army; if they're with competent, confident, capable leaders they turn into good soldiers.

Sergeant Major of the Army Robert E. Hall

5-50. Executing means acting to accomplish the mission, moving to achieve the leader's goals as expressed in the leader's vision—to standard and on time—while taking care of your people.

5-51. Execution, the payoff, is based on all the work that has gone before. But planning and preparation alone can't guarantee success. Things will go wrong. Large chunks of the plan will go flying out the window. At times, it will seem as if everything is working against you. Then you must have the will to fight through, keeping in mind your higher leaders' intent and the mission's ultimate goal. You must adapt and improvise.

5-52. In a tactical setting, all leaders must know the intent of commanders two levels up. During execution, position yourself to best lead your people, initiate and control the action, get others to follow the plan, react to changes, keep your people focused, and work the team to accomplish the goal to standard. A well-trained organization accomplishes the mission, even when things go wrong.

5-53. Finally, leaders ensure they and their subordinate leaders are doing the right jobs. This goes hand in hand with empowerment. A company commander doesn't do a squad leader's job. A division chief doesn't do a branch chief's job. A supervisor doesn't do a team leader's job.

Maintaining Standards

5-54. The Army has established standards for all military activities. Standards are formal, detailed instructions that can be stated, measured, and achieved. They provide a performance baseline to evaluate how well a specific task has been executed. You must know, communicate and enforce standards. Explain the ones that apply to your organization and give your subordinate leaders the authority to enforce them. Then hold your subordinates responsible for achieving them.

5-55. Army leaders don't set the minimum standards as goals. However, everything can't be a number one priority. As an Army leader, you must exercise judgment concerning which tasks are most important. Organizations are required to perform many tasks that are not mission-related. While some of these are extremely important, others require only a minimum effort. Striving for excellence in every area, regardless of how trivial, quickly works an organization to death. On the other hand, the fact that a task isn't a first priority doesn't excuse a sloppy performance. Professional soldiers accomplish all tasks to standard. Competent leaders make sure the standard fits the task's importance.

Setting Goals

5-56. The leader's ultimate goal—your ultimate goal—is to train the organization to succeed in its wartime mission. Your daily work includes setting intermediate goals to get the organization ready. Involve your subordinates in goal setting. This kind of cooperation fosters trust and makes the best use of subordinates' talents. When developing goals, consider these points:

- Goals must be realistic, challenging, and attainable.
- Goals should lead to improved combat readiness.
- Subordinates ought to be involved in the goal setting.
- Leaders develop a plan of action to achieve each goal.

ASSESSING

Schools and their training offer better ways to do things, but only through experience are we able to capitalize on this learning. The process of profiting from mistakes becomes a milestone in learning to become a more efficient soldier.

Former Sergeant Major of the Army
William G. Bainbridge

5-57. Setting goals and maintaining standards are central to assessing mission accomplishment. Whenever you talk about accomplishing the mission, always include the phrase "to standard." When you set goals for your subordinates, make sure they know what the standards are. To use a simple example, the goal might be "All unit members will pass the APFT." The APFT standard tells you, for each exercise, how many repetitions are required in how much time, as well as describing a proper way to do the exercise.

5-58. Also central to assessing is spot checking. Army leaders check things: people, performance, equipment, resources. They check things to ensure the organization is meeting standards and moving toward the goals the leader has established. Look closely; do it early and often; do it both before and after the fact. Praise good performance and figure out how to fix poor performance. Watch good first sergeants or command sergeants major as they go through the mess line at the organizational dining facility. They pick up the silverware and run their fingers over it—almost unconsciously—checking for cleanliness. Good leaders supervise, inspect, and correct their subordinates. They don't waste time; they're always on duty.

5-59. Some assessments you make yourself. For others, you may want to involve subordinates. Involving subordinates in assessments and obtaining straightforward feedback from them become more important as your span of authority increases. Two techniques that involve your subordinates in assessing are in-process reviews (IPRs) and after-action reviews (AARs).

In-Process Reviews

5-60. Successful assessment begins with forming a picture of the organization's performance early. Anticipate which areas the organization might have trouble in; that way you know which areas to watch closely. Once the organization begins the mission, use IPRs to evaluate performance and give feedback. Think of an IPR as a checkpoint on the way to mission accomplishment.

5-61. Say you tell your driver to take you to division headquarters. If you recognize the landmarks, you decide your driver knows the way and probably say nothing. If you don't recognize the landmarks, you might ask where you are. And if you determine that the driver is lost or has made a wrong turn, you give instructions to get back to where you need to be. In more complex missions, IPRs give leaders and subordinates a chance to talk about what's going on. They can catch problems early and take steps to correct or avoid them.

After-Action Reviews

5-62. AARs fill a similar role at the end of the mission. Army leaders use AARs as opportunities to develop subordinates. During an AAR, give subordinates a chance to talk about how they saw things. Teach them how to look past a problem's symptoms to its root cause. Teach them how to give constructive, useful feedback. ("Here's what we did well; here's what we can do better.") When subordinates share in identifying reasons for success and failure, they become owners of a stake in how things get done. AARs also give you a chance to hear what's on your subordinates' minds—and

good leaders listen closely. (FM 25-101 and TC 25-20 discuss how to prepare, conduct, and follow up after AARs.)

5-63. Leaders base reviews on accurate observations and correct recording of those observations. If you're evaluating a ten-day field exercise, take good notes because you won't remember everything. Look at things in a systematic way; get out and see things firsthand. Don't neglect tasks that call for subjective judgment: evaluate unit cohesion, discipline, and morale. (FM 25-100 and FM 25-101 discuss training assessment.)

Initial Leader Assessments

5-64. Leaders often conduct an initial assessment before they take over a new position. How competent are your new subordinates? What's expected of you in your new job? Watch how people operate; this will give you clues about the organizational climate. (Remember SSG Withers and the vehicle inspection in Chapter 3?) Review the organization's SOP and any regulations that apply. Meet with the outgoing leader and listen to his assessment. (But don't take it as the absolute truth; everyone sees things through filters.) Review status reports and recent inspection results. Identify the key people outside the organization whose help you'll need to be successful. However, remember that your initial impression may be off-base. After you've been in the position for a while, take the necessary time to make an in-depth assessment.

5-65. And in the midst of all this checking and rechecking, don't forget to take a look at yourself. What kind of leader are you? Do you oversupervise? Undersupervise? How can you improve? What's your plan for working on your weak areas? What's the best way to make use of your strengths? Get feedback on yourself from as many sources as possible: your boss, your peers, even your subordinates. As Chapter 1 said in the discussion of character, make sure your own house is in order.

Assessment of Subordinates

5-66. Good leaders provide straightforward feedback to subordinates. Tell them where you see their strengths; let them know where they can improve. Have them come up with a plan of action for self-improvement; offer your help. Leader assessment should be a positive experience that your subordinates see as a chance for them to improve. They should see it as an opportunity to tap into your experience and knowledge for their benefit.

5-67. To assess your subordinate leaders, you must—

- Observe and record leadership actions. Figure 1-1 is a handy guide for organizing your thoughts.

- Compare what you see to the performance indicators in Appendix B or the appropriate reference.

- Determine if the performance meets, exceeds, or falls below standard.

- Tell your subordinates what you saw; give them a chance to assess themselves.

- Help your subordinate develop a plan of action to improve performance.

Leader Assessments and Plans of Action

5-68. Leader assessment won't help anyone improve unless it includes a plan of action designed to correct weaknesses and sustain strengths. Not only that, you and the subordinate must use the plan; it doesn't do anyone any good if you stick it in a drawer or file cabinet and never think about it again. Here is what you must do:

- Design the plan of action together; let your subordinate take the lead as much as possible.

- Agree on the actions necessary to improve leader performance; your subordinate must buy into this plan if it's going to work.

- Review the plan frequently, check progress, and change the plan if necessary.

(Appendix C discusses the relationship between a leader plan of action and developmental counseling.)

IMPROVING ACTIONS

How can you know if you've made a difference? Sometimes—rarely—the results are instant. Usually it takes much longer. You may see a soldier again as a seasoned NCO; you may get a call or a letter or see a name in the Army Times. *In most cases, you will never be sure how well you succeeded, but don't let that stop you.*

Command Sergeant Major John D. Woodyard, 1993

5-69. Improving actions are things leaders do to leave their organizations better than they found them. Improving actions fall into the categories highlighted in Figure 5-3: developing, building, and learning.

5-70. Developing refers to people: you improve your organization and the Army as an institution when you develop your subordinates.

5-71. Building refers to team building: as a direct leader, you improve your organization by building strong, cohesive teams that perform to standard, even in your absence.

5-72. Learning refers to you, your subordinates, and your organization as a whole. As a leader, you must model self-development for your people; you must constantly be learning. In addition, you must also encourage your subordinates to learn and reward their self-development efforts. Finally, you must establish an organizational climate that rewards collective learning and act to ensure your organization learns from its experiences.

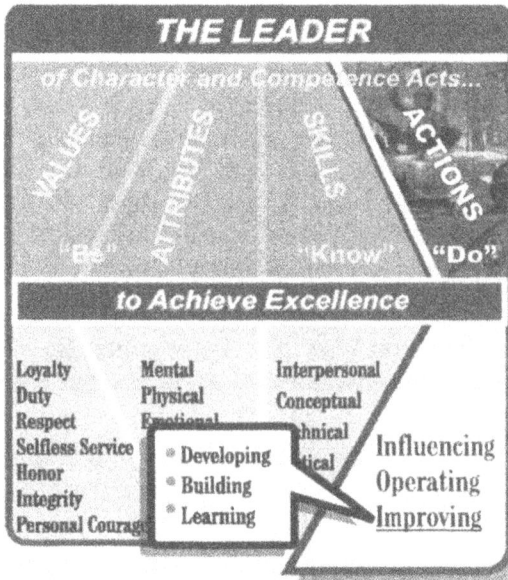

Figure 5-3. Direct Leader Actions —Improving

DEVELOPING

I've reminded many NCOs that they wouldn't be where they are today if someone hadn't given them a little extra time. I know I wouldn't be where I am.

Former Sergeant Major of the Army Glenn E. Morrell

5-73. In the Army, developing means developing people. Your subordinates are the leaders of tomorrow's Army. You have a responsibility to train them, to be the kind of leader they deserve so that they'll see how leading is done. It's your duty to invest the time and energy it takes to help them reach their fullest potential. The driving principle behind Army leader development is that leaders must be prepared before assuming leadership positions; they must be competent and confident in their abilities. This principle applies to all ranks and levels, to soldiers and DA civilians, and to both the active and reserve components.

5-74. As Figure 5-4 shows, a trained and ready Army rests on effective leader development. In turn, leader development rests on a foundation of training and education, expectations and standards, and values and ethics. This foundation supports the three leader development

pillars: institutional training (schooling), operational assignments, and self-development. (DA Pam 350-58 discusses Army leader development.)

Institutional Training

5-75. The Army school system provides formal education and training for job-related and leadership skills. The American public education system is progressive; that is, children attend primary school before middle school or junior high and then go on to high school. Likewise, the Army school system is progressive. The main difference is that you can expect to go out and use your skills in an assignment before being considered for the next level of schooling. Institutional training is critical in developing leaders and preparing them for increased positions of responsibility throughout the Army.

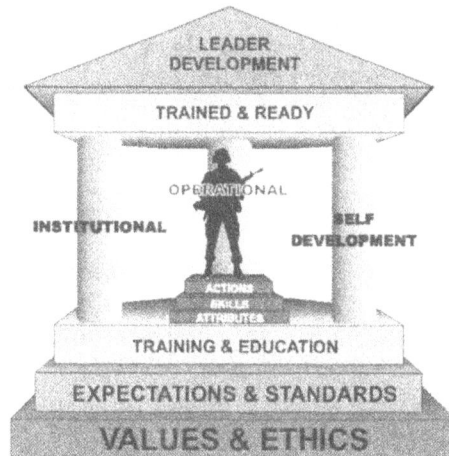

Figure 5-4. Leader Development

Operational Assignments

5-76. When you take what you've learned in school into the field Army, you continue to learn through on-the-job experience and by watching your leaders, peers, and subordinates. Operational assignments provide opportunities to broaden your knowledge and refine skills you gain during institutional training and previous assignments. You gain and expand your experience base by performing a wide range of duties and tasks under a variety of frequently changing conditions and situations. Operational assignments provide a powerful resource for leader development—an opportunity to learn by doing.

Self-Development

5-77. Self-development is a process you should use to enhance previously acquired skills, knowledge, and experience. Its goal is to increase your readiness and potential for positions of greater responsibility. Effective self-development focuses on aspects of your character, knowledge, and capabilities that you believe need developing or improving. You can use the dimensions of the Army leadership framework to help you determine what areas to work on. Self-development is continuous: it takes place during institutional training and operational assignments.

5-78. Self-development is a joint effort involving you, your first-line leader, and your commander. Commanders establish and monitor self-development programs for their organizations. You and your first-line leader together establish goals to meet your individual needs and plan the actions you must take to meet them. You do this as part of developmental counseling, which is discussed below and in Appendix C. Finally, you must execute your plan of action. If you have subordinates, you monitor how well they're acting on their plans of action. You can't execute their plans for them, but you can give them advice, encouragement, and—when necessary and mission permits—time.

5-79. Self-development for junior personnel is very structured and generally narrow in focus. The focus broadens as individuals learn their strengths and weaknesses, determine their individual needs, and become more independent. Everyone's knowledge and perspective increases with age, experience, institutional training, and operational assignments. Specific, goal-oriented self-development actions can accelerate and broaden a person's skills and knowledge. As a member of the Army, you're obligated to develop your abilities to the greatest extent possible. As an Army leader, you're responsible to assist your subordinates in their self-development.

5-80. Civilian and military education is part of self-development. Army leaders never stop learning. They seek to educate and train themselves beyond what's offered in formal schooling or even in their duty assignments. Leaders look for educational opportunities to prepare themselves for their next job and future responsibilities. Look for Army off-duty education that interests you and will give you useful skills. Seek civilian education to broaden your outlook on life. Look for things to read that will develop your mind and help you build skills. Challenge yourself and apply the same initiative here as you do in your day-to-day duties.

5-81. Remember that Army leaders challenge themselves and take advantage of work done by others in such fields as leadership and military history as well as in their off-duty areas of interest. In the leadership area, you can begin with some of the books listed in the bibliography or go to any bookstore or library. You'll find hundreds of titles under the heading of leadership. The libraries of your post, nearby civilian communities, and colleges contain works on these topics. In addition, the Internet can also be a useful place for obtaining information on some areas. However, be careful. Some books contain more reliable and useful information than others; the same is true of Internet sites.

5-82. Figure 5-4 also shows that actions, skills, and attributes form the foundation of success in operational assignments. This is where you, the leader, fit into Army leader development. As a leader, you help your subordinates internalize Army values. You also assist them in developing the individual attributes, learning the skills, and mastering the actions required to become leaders of character and competence themselves. You do this through the action of mentoring.

Mentoring

Good NCOs are not just born—they are groomed and grown through a lot of hard work and strong leadership by senior NCOs.

Former Sergeant Major of the Army
William A. Connelly

> **Mentoring** (in the Army) is the proactive development of each subordinate through observing, assessing, coaching, teaching, developmental counseling, and evaluating that results in people being treated with fairness and equal opportunity. Mentoring is an inclusive process (not an exclusive one) for everyone under a leader's charge.

5-83. Mentoring is totally inclusive, real-life leader development for every subordinate. Because leaders don't know which of their subordinates today will be the most significant contributors and leaders in the future, they strive to provide all their subordinates with the knowledge and skills necessary to become the best they can be—for the Army and for themselves.

5-84. Mentoring begins with the leader setting the right example. As an Army leader, you mentor people every day in a positive or negative way, depending on how you live Army values and perform leader actions. Mentoring shows your subordinates a mature example of values, attributes, and skills in action. It encourages them to develop their own character and leader attributes accordingly.

5-85. Mentoring links operating leader actions to improving leader actions. When you mentor, you take the observing, assessing, and evaluating you do when you operate and apply these actions to developing individual subordinates. Mentoring techniques include teaching, developmental counseling, and coaching.

> **Teaching** gives knowledge or provides skills to others, causing them to learn by example or experience.

5-86. Teaching is passing on knowledge and skills to subordinates. It's a primary task for first-line leaders. Teaching focuses primarily on technical and tactical skills. Developmental counseling is better for improving interpersonal and conceptual skills. Technical competence is critical to effective teaching. In order to develop subordinates, you must be able to demonstrate the technical and tactical skills you expect them to perform; otherwise they won't listen to you.

5-87. To be an Army leader, you must be a teacher. You give your subordinates knowledge and skills all the time, whether in formal, classroom settings or through your example. To be an effective teacher, you must first be professionally competent; then you must create conditions in which your subordinates can learn.

> *Soldiers learn to be good leaders from good leaders.*
>
> Former Sergeant Major of the Army Richard A. Kidd

5-88. The measure of how well you teach is how well your people learn. In most cases, your people will learn more by performing a skill than they will by watching you do it or by hearing you talk about how to do it. However, it's up to you to choose the teaching method that best fits the material. To make this choice, you need to understand the different ways people learn. People learn—

- Through the example of others (observing).
- By forming a picture in their minds of what they're trying to learn (thinking).
- By absorbing information (thinking).
- Through practice (hands-on experience).

5-89. Teaching is a complex art, one that you must learn in addition to the competencies you seek to teach. Just because you can pull the engine out of a tank doesn't mean you would be any good at teaching other people to do it. There are techniques and methods involved in teaching that have nothing to do with how good you are on the job; you must know both the skills related to the subject and another set of teaching skills. As an Army leader, you must develop these teaching skills as well. A subject matter expert who has acquired technical knowledge but is unable to teach that knowledge to others isn't improving the organization or the Army. (FM 25-101 addresses these and other areas related to conducting training.)

> **Developmental Counseling** is subordinate-centered communication that produces a plan outlining actions necessary for subordinates to achieve individual or organizational goals.

5-90. Developmental counseling is central to leader development. It's the means by which you prepare your subordinates of today to be the leaders of tomorrow. (Appendix C contains more details on developmental counseling.)

5-91. Developmental counseling isn't a time for war stories or for tales of how things were done way back when. It should focus on today's performance and problems and tomorrow's plans and solutions. Effective developmental counseling is centered on the subordinate, who is actively involved—listening, asking for more feedback, seeking elaboration of what the counselor has to say.

5-92. Developmental counseling isn't an occasional event that you do when you feel like it. It needs to be part of your program to develop your subordinates. It requires you to use all your counseling tools and skills. This means using counseling requirements such as those prescribed in the NCO Evaluation Reporting System (NCOERS), Officer Evaluation Reporting System (OERS), and Total Army Performance Evaluation System (TAPES, which is used to evaluate DA civilians) as more than paper drills. It means face-to-face counseling of individuals you rate. But more important, it means making time throughout the rating period to discuss performance objectives and provide meaningful assessments and feedback. No evaluation report—positive or negative—should be a surprise. A consistent developmental counseling program ensures your people know where they stand and what they should be doing to improve their performance and develop themselves. Your program should include all your people, not just the ones you think have the most potential. (The bibliography lists the evaluation and support forms prescribed by the OERS, NCOERS, and TAPES. Appendix C discusses

how to use support forms to assist you with developmental counseling.)

5-93. New direct leaders are sometimes uncomfortable confronting a subordinate who isn't performing to standard. However, remember that counseling isn't about how comfortable or uncomfortable you are; counseling is about correcting the performance or developing the character of a subordinate. Therefore, be honest and frank with your subordinates during developmental counseling. If you let your people get away with substandard behavior because you want them to like you or because you're afraid to make a hard call, you're sacrificing Army standards for your personal wellbeing—and you're not developing your subordinates.

5-94. This manual has emphasized throughout the importance of the example that you, as an Army leader, set for your subordinates. Your people look to you to see what kind of leader they want to be. The example you set in counseling is especially important. Army leaders at every level must ensure their subordinate leaders use counseling to develop their own subordinates. Setting the example is a powerful leadership tool: if you counsel your subordinates, your subordinate leaders will counsel theirs as well. The way you counsel is the way they'll counsel. Your people copy your behavior. The significance of your position as a role model can't be understated. It's a powerful teaching tool, for developmental counseling as well as other behaviors.

5-95. Although you're responsible for developing your subordinates, no leader can be all things to all people. In addition, the Army is already culturally diverse and is becoming increasingly technologically complex. In this environment, some of your subordinates may seek advice and counsel from informal relationships in addition to their leadership chain. Such relationships can be particularly important for women, minorities, and those in low-density specialties who have relatively few role models nearby.

5-96. This situation in no way relieves you, the leader, of any of your responsibilities regarding caring for and developing your people. Rather, being sensitive to your subordinates' professional development and cultural needs is part of the cultural awareness leader attribute. As an Army leader, you must know your people and take advantage of every resource available to help your subordinates develop as leaders. This includes other leaders who have skills or attributes different from your own.

> **Coaching** involves a leader's assessing performance based on observations, helping the subordinate develop an effective plan of action to sustain strengths and overcome weaknesses, and supporting the subordinate and the plan.

5-97. You can consider coaching to be both an operating and an improving leader action. It's less formal than teaching. When you're dealing with individuals, coaching is a form of specific instance counseling (which Appendix C discusses). When you're dealing with all or part of a team, it's generally associated with AARs (which you read about earlier in this chapter).

5-98. Coaching follows naturally from the assessing leader action. As you observe your subordinates at work, you'll see them perform some tasks to standard and some not to standard. Some of their plans will work; some won't. Your subordinates know when you're watching them. They expect you to tell them what they need to do to meet the standard, improve the team's performance, or develop themselves. You provide this sort of feedback through coaching. And don't limit your coaching to formal sessions. Use every opportunity to teach, counsel or coach from quarterly training briefings to AARs. Teaching moments and coaching opportunities occur all the time when you concentrate on developing leaders.

Mentoring and Developing Tomorrow's Army

5-99. Mentoring is demanding business, but the future of the Army depends on the trained and effective leaders whom you leave behind. Sometimes it requires you to set priorities, to balance short-term readiness with long-term leader development. The commitment to mentoring future leaders may require you to take risks. It requires you to give subordinates the opportunity to learn and develop themselves while using your experience to guide them without micromanaging. Mentoring will lead your subordinates to successes that build their confidence and skills for the future.

5-100. Mentoring isn't something new for the Army. Past successes and failures can often be traced to how seriously those in charge took the challenge of developing future leaders. As you consider the rapid pace of change in today's world, it's critical that you take the time to develop leaders capable of responding to that change. The success of the next generation of Army leaders depends on how well you accept the responsibility of mentoring your subordinates. Competent and confident leaders trained to meet tomorrow's challenges and fight and win future conflicts will be your legacy.

5-101. As you assume positions of greater responsibility, as the number of people for whom you are responsible increases, you need to do even more to develop your subordinates. More, in this case, means establishing a leader development program for your organization. It also means encouraging your subordinates to take actions to develop themselves personally and professionally. In addition, you may have to provide time for them to pursue self-development. (FM 25-101 discusses leader development programs.)

> What have YOU done TODAY to develop the leaders of tomorrow's Army?

BUILDING
Building Teams

5-102. You've heard—no doubt countless times—that the Army is a team. Just how important is it that people have a sense of the team? Very important. The national cause, the purpose of the mission, and all the larger concerns may not be visible from the battlefield.

Regardless of other issues, soldiers perform for the other people in the squad or section, for others in the team or crew, for the person on their right or left. This is a fundamental truth: soldiers perform because they don't want to let their buddies down.

5-103. If the leaders of the small teams that make up the Army are competent, and if their members trust one another, those teams and the larger team of teams will hang together and get the job done. People who belong to a successful team look at nearly everything in a positive light; their winners' attitudes are infectious, and they see problems as challenges rather than obstacles. Additionally, a cohesive team accomplishes the mission much more efficiently than a group of individuals. Just as a football team practices to win on the gridiron, so must a team of soldiers practice to be effective on the battlefield.

5-104. Training together builds collective competence; trust is a product of that competence. Subordinates learn to trust their leaders if the leaders know how to do their jobs and act consistently—if they say what they mean and mean what they say. Trust also springs from the collective competence of the team. As the team becomes more experienced and enjoys more successes, it becomes more cohesive.

Trust Earned

In a 1976 interview, Congressman Hamilton Fish of New York told of his experiences as a white officer with the 369th Infantry Regiment, an all-black unit in the segregated Army of 1917. Fish knew that his unit would function only if his soldiers trusted him; his soldiers, all of whom had volunteered for combat duty, deserved nothing less than a trustworthy leader. When a white regiment threatened to attack the black soldiers in training camp, Fish, his pistol drawn, alerted the leaders of that regiment and headed off a disaster.

"There was one thing they wanted above all from a white officer," [Fish recalled in an interview nearly 60 years later] "and that was fair treatment. You see, even in New York City [home of most of his soldiers] they really did not get a square deal most of the time. But if they felt you were on the level with them, they would go all out for you. And they seemed to have a sixth sense in realizing just how you felt. I sincerely wanted to lead them as real soldiers, and they knew it."

5-105. Developing teams takes hard work, patience, and quite a bit of interpersonal skill on the part of the leader, but it's a worthwhile investment. Good teams get the job done. People who are part of a good team complete the mission on time with the resources given them and a minimum of wasted effort; in combat, good teams are the most effective and take the fewest casualties.

5-106. Good teams—

- Work together to accomplish the mission.
- Execute tasks thoroughly and quickly.
- Meet or exceed the standard.
- Thrive on demanding challenges.
- Learn from their experiences and are proud of their accomplishments.

5-107. The Army is a team that includes members who are not soldiers but whose contributions are essential to mission success. The contributions made by almost 1,600 DA civilians in the Persian Gulf region were all but lost in the celebrations surrounding the military victory against Iraq and the homecoming celebration for the soldiers that followed. However, one safety specialist noted that these deployed DA civilians recognized the need for a team effort:

Patriotism was their drawing force for being there....We were part of the team supporting our soldiers! The focus is where it should be—on the military. They're here to do the job; we're here to help them.

5-108. People will do the most extraordinary things for their buddies. It's your job as an Army leader to pull each member into the team because you may someday ask that person for extraordinary effort. Team building involves applying interpersonal leader skills that transform individuals into productive teams. If you've done your work, the team member won't let you down.

5-109. Within a larger team, smaller teams may be at different stages of development. For instance, members of First Squad may be used to working together. They trust one another and get the job done—usually exceeding the standard—with no wasted motion. Second Squad in the same platoon just received three new soldiers and a team leader from another company. As a team, Second Squad is less mature; it will take them some time to get up to the level of First Squad. New team members have to learn how things work: they have to be brought on board and made to feel members of the team; they must learn the standards and the climate of their new unit; they'll have to demonstrate some competence before other members really accept them; and finally, they must practice working together. Leaders, who must oversee all this, are better equipped if they know what to expect. Make use of the information on the next few pages; learn what to look for—and stay flexible.

5-110. Figure 5-5 lists things you must do to pull a team together, get it going in the right direction, and keep it moving. And that list only hints at the work that lies ahead as you get your team to work together. Your subordinates must know—must truly believe—that they're a part of the team, that their contribution is important and valued. They must know that you'll train them and listen to them. They don't want you to let them get away with shoddy work or half-baked efforts; there's no pride in loafing. You must constantly observe, counsel, develop, listen; you must be every bit the team player you want your subordinates to be—and more.

5-111. Teams don't come together by accident; leaders must build and guide them through a series of developmental stages: formation, enrichment, and sustainment. This discussion may make the process seem more orderly than it actually is; as with so many things leaders do, the reality is more complicated than the explanation. Each team develops differently: the boundaries between stages are not hard and fast. As a leader, you must be sensitive to the characteristics of the team you're building and of its individual members—your people. Compare the characteristics of your team with the team building stage descriptions. The information that results can help you determine what to expect of your team and what you need to do to improve its capabilities.

Stages of Team Building

5-112. Teams, like individuals, have different personalities. As with individuals, the leader's job isn't to make teams that are clones of one another; the job is to make best use of the peculiar talents of the team, maximize the potential of the unit climate, and motivate aggressive execution.

5-113. **Formation stage.** Teams work best when new members are brought on board quickly, when they're made to feel a part of the team. The two steps—reception and orientation—are dramatically different in peace and war. In combat, this sponsorship process can literally mean life or death to new members and to the team.

5-114. Reception is the leader's welcome: the orientation begins with meeting other team members, learning the layout of the workplace, learning the schedule and other requirements, and generally getting to know the lay of the land. In combat, leaders may not have time to spend with new members. In this case, new arrivals are often assigned a buddy who will help them get oriented and keep them out of trouble until they learn their way around. Whatever technique you use, your soldiers should never encounter a situation similar to the one in the next example.

TEAM BUILDING STAGES

	SUBORDINATE CHALLENGES	LEADER & UNIT/ORGANIZATION ACTIONS
FORMATION STAGE GENERIC	• Achieve belonging and acceptance • Set personal & family concerns • Learn about leaders and other members	• Listen to and care for subordinates • Design effective reception and orientation • Communicate • Reward positive contributions • Set example
SOLDIER CRITICAL	• Face the uncertainty of war • Cope with fear of unknown injury and death • Adjust to sights and sounds of war • Adjust to separation from home and family	• Talk with each soldier • Reassure with calm presence • Communicate vital safety tips • Provide stable situation • Establish buddy system • Assist soldiers to deal with immediate problems
ENRICHMENT STAGE GENERIC	• Trust leaders & other members • Find close friends • Learn who is in charge • Accept the way things are done • Adjust to feelings about how things ought to be done • Overcome family-versus-unit conflict	• Trust and encourage trust • Allow growth while keeping control • Identify and channel emerging leaders • Establish clear lines of authority • Establish individual and unit goals • Train as a unit for mission • Build pride through accomplishment • Acquire self-evaluation/self-assessment habits • Be fair and give responsibility
SOLDIER CRITICAL	• Survive • Demonstrate competence • Become a team member quickly • Learn about the enemy • Learn about the battlefield • Avoid life-threating mistakes	• Train as a unit for combat • Demonstrate competence • Know the soldiers • Pace subordinate battlefield integration • Provide stable unit climate • Emphasize safety awareness for improved readiness
SUSTAINMENT STAGE GENERIC	• Trust others • Share ideas and feelings freely • Assist other team members • Sustain trust and confidence • Share mission and values	• Demonstrate trust • Focus on teamwork, training & maintaining • Respond to subordinate problems • Devise more challenging training • Build pride and spirit through unit sports, social & spiritual activities.
SOLDIER CRITICAL	• Adjust to continuous operations • Cope with casualties • Adjust to enemy actions • Overcome boredom • Avoid rumors • Control fear, anger, despair, and panic	• Observe and enforce sleep discipline • Sustain safety awareness • Inform soldiers • Know and deal with soldiers' perceptions • Keep soldiers productively busy • Use in-process reviews (IPRs) and after-action reviews (AARs) • Act decisively in face of panic

Figure 5-5. Team Building Stages

Replacements in the ETO

Most historians writing about World War II agree that the replacement system that fed new soldiers into the line units was seriously flawed, especially in the ETO, and did tremendous harm to the soldiers and the Army. Troops fresh from stateside posts were shuffled about in tent cities where they were just numbers. 1LT George Wilson, an infantry company commander who received one hundred replacements on December 29, 1944, in the midst of the Battle of the Bulge, remembers the results: "We discovered that these men had been on a rifle range only once; they had never thrown a grenade or fired a bazooka [antitank rocket], mortar or machine gun."

PVT Morris Dunn, another soldier who ended up with the 84th Division after weeks in a replacement depot recalls how the new soldiers felt: "We were just numbers, we didn't know anybody, and I've never felt so alone and miserable and helpless in my entire life—we'd been herded around like cattle at roundup time....On the ride to the front it was cold and raining with the artillery fire louder every mile, and finally we were dumped out in the middle of a heavily damaged town."

5-115. In combat, Army leaders have countless things to worry about; the mental state of new arrivals might seem low on the list. But if those soldiers can't fight, the unit will suffer needless casualties and may fail to complete the mission.

5-116. Discipline and shared hardship pull people together in powerful ways. SGT Alvin C. York, who won the Medal of Honor in an action you'll read about later in this chapter, talked about cohesion this way:

War brings out the worst in you. It turns you into a mad, fighting animal, but it also brings out something else, something I just don't know how to describe, a sort of tenderness and love for the fellow fighting with you.

5-117. However, the emotions SGT York mentions don't emerge automatically in combat. One way to ensure cohesion is to build it during peacetime. Team building begins with receiving new members; you know how important first impressions are when you meet someone new. The same thing is true of teams; the new member's reception and orientation creates that crucial first impression that colors the person's opinion of the team for a long time. A good experience joining the organization will make it easier for the new member to fit in and contribute. Even in peacetime, the way a person is received into an organization can have long-lasting effects—good or bad—on the individual and the team. (Appendix C discusses reception and integration counseling.)

Reception on Christmas Eve

An assistant division commander of the 25th Infantry Division told this story as part of his farewell speech:

"I ran across some new soldiers and asked them about their arrival on the island [of Oahu]. They said they got in on Christmas Eve, and I thought to myself, 'Can't we do a better job when we ship these kids out, so they're not sitting in some airport on their first big holiday away from home?' I mean, I really felt sorry for them. So I said, 'Must have been pretty lonesome sitting in a new barracks where you didn't know anyone.' And one of them said, 'No, sir. We weren't there a half-hour before the CQ [charge of quarters] came up and told us to get into class B's and be standing out front of the company in 15 minutes. Then this civilian drives up, a teenager, and the CQ orders us into the car. Turns out the kid was the first sergeant's son; his father had sent him over to police up anybody who was hanging around the barracks. We went over to the first sergeant's house to a big luau [party] with his family and a bunch of their neighbors and friends.'

5-118. **Enrichment stage.** New teams and new team members gradually move from questioning everything to trusting themselves, their peers, and their leaders. Leaders earn that trust by listening, following up on what they hear, establishing clear lines of authority, and setting standards. By far the most important thing a leader does to strengthen the team is training. Training takes a group of individuals and molds them into a team while preparing them to accomplish their missions. Training occurs during all three team building stages, but is particularly important during enrichment; it's at this point that the team is building collective proficiency.

5-119. **Sustainment stage.** When a team reaches this stage, its members think of the team as "their team." They own it, have pride in it, and want the team to succeed. At this stage, team members will do what needs to be done without being told. Every new mission gives the leader a chance to make the bonds even stronger, to challenge the team to reach for new heights. The leader develops his subordinates because they're tomorrow's team leaders. He continues to train the team so that it maintains proficiency in the collective and individual tasks it must perform to accomplish its missions. Finally, the leader works to keep the team going in spite of the stresses and losses of combat.

Building the Ethical Climate

5-120. As an Army leader, you are the ethical standard bearer for your organization. You're responsible for building an ethical climate that demands and rewards behavior consistent with Army values. The primary factor affecting an organization's ethical climate is its leader's ethical standard. Leaders can look to other organizational or installation personnel—for example, the chaplain, staff judge advocate, inspector general, and equal employment opportunity manager—to assist them in building and assessing their organization's ethical climate, but the ultimate responsibility belongs to the leader—period.

5-121. Setting a good ethical example doesn't necessarily mean subordinates will follow it. Some of them may feel that circumstances justify unethical behavior. (See, for example, the situation portrayed in Appendix D.) Therefore, you must constantly seek to maintain a feel for your organization's current ethical climate and take prompt action to correct any discrepancies between the climate and the standard. One tool to help you is the Ethical Climate Assessment Survey (ECAS), which is discussed in Appendix D. You can also use some of the resources listed above to help you get a feel for your organization's ethical climate. After analyzing the information gathered from the survey or other sources, a focus group may be a part of your plan of action to improve the ethical climate. Your abilities to listen and decide are the most important tools you have for this job.

5-122. It's important for subordinates to have confidence in the organization's ethical environment because much of what is necessary in war goes against the grain of the societal values individuals bring into the Army. You read in the part of Chapter 4 that discusses ethical reasoning that a soldier's conscience may tell him it's wrong to take human life while the mission of the unit calls for exactly that. Unless you've established a strong ethical climate that lets that soldier know his duty, the conflict of values may sap the soldier's will to fight.

SGT York

A conscientious objector from the Tennessee hills, Alvin C. York was drafted after America's entry into World War I and assigned to the 328th Infantry Regiment of the 82d Division, the "All Americans." PVT York, a devout Christian, told his commander, CPT E. C. B. Danforth, that he would bear arms against the enemy but didn't believe in killing. Recognizing PVT York as a potential leader but unable to sway him from his convictions, CPT Danforth consulted his battalion commander, MAJ George E. Buxton, about how to handle the situation.

MAJ Buxton was also deeply religious and knew the Bible as well as PVT York did. He had CPT Danforth bring PVT York to him, and they talked at length about the Scriptures, about God's teachings, about right and wrong, about just wars. Then MAJ Buxton sent PVT York home on leave to ponder and pray over the dilemma. The battalion commander promised to release him from the Army if PVT York decided he could not serve his country without sacrificing his integrity. After two weeks of reflection and deep soul-searching, PVT York returned, having reconciled his personal values with those of the Army. PVT York's decision had great consequences for both himself and his unit.

Alvin York performed an exploit of almost unbelievable heroism in the morning hours of 8 October 1918 in France's Argonne Forest. He was now a corporal (CPL), having won his stripes during combat in the Lorraine. That morning CPL York's battalion was moving across a valley to seize a German-held rail point when a German infantry battalion, hidden on a wooded ridge overlooking the valley, opened up with machine gun fire. The American battalion dived for cover, and the attack stalled. CPL York's platoon, already reduced to 16 men, was sent to flank the enemy machine guns.

As the platoon advanced through the woods to the rear of the German outfit, it surprised a group of about 25 German soldiers. The shocked enemy offered only token resistance, but then more hidden machine guns swept the clearing with fire. The Germans dropped safely to the ground, but nine Americans, including the platoon leader and the other two corporals, fell dead or wounded. CPL York was the only unwounded leader remaining.

CPL York found his platoon trapped and under fire within 25 yards of the enemy's machine gun pits. Nonetheless, he didn't panic. Instead, he began firing into the nearest enemy position, aware that the Germans would have to expose themselves to get an aimed shot at him. An expert marksman, CPL York was able to hit every enemy soldier who popped his head over the parapet.

After he had shot more than a dozen enemy, six German soldiers charged him with fixed bayonets. As the Germans ran toward him, CPL York once again drew on the instincts of a Tennessee hunter and shot the last man first (so the ones in front wouldn't see the ones he shot fall), then the fifth, and so on. After he had shot all the assaulting Germans, CPL York again turned his attention to the machine gun pits. In between shots, he called for the Germans to give up. It may have initially seemed ludicrous for a lone soldier in the open to call on a well-entrenched enemy to surrender, but their situation looked desperate to the German battalion commander, who had seen over 20 of his soldiers killed by this one American. The commander advanced and offered to surrender if CPL York would stop shooting.

CPL York now faced a daunting task. His platoon, now numbering seven unwounded soldiers, was isolated behind enemy lines with several dozen prisoners. However, when one American said their predicament was hopeless, CPL York told him to be quiet and began organizing the prisoners for a movement. CPL York moved his unit and prisoners toward American lines, encountering other German positions and forcing their surrender. By the time the platoon reached the edge of the valley they had left just a few hours before, the hill was clear of German machine guns. The fire on the Americans in the valley was substantially reduced and their advance began again.

SGT York (continued)

CPL York returned to American lines, having taken a total of 132 prisoners and putting 35 machine guns out of action. He left the prisoners and headed back to his own outfit. Intelligence officers questioned the prisoners and learned from their testimony the incredible story of how a fighting battalion was destroyed by one determined soldier armed only with a rifle and pistol. Alvin C. York was promoted to sergeant and awarded the Medal of Honor for this action. His character, physical courage, technical competence, and leadership enabled him to destroy the morale and effectiveness of an entire enemy infantry battalion.

5-123. CPT Danforth and MAJ Buxton could have ordered SGT York to go to war, or they might have shipped him out to a job that would take him away from the fight. Instead, these leaders carefully addressed the soldier's ethical concerns. MAJ Buxton, in particular, established the ethical climate by showing that he, too, had wrestled with the very questions that troubled SGT York. The climate these leaders created held that every person's beliefs are important and should be considered. MAJ Buxton demonstrated that a soldier's duties could be consistent with the ethical framework established by his religious beliefs. Leaders who create a healthy ethical environment inspire confidence in their subordinates; that confidence and the trust it engenders builds the unit's will. They create an environment where soldiers can truly "be all they can be."

LEARNING

For most men, the matter of learning is one of personal preference. But for Army [leaders], the obligation to learn, to grow in their profession, is clearly a public duty.

General of the Army Omar N. Bradley

5-124. The Army is a learning organization, one that harnesses the experience of its people and organizations to improve the way it does business. Based on their experiences, learning organizations adopt new techniques and procedures that get the job done more efficiently or effectively. Likewise, they discard techniques and procedures that have outlived their purpose. However, you must remain flexible when trying to make sense of your experiences. The leader who works day after day after day and never stops to ask "How can I do this better?" is never going to learn and won't improve the team.

5-125. Leaders who learn look at their experience and find better ways of doing things. Don't be afraid to challenge how you and your subordinates operate. When you ask "Why do we do it that way?" and the only answer you get is "Because we've always done it that way," it's time for a closer look. Teams that have found a way that works still may not be doing things the best way. Unless leaders are willing to question how things are, no one will ever know what can be.

"Zero Defects" and Learning

5-126. There's no room for the "zero-defects" mentality in a learning organization. Leaders willing to learn welcome new ways of looking at things, examine what's going well, and are not afraid to look at what's going poorly. When direct leaders stop receiving feedback from subordinates, it's a good indication that something is wrong. If the message you hammer home is "There will be no mistakes," or if you lose your temper and "shoot the messenger" every time there's bad news, eventually your people will just stop telling you when things go wrong or suggesting how to make things go right. Then there will be some unpleasant surprises in store. Any time you have human beings in a complex organization doing difficult jobs, often under pressure, there are going to be problems. Effective leaders use those mistakes to figure out how to do things better and share what they have learned with other leaders in the organization, both peers and superiors.

5-127. That being said, all environments are not learning environments; a standard of "zero-defects" is acceptable, if not mandatory, in some circumstances. A parachute rigger is charged with a "zero-defect" standard. If a rigger makes a mistake, a parachutist will die. Helicopter repairers live in a "zero-defect" environment as well. They can't allow aircraft to be mechanically unstable during flight. In these and similar work environments, safety concerns mandate a "zero-defects" mentality. Of course, organizations and people make mistakes; mistakes are part of training and may be the price of taking action. Leaders must make their intent clear and ensure their people understand the sorts of mistakes that are acceptable and those that are not.

5-128. Leaders can create a "zero-defects" environment without realizing it. Good leaders want their organizations to excel. But an organizational "standard" of excellence can quickly slide into "zero defects" if the leader isn't careful. For example, the published minimum standard for passing the APFT is 180 points—60 points per event. However, in units that are routinely assigned missions requiring highly strenuous physical activities, leaders need to train their people to a higher-than-average level of physical fitness. If leaders use APFT scores as the primary means of gauging physical fitness, their soldiers will focus on the test rather than the need for physical fitness. A better course would be for leaders to train their people on mission-related skills that require the higher level of physical readiness while at the same time motivating them to strive for their personal best on the APFT.

Barriers to Learning

5-129. Fear of mistakes isn't the only thing that can get in the way of learning; so can rigid, lockstep thinking and plain mental laziness. These habits can become learning barriers leaders are so used to that they don't even notice them. Fight this tendency. Challenge yourself. Use your imagination. Ask how other people do things. Listen to subordinates.

Helping People Learn

5-130. Certain conditions help people learn. First, you must motivate the person to learn. Explain to the subordinate why the subject is important or show how it will help the individual perform better. Second, involve the subordinate in the learning process; make it active. For instance, you would never try to teach someone how to drive a vehicle with classroom instruction alone; you have to get the person behind the wheel. That same approach applies to much more complex tasks; keep the lecture to a minimum and maximize the hands-on time.

5-131. Learning from experience isn't enough; you can't have every kind of experience. But if you take advantage of what others have learned, you get the benefit without having the experience. An obvious example is when combat veterans in a unit share their experiences with soldiers who haven't been to war. A less obvious, but no less important, example is when leaders share their experience with subordinates during developmental counseling.

After-Action Reviews and Learning

5-132. Individuals benefit when the group learns together. The AAR is one tool good leaders use to help their organizations learn as a group. Properly conducted, an AAR is a professional discussion of an event, focused on performance standards, that enables people to discover for themselves what happened, why it happened, and how to sustain strengths and improve on weaknesses. Like warning orders and rehearsals, the AAR is a technique that all leaders—military or DA civilian—can use in garrison as well as field environments. (FM 25-101 and TC 25-20 discuss how to prepare for, conduct, and follow up after an AAR.) When your team sits down for an AAR, make sure everyone participates and all understand what's being said. With input from the whole team, your people will learn more than if they just think about the experience by themselves.

Organizational Climate and Learning

5-133. It takes courage to create a learning environment. When you try new things or try things in different ways, you're bound to make mistakes. Learn from your mistakes and the mistakes of others. Pick your team and yourself up, determine what went right and wrong, and continue the mission. Be confident in your abilities. Theodore Roosevelt, a colonel during the Spanish-American War and twenty-sixth President of the United States, put it this way:

Whenever you are asked if you can do a job, tell 'em, Certainly I can!—and get busy and find out how to do it.

5-134. Your actions as a direct leader move the Army forward. How you influence your subordinates and the people you work for, how you operate to get the job done, how you improve the organization for a better future, all determine the Army's success or failure.

SUMMARY

5-135. Direct leaders influence their subordinates face-to-face as they operate to accomplish the mission and improve the organization. Because their leadership is face-to-face, direct leaders see the outcomes of their actions almost immediately. This is partly because they receive immediate feedback on the results of their actions.

5-136. Direct leaders influence by determining their purpose and direction from the boss's intent and concept of the operation. They motivate subordinates by completing tasks that reinforce this intent and concept. They continually acquire and assess outcomes and motivate

their subordinates through face-to-face contact and personal example.

5-137. Direct leaders operate by focusing their subordinates' activities toward the organization's objective and achieving it. Direct leaders plan, prepare, execute, and assess as they operate. These functions sometimes occur simultaneously.

5-138. Direct leaders improve by living Army values and providing the proper role model for subordinates. Leaders must develop all subordinates as they build strong, cohesive teams and establish an effective learning environment.

PART THREE

Organizational and Strategic Leadership

As they mature and assume greater responsibilities, Army leaders must also learn new skills, develop new abilities, and act in more complex environments. Organizational and strategic leaders maintain their own personalities and propensies, but they also expand what they know and refine what they do.

Chapters 6 and 7 describe (rather than prescribe or mandate) skills and actions required of organizational and strategic leaders. The chapters discuss much of what developing leaders often sense and explore some concepts that may seem foreign to them. Neither chapter outlines exhaustively what leaders know and do at higher levels; they simply introduce what's different.

The audience for Chapters 6 and 7 is only in part organizational and strategic leaders, who have prepared to serve in those positions by career-long experience and study. Primarily, these chapters offer staffs and subordinates who work for those leaders insight into the additional concerns and activities of organizational and strategic leadership.

Chapter 6

Organizational Leadership

6-1. During the Battle of the Bulge, with the Germans bearing down on retreating US forces, PFC Vernon L. Haught dug in and told a sergeant in a tank destroyer, "Just pull your vehicle behind me . . . I'm the 82d Airborne, and this is as far as the bastards are going." He knew his division commander's intent. Despite desperate odds, he had confidence in himself and his unit and knew they would make the difference. Faced with a fluid situation, he knew where the line had to be drawn; he had the will to act and he didn't hesitate to do what he thought was right.

6-2. Whether for key terrain in combat or for results in peacetime training, leaders in units and organizations translate strategy into policy and practice. They develop programs, plans, and systems that allow soldiers in teams, like the infantryman in the All-American Division, to turn plans and orders into fire and maneuver that seize victory at the least possible cost in sweat and blood. By force of will and application of their leadership skills, organizational leaders build teams with discipline, cohesion, trust, and proficiency. They clarify missions throughout the ranks by producing an intent, concept, and systematic approach to execution.

6-3. Organizational leadership builds on direct leader actions. Organizational leaders apply direct leader skills in their daily work with their command and staff teams and, with soldiers and subordinate leaders, they influence during their contacts with units. But to lead complex organizations like brigades, divisions, and corps at today's OPTEMPO and under the stresses of training, contingency operations, and combat, organizational leaders must add a whole new set of skills and actions to their leadership arsenal. They must practice direct and organizational leadership simultaneously.

6-4. Communicating to NCOs, like the airborne soldier at the Battle of the Bulge, occurs through individual subordinates, the staff, and the chain of command. Organizational leaders divide their attention between the concerns of the larger organization and their staffs and those of their subordinate leaders, units, and individuals. This tradeoff requires them to apply interpersonal and conceptual skills differently when exercising organizational leadership than when exercising direct leadership.

6-5. Organizational leaders rely heavily on mentoring subordinates and empowering them to execute their assigned responsibilities and missions. They stay mentally and emotionally detached from their immediate surroundings so they can visualize the larger impact on the organization and mission. Soldiers and subordinate leaders look to their organizational leaders

to establish standards for mission accomplishment and provide resources (conditions) to achieve that goal. Organizational leaders provide direction and programs for training and execution that focus efforts on mission success.

6-6. Due to the indirect nature of their influence, organizational leaders assess interrelated systems and design long-term plans to accomplish the mission. They must sharpen their abilities to assess their environments, their organization, and their subordinates. Organizational leaders determine the cause and effect of shortcomings, translate these new understandings into plans and programs, and allow their subordinate leaders latitude to execute and get the job done.

6-7. Organizational demands also differ as leaders develop a systems perspective. At the strategic level, the Army has identified six imperatives: quality people, training, force mix, doctrine, modern equipment, and leader development. In organizations these imperatives translate into doctrine, training, leader development, organization, materiel, and soldiers—commonly called DTLOMS. Together with Army values, these systems provide the framework for influencing people and organizations at all levels, conducting a wide variety of operations, and continually improving the force. Doctrine includes techniques to drive the functional systems in Army organizations. FMs 25-100, 25-101, and 101-5 lay out procedures for training management and military decision making that enable and focus execution. The training management and military decision-making processes provide a ready-made, systemic approach to planning, preparing, executing, and assessing.

SECTION I

WHAT IT TAKES TO LEAD ORGANIZATIONS—SKILLS

6-8. Organizational leaders continue to use the direct leader skills discussed in Chapter 4. However their larger organizations and spans of authority require them to master additional skills. As with direct leader skills, these span four areas: interpersonal, conceptual, technical, and tactical.

INTERPERSONAL SKILLS

To get the best out of your men, they must feel that you are their real leader and must know that they can depend upon you.

General of the Armies John J. Pershing

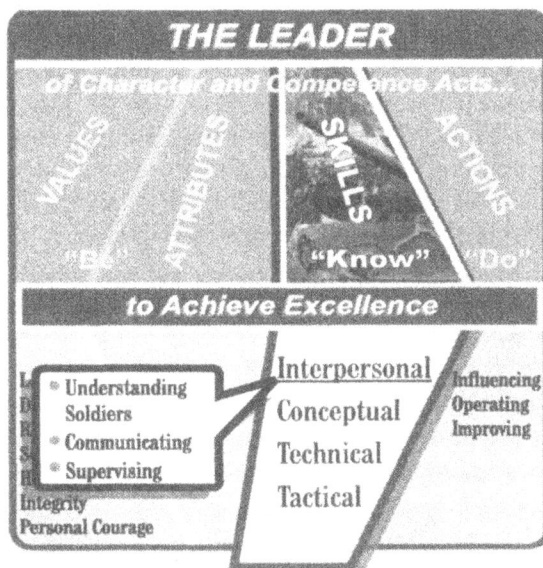

Figure 6-1. Organizational Leader Skills—Interpersonal

UNDERSTANDING SOLDIERS

6-9. Good organizational leaders understand the human dimension, which Chapter 3 discussed. They use that understanding to motivate subordinates and to encourage initiative. Chapter 5 explained that motivation means much more than an individual willingness to do what's directed. It imparts a desire on the part of individuals and organizations to do what's needed *without* being directed. This collective desire to accomplish the mission underlies good organizational discipline: good soldiers and competent DA civilians adhere to standards because they understand that doing so, even when it's a nuisance or hardship, leads to success.

6-10. This understanding, along with Army values, forms the foundation of great units. Units that have solid discipline can take tremendous stress and friction yet persevere, fight through, and win. Fostering initiative builds on motivation and discipline. It requires subordinates' confidence that in an uncertain situation, when they know the commander's intent and develop a competent solution, the commander will underwrite the risk they take. While this principle applies to both direct and organizational leaders, the stakes are usually higher in larger, more complex organizations. Additionally, organizational leaders may be more remote in time and distance and subordinates' ability to check back with them is diminished. Therefore, organizational leaders' understanding must develop beyond what they can immediately and personally observe.

COMMUNICATING

6-11. Persuasion is a communication skill important to organizational leaders. Well-developed skills of persuasion and an openness to working through controversy in a positive way help organizational leaders overcome resistance and build support. These characteristics are particularly important in dealing with other organizational leaders. By reducing grounds for misunderstanding, persuasion reduces time wasted in overcoming unimportant issues. It also ensures involvement of others, opens communication with them, and places value on their opinions—all team-building actions. Openness to discussing one's position and a positive attitude toward a dissenting view often diffuses tension and saves time and resistance in the long run. By demonstrating these traits, organizational leaders also provide an example that subordinates can use in self-development.

6-12. In some circumstances, persuasion may be inappropriate. In combat, all leaders make decisions quickly, modifying the decision-making process to fit the circumstances. But this practice of using the directing leadership styles as opposed to more participatory ones should occur when situations are in doubt, risks are high, and time is short—circumstances that often appear in combat. No exact blueprints exist for success in every context; leadership and the ability to adapt to the situation will carry the day. Appropriate style, seasoned instinct, and the realities of the situation must prevail.

SUPERVISING

6-13. Organizations pay attention to things leaders check. Feedback and coaching enhance motivation and improve performance by showing subordinates how to succeed. But how much should you check and how much is too much? When are statistics and reports adequate indicators and when must you visit your front-line organizations, talk to your soldiers and DA civilians and see what's going on yourself?

6-14. Overcentralized authority and oversupervising undermine trust and empowerment. Undersupervising can lead to failure, especially in cases where the leader's intent wasn't fully understood or where subordinate organizations lack the training for the task. Different subordinate commanders need different levels of supervision: some need a great deal of coaching and encouragement, though most would just as soon be left alone. As always, a good leader knows his subordinates and has the skill to supervise at the appropriate level.

Knowing Your People

This General said, "Each of our three regimental commanders must be handled differently. Colonel 'A' does not want an order. He wants to do everything himself and always does well. Colonel 'B' executes every order, but has no initiative. Colonel 'C' opposes everything he is told to do and wants to do the contrary."

A few days later the troops confronted a well-entrenched enemy whose position would have to be attacked. The General issued the following orders:

To Colonel "A" (who wants to do everything himself): "My dear Colonel "A", I think we will attack. Your regiment will have to carry the burden of the attack. I have, however, selected you for this reason. The boundaries of your regiment are so-and-so. Attack at X-hour. I don't have to tell you anything more."

To Colonel "C" (who opposes everything): "We have met a very strong enemy. I am afraid we will not be able to attack with the forces at our disposal." "Oh, General, certainly we will attack. Just give my regiment the time of attack and you will see that we are successful," replied Colonel "C." "Go then, we will try it," said the General, giving him the order for the attack, which he had prepared some time previously.

To Colonel "B" (who always must have detailed orders) the attack order was merely sent with additional details.

All three regiments attacked splendidly.

Adolph von Schell
German liaison to the Infantry School between the World Wars

CONCEPTUAL SKILLS

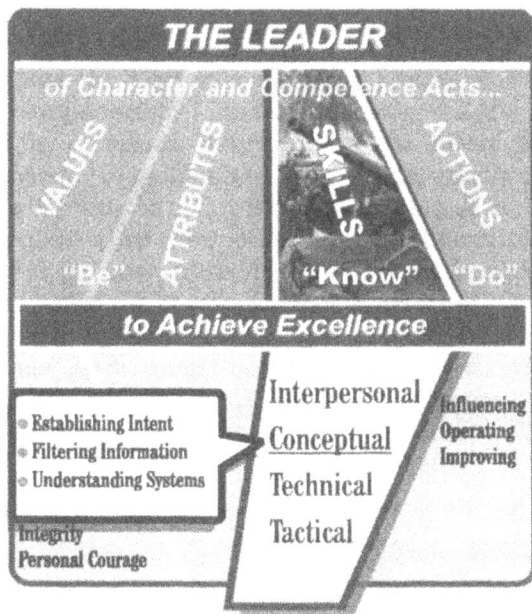

Figure 6-2. Organizational Leader Skills— Conceptual

6-15. The complexity of the organizational leader's environment requires patience, the willingness to think before acting. Furthermore, the importance of conceptual and analytical skills increases as an organizational leader moves into positions of greater responsibility. Organizational environments with multiple dimensions offer problems that become more abstract, complex, and uncertain.

6-16. Figure 6-2 identifies the conceptual skills required of organizational leaders. For organizational leaders, reasoning skills are crucial for developing intent and direction toward common goals. Critical thinking at the organizational level requires understanding systems and an increased ability to filter information, that is, to identify quickly information that applies to the task at hand and separate the important from the unimportant. Organizational leaders use this analytical ability to assess ambiguous environments and to calculate and manage risk. Their experience may allow them to see and define problems more easily—but not necessarily *fix* them quickly. Therefore, they

also dedicate time to think and generate alternative ways of organizing their organizations and resources for maximum effect. It's important for organizational leaders to encourage critical thinking in subordinates because subordinates also assess organizational challenges, analyze indicators, and recommend courses of action. It's also important, time and mission permitting, to allow subordinates' solutions to bear fruit.

ESTABLISHING INTENT

In an organization like ours, you have to think through what it is that you are becoming. Like a marathon runner, you have to get out in front, mentally, and pull the organization to you. You have to visualize the finish line—to see yourself there—and pull yourself along—not push—pull yourself to the future.

General Gordon R. Sullivan
Former Army Chief of Staff

6-17. Intent is the leader's personal expression of a mission's end state and the key tasks the organization must accomplish to achieve it. During operations and field training, it's a clear, concise statement of what the force must do to succeed with respect to the enemy and the terrain and to the desired end state. It provides the link between the mission and the concept of operations. By describing their intent, organizational leaders highlight the key tasks that, along with the mission, are the basis for subordinates to exercise initiative when unanticipated opportunities arise or when the original concept of operations no longer applies. Clear and concise, the leader's intent includes a mission's overall purpose and expected results. It provides purpose, motivation, and direction, whether the leader is commanding a division or running a staff directorate. An organizational leader visualizes the sequence of activities that will move the organization from its current state to the desired end state and expresses it as simply and clearly as possible. (FM 101-5-1 contains a complete definition of commander's intent. FM 100-34 discusses the relationship of intent and visualization to command and control. FM 101-5 discusses the development of intent in the MDMP.)

6-18. After establishing a clear and valid intent, the art of organizational leadership lies in having subordinates take actions on their own to transform that intent into reality. Since organizational leaders are likely to be farther away from the point of execution in time and space, they must describe the collective goal rather than list tasks for individual subordinates. With clearly communicated purpose and direction, subordinates can then determine what they must do and why. Within that broad framework, leaders empower subordinates, delegating authority to act within the intent: "Here's where we're headed, why we're going there, and how we're going to get there." Purpose and direction align the efforts of subordinates working toward common goals.

6-19. A former division commander has said, "You must be seen to be heard." There's a great temptation for organizational leaders to rely exclusively on indirect leadership, to spread intent by passing orders through subordinates or communicating electronically with troops scattered far and wide. However, nothing can take the place of face-to-face contact. Organizational leaders make every effort to get out among the troops. There they can spot-check intent to see that it's disseminated and understood among those who must execute it.

GEN Grant and the End of the Civil War

I propose to fight it out on this line if it takes all summer.

General Ulysses S. Grant
Dispatch, May 11, 1864

GEN Ulysses S. Grant penned those words at Spotsylvania, Virginia, after being appointed general in chief of the Union Army and stationing himself forward with the Army of the Potomac. After fighting a bloody draw at the Wilderness, the Army of the Potomac had moved aggressively to outflank GEN Robert E. Lee's Army of Northern Virginia but once more faced a dug-in Confederate Army. However, where previous Union commanders had turned away, GEN Grant would not relent. His intent was clear. It was reinforced in the sentiment of his army, which wanted to finish the war.

In a series of determined attacks, the new Union commander broke GEN Lee's defense and used a turning movement to force his opponent out of position. Again they met at Cold Harbor, where GEN Grant attacked frontally and failed. Prior to his attack, Union soldiers, literally days from the end of their enlistment, were seen writing notes to their families and pinning them on the backs of their shirts so one last message would get home if they were killed. Their resolution demonstrates the power of their commitment to a shared intent.

After his bloody repulse at Cold Harbor, GEN Grant again moved south, maintaining the initiative, always pressing, always threatening to turn the Confederate flank and expose Richmond, the capital of Virginia. GEN Lee, in turn, was forced to block and defend Petersburg, Richmond's railroad hub. Uncovering it would have isolated Richmond as well as his army's rail-based lines of communication. GEN Grant had his opponent pinned to a critical strategic resource. Doing this denied the Army of Northern Virginia its greatest asset, its excellent ability to maneuver. It would not escape.

GEN Grant was unstinting in his resolve to totally defeat GEN Lee. He was a familiar figure to his soldiers, riding among them with his slouch hat in his private's uniform with general's rank. He drove his subordinates, who themselves wanted to finish off their old foe; despite casualties, he unflinchingly resisted pressure to back away from his intent.

6-20. Having established an azimuth, organizational leaders assist their subordinates' efforts to build and train their organizations on those tasks necessary for success. Finally, they act to motivate subordinate leaders and organizations to meet the operational standards upon which discipline depends.

FILTERING INFORMATION

Leaders at all levels, but particularly those at higher levels who lack recent personal observations, can only make decisions based on the information given to them. What sets senior leaders apart is their ability to sort through great amounts of information, key in on what is significant, and then make decisions. But, these decisions are only as good as the information provided.

A Former Battalion Commander

6-21. Organizational leaders deal with a tremendous amount of information. Some information will make sense only to someone with a broad perspective and an understanding of the entire situation. Organizational leaders communicate clearly to their staffs what information they need and then hold the staff accountable for providing it. Then, they judge—based on their education, training, and experience—what's important and make well-informed, timely decisions.

6-22. Analysis and synthesis are essential to effective decision-making and program development. Analysis breaks a problem into its component parts. Synthesis assembles complex and disorganized data into a solution. Often, data must be processed before it fits into place.

6-23. Commander's critical information requirements (CCIR) are the commander's most important information filters. Commanders must know the environment, the situation, their organizations, and themselves well enough to articulate what they need to know to control their organizations and accomplish their missions. They must also ensure they have thought through the feedback systems necessary to supervise execution. Organizational-level commanders must not only establish CCIR but also train their staffs to battle drill proficiency in information filtering. (FM 101-5 discusses CCIR, mission analysis, information management, and other staff operations.)

UNDERSTANDING SYSTEMS

6-24. Organizational leaders think about systems in their organization: how they work together, how using one affects the others, and how to get the best performance from the whole. They think beyond their own organizations to how what their organization does affects other organizations and the team as a whole. Whether coordinating fires among different units or improving sponsorship of new personnel, organizational leaders use a systems perspective. While direct leaders think about tasks, organizational leaders integrate, synchronize, and fine-tune systems and monitor outcomes. If organizational leaders can't get something done, the flaw or failure is more likely systemic than human. Being able to understand and leverage systems increases a leader's ability to achieve organizational goals and objectives.

6-25. Organizational leaders also know how effectively apply all available systems to achieve mission success. They constantly make sure that the systems for personnel, administration, logistical support, resourcing, and training work effectively. They know where to look to see if the critical parts of the system are functioning properly.

DA Civilian Support to Desert Shield

During Operation Desert Shield, a contingent of DA civilians deployed to a depot in the combat theater to provide warfighting supplies and operational equipment to Third (US) Army. These DA civilians were under the supervision of DA civilian supervisors, who motivated their employees in spite of the harsh conditions in the region: hot weather, a dismal environment, and the constant threat of Iraqi missile and chemical attacks. It turned out that the uplifting organizational climate these leaders provided overcame the physical deprivation.

Two senior DA civilian leaders, the depot's deputy director of maintenance and the chief of the vehicle branch, developed a plan to replace arriving units' M1 tanks with M1A1s, which boasted greater firepower, better armor, and a more advanced nuclear, biological, and chemical protective system. They also developed systems for performing semiannual and annual maintenance checks, quickly resolving problems, applying modifications such as additional armor, and repainting the tanks in the desert camouflage pattern.

Although similar programs normally take 18 to 24 months to complete, the two leaders set an ambitious objective of 6 months. Many experts thought the goal could not be met, but the tenacious leaders never wavered in their resolve. After 24-hour-a-day, 7-day-a week operations, their inspired team of teams completed the project in 2 months. These DA civilian leaders with clear intent, firm objectives, and unrelenting will motivated their team and provided modern, lethal weapons to the soldiers who needed them when they were needed.

6-26. Organizational leaders analyze systems and results to determine why things happened the way they did. Performance indicators and standards for systems assist them in their analysis. Equipment failure rates, unit status reports (USR) Standard Installation/Division Personnel System (SIDPERS), Defense Civilian Personnel Data System (DCPDS) data, and evaluation report timeliness all show the health of systems. Once an assessment is complete and causes of a problem known, organizational leaders develop appropriate solutions that address the problem's root cause.

6-27. Isolating why things go wrong and where systems break down usually requires giving subordinates time and encouragement to ferret out what's really happening. The dilemma for organizational leaders occurs when circumstances and mission pressures require immediate remedial action and preclude gathering more data. It's then that they must fall back on their experience and that of their subordinates, make a judgment, and act.

Innovative Reorganization

Facing a long-term downsizing of his organization, a DA civilian director didn't simply shrink its size. Instead, the director creatively flattened the organization by reducing the number of deputy executives, managers, and supervisors. The director increased responsibilities of those in leadership positions and returned to a technical focus those managers and supervisors with dominant mission skills. The result was a better leader-to-led ratio, a reduced number of administrative and clerical positions, and a smoother transition to multidisciplined team operations. The director's systems understanding led him to tailor inputs that maintained healthy systems and improved outputs.

TECHNICAL SKILLS

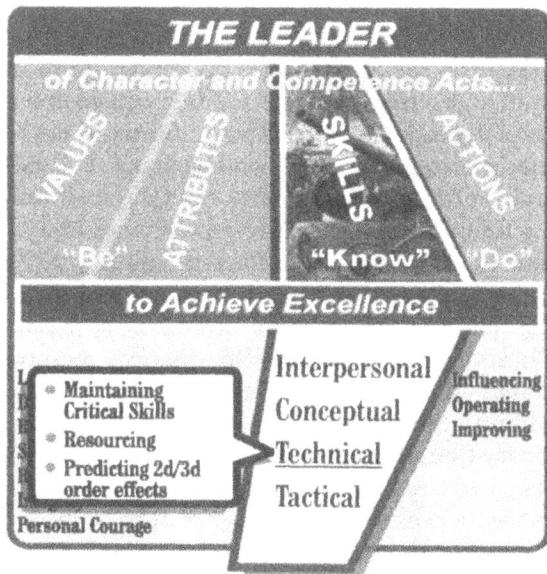

Figure 6-3. Organizational Leader Skills—Technical

6-28. The external responsibilities of organizational leaders are greater than those of direct leaders, both vertically and horizontally. Their organizations have more levels than direct-level organizations and, depending on the organization's role, command interest may reach to the CINC or national command authority. There are more requirements to coordinate with other organizations, which may include agencies outside the Department of Defense (DOD). To make full use of their organizations' capabilities, organizational leaders must continue to master technical skills outside their original area of expertise.

MAINTAINING CRITICAL SKILLS

6-29. Organizational leaders have fewer opportunities to practice many of the technical skills they mastered as direct leaders. However, that doesn't mean they can forget about them. In every organization there are certain skills in which all members must be proficient. Soldiers know what they are and expect their leaders to be able to perform them. This doesn't mean that organizational leaders must be able to perform every specialty-related skill as well as an individual holding that specialty. The Army is

too complex for that. It does, however, mean that organizational leaders must identify and be proficient in those critical, direct-leader skills they need to assess tactical training and set the example.

6-30. One organizational leader who set the example by drawing on deeply embedded technical skills was COL Marian Tierney. In her final military assignment, COL Tierney was responsible for nursing operations at 38 hospitals with 2500 nurses in the Republic of Vietnam. In 1966 she had to call on the basic medical skills and personal character she had honed throughout a career in places like Omaha Beach during the Normandy invasion. That day, 22 years after D-Day, the aircraft on which she was a passenger crashed, leaving many injured and panicked survivors. Ignoring her own injuries COL Tierney treated her comrades and took charge of evacuating the scene. For her heroism she received the Soldier's Medal. Her actions demonstrate that courageous leaders of character and competence serve at all levels.

RESOURCING

6-31. In addition to using the technical skills they learned as direct leaders, organizational leaders must also master the skill of resourcing. Resources—which include time, equipment, facilities, budgets, and people—are required to achieve organizational goals. Organizational leaders must aggressively manage the resources at their disposal to ensure their organizations' readiness. The leader's job grows more difficult when unprogrammed costs—such as an emergency deployment—shift priorities.

6-32. Organizational leaders are stewards of their people's time and energy and their own will and tenacity. They don't waste these resources but skillfully evaluate objectives, anticipate resource requirements, and efficiently allocate what's available. They balance available resources with organizational requirements and distribute them in a way that best achieves organizational goals—in combat as well as peacetime. For instance, when a cavalry squadron acting as the division flank guard makes contact, its commander asks for priority of fires. The division commander considers the needs of the squadron but must weigh it against the overall requirements of the current and future missions.

PREDICTING SECOND- AND THIRD-ORDER EFFECTS

6-33. Because the decisions of organizational leaders have wider-ranging effects than those of direct leaders, organizational leaders must be more sensitive to how their own actions affect the organization's climate. These actions may be conscious, as in the case of orders and policies, or unconscious, such as requirements for routine or unscheduled reports and meetings. The ability to discern and predict second- and third-order effects helps organizational leaders assess the health of the organizational climate and provide constructive feedback to subordinates. It can also result in identifying resource requirements and changes to organizations and procedures. (The ECAS process illustrated in Appendix D or a similar one can be applied by organizational as well as direct leaders.)

6-34. For instance, when the Army Chief of Staff approved a separate military occupational specialty code for mechanized infantry soldiers, the consequences were wide-ranging. Second-order effects included more specialized schooling for infantry NCOs, a revised promotion system to accommodate different infantry NCO career patterns, and more doctrinal and training material to support the new specialty. Third-order effects included resource requirements for developing the training material and adding additional instructor positions at the Infantry Center and School. Organizational leaders are responsible for anticipating the consequences of any action they take or direct. Requiring thorough staff work can help. However, proper anticipation also requires imagination and vision as well as an appreciation for other people and organizations.

TACTICAL SKILLS

Soldiers need leaders who know how to fight and how to make the right decisions.

General Carl F. Vuono
Former Army Chief of Staff

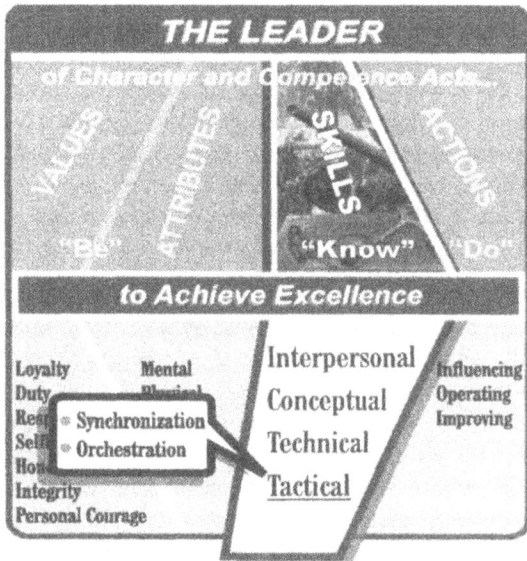

Figure 6-4. Organizational Leader Skills—Tactical

6-35. Organizational leaders must master the tactical skills of synchronization and orchestration. Synchronization applies at the tactical level of war; orchestration is an operational-level term. Synchronization arranges activities in time, space, and purpose to focus maximum relative military power at a decisive point in space and time. Organizational leaders synchronize battles, each of which may comprise several synchronized engagements. (FM 100-40 discusses synchronization. FM 100-5 discusses orchestration.)

6-36. Organizational leaders at corps and higher levels orchestrate by applying the complementary and reinforcing effects of all military and nonmilitary assets to overwhelm opponents at one or more decisive points. Both synchronization and orchestration require leaders to put together technical, interpersonal, and conceptual skills and apply them to war-fighting tasks.

6-37. Tactical skill for direct leaders involves employing individuals and teams of company size and smaller. In contrast, tactical skill for organizational leaders entails employing *units* of battalion size and larger. Organizational leaders get divisions, brigades, and battalions to the right place, at the right time, and in the right combination to fight and win battles and engagements. (FM 100-40 discusses battles and engagements.) They project the effects of their decisions further out—in time and distance—than do direct leaders.

6-38. The operational skill of orchestrating a series of tactical events is also more demanding and far-reaching. Time horizons are longer. Effects take more time to unfold. Decision sets are more intricate. GEN Grant's Vicksburg campaign in the spring of 1863, which split the Confederacy and opened the Mississippi River to Union use, is a classic example of an organizational leader orchestrating the efforts of subordinate forces.

GEN Grant at Vicksburg

After failing to capture Vicksburg by attacking from the north, GEN Ulysses S. Grant moved along the west bank of the Mississippi River to a point south of the city. He masked his movement and intentions by sending COL Benjamin Grierson's cavalry deep into Mississippi to conduct a series of raids. The Union commander also synchronized the daring dispatch of US Navy gunboats through Confederate shore batteries to link up with his army south of Vicksburg. Using Admiral (ADM) David D. Porter's gunboats, the Union Army crossed to the east bank of the Mississippi while MG William T. Sherman conducted a diversionary attack on the northern approaches to Vicksburg.

Once across the Mississippi, GEN Grant bypassed Vicksburg, used the Big Black River to protect his flank, and maneuvered east toward Jackson, Mississippi. By threatening both Jackson and Vicksburg, GEN Grant prevented Confederate forces from uniting against him. After a rapid series of engagements, the Union Army forced the enemy out of Jackson, blocking Vicksburg's main line of supply. It then turned west for an assault of Vicksburg, the key to control of the Mississippi. With supply lines severed and Union forces surrounding the city, Confederate forces at Vicksburg capitulated on 4 July 1863.

GEN Grant's Vicksburg campaign demonstrates the orchestration of a series of subordinate unit actions. In a succession of calculated moves, he defeated the Confederate forces under the command of Generals Joseph E. Johnston and John C. Pemberton, gained control of the Mississippi River, and divided the Confederacy.

6-39. Organizational leaders know doctrine, tactics, techniques, and procedures. Their refined tactical skills allow them to understand, integrate, and synchronize the activities of systems, bringing all resources and systems to bear on warfighting tasks.

SECTION II

WHAT IT TAKES TO LEAD ORGANIZATIONS—ACTIONS

Making decisions, exercising command, managing, administering—those are the dynamics of our calling. Responsibility is its core.

General Harold K. Johnson
Former Army Chief of Staff

6-40. Actions by organizational leaders have far greater consequences for more people over a longer time than those of direct leaders. Because the connections between action and effect are sometimes more remote and difficult to see, organizational leaders spend more time thinking about what they're doing and how they're doing it than direct leaders do. When organizational leaders act, they must translate their intent into action through the larger number of people working for them.

6-41. Knowledge of subordinates is crucial to success. To maximize and focus the energy of their staffs, organizational leaders ensure that subordinates know what must be done and why. In addition, they ensure that work being done is moving the organization in the right direction. They develop concepts for operations and policies and procedures to control and monitor their execution. Since the challenges they face are varied and complicated, no manual can possibly address them all. However, the following section provides a framework for examining, explaining, and reflecting on organizational leader actions.

INFLUENCING ACTIONS

A soldier may not always believe what you say, but he will never doubt what you do.

The Battalion Commander's Handbook

6-42. As Figure 6-5 shows, influencing is achieved through communicating, decision making, and motivating. At the organizational level, influencing means not only getting the order or concept out; it means marshaling the activities of the staff and subordinate leaders to move towards the organization's objective. Influencing involves continuing to reinforce the intent and concept, continually acquiring and assessing available feedback, and inspiring subordinates with the leader's own presence and encouragement.

6-43. The chain of command provides the initial tool for getting the word out from, and returning feedback to, the commander. In training, commanders must constantly improve its functioning. They must stress it in training situations, pushing it to the point of failure. Combat training centers (CTCs) offer tremendous opportunities to exercise and assess the chain of command in their

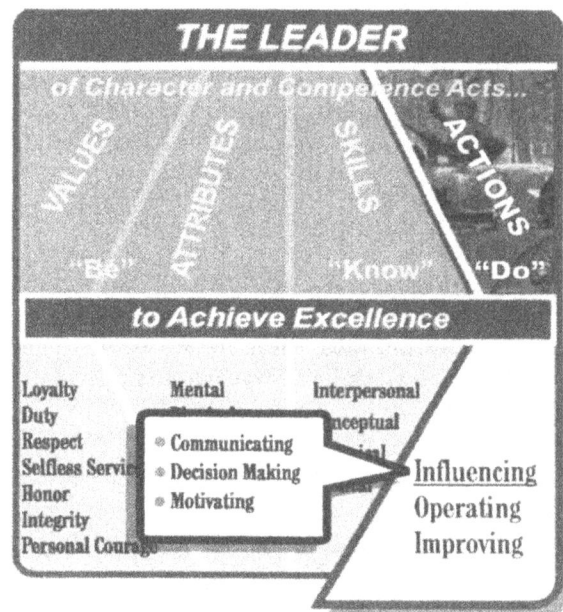

Figure 6-5. Organizational Leader Actions—Influencing

communicating and monitoring tasks. Programs for officer and NCO professional development based on either terrain walks or seminars can reinforce chain of command functioning. Checking organizational functions daily ("leading by walking around") can reveal whether the commander's intent is getting to the lowest level.

6-44. Communication becomes more complex for organizational leaders because of their increased span of control and separation from elements actually executing the mission. Whatever organizational leaders ask for, explicitly or implicitly, causes ripples throughout the organization. Therefore, they must consider how subordinates might interpret their wishes. Directives and actions must be clear and issued in a manner that discourages overreaction. The installation commander who remarks out loud about bland walls may cause an entire organization's soldiers to paint all weekend (it has happened).

6-45. Organizational leaders also lose the right to complain in public; for example, their opinion of a support agency affects the attitude of hundreds or thousands of people. Where the leader is, how the leader looks, and what the leader does and says influence routine leadership actions throughout the organization. Like direct leaders, organizational leaders are always on display, and their demeanor and presence set the tone and climate for subordinate organizations. However, the position of organizational leaders makes them more prominent, and they must remain aware of how their behavior affects their organization. A bad day for the leader should not have to be a bad day for everyone else.

COMMUNICATING

Too often we place the burden of comprehension on [those at a different level from] us, assuming both the existence of a common language and motivation.

General Edward C. Meyer
Former Army Chief of Staff

6-46. Ironically, organizational leaders' face-to-face communication must be more powerful,

more focused, and more unequivocal than direct leaders' communication. Because organizational leaders move quickly from one project to another and one part of the organization to the other, they must be careful that the right message goes out the first time. Poor communication can have tremendously negative consequences.

Know Yourself

6-47. Even before assuming an organizational, leadership position, leaders must assess themselves, understand their strengths and weaknesses, and commit to an appropriate leadership philosophy. Organizational leaders must realize that some techniques that worked in direct-level positions may no longer work at the organizational level. They must resist the temptation to revert to their old role and thus preempt their subordinates by making decisions for them.

6-48. That said, personal qualities that contributed to their previous success are still important for organizational leaders. They must be themselves. They must know their biases, frustrations, and desires and try to keep these factors from negatively influencing their communication. It's not enough to be careful about what they say. Nonverbal communication is so powerful that organizational leaders need to be aware of personal mannerisms, behavioral quirks, and demeanor that reinforce or contradict a spoken message.

Know the Purpose

6-49. Organizational leaders know themselves, the mission, and the message. They owe it to their organization and their people to share as much as possible. People have to know what to do and why. At the most basic level, communication provides the primary way that organizational leaders show they care. If subordinates are to succeed and the organization is to move forward, then the organizational leader must work hard at maintaining positive communication. Encouraging open dialogue, actively listening to all perspectives, and ensuring that subordinate leaders and staffs can have a forthright, open, and honest voice in the organization without fear of negative consequences greatly fosters communication at all levels. Organizational leaders who communicate openly and genuinely reinforce team values, send a message of trust to subordinates, and benefit from subordinates' good ideas.

The Commander's Notebook

A brigade commander met with his subordinate leaders and outlined his goals for an upcoming training exercise. In the days following, while the brigade staff worked on the formal orders and requirements, the commander spent time visiting subordinate units as they trained. As a part of each visit, he asked his subordinate leaders for specific feedback on his intent. Was it clear? Could they repeat the three main points he had tried to make? What would they add to the unit's goals for the training? He listened, asked his own questions, and allowed them to question him. It turned out that most of the people he spoke to missed one of his three main points, which led the commander to believe that he hadn't made himself clear the first time. Eventually, he started the conversation by saying, "There are a couple of points I tried to make in my talk; apparently, I dropped the ball on at least one of them. Let me take another shot at it." Then he explained the point again.

Whenever subordinate leaders offered suggestions about the upcoming exercise, the brigade commander took out a pocket notebook and wrote some notes. Even when suggestions sounded lame, he wrote them down. That way, he signaled to the speaker, "Yes, your opinion counts, too." Secondly, by writing down the ideas, the commander guaranteed himself a chance to look at the comments later. He knew from experience that sometimes the ones that don't seem to make sense at first turn out to be quite useful later. Many of the direct leaders remarked that they had never seen a brigade commander do anything like that before. They were even more astonished when they got feedback on the suggestion. The brigade adjutant even explained to one company commander why his suggestion wasn't implemented. On a Saturday morning the brigade commander was standing in line at the PX when a platoon sergeant engaged him in conversation. "I wasn't around the day you visited my company last week, sir," the NCO said, "but I heard the other folks had a few suggestions for you. I wonder if I could add something?"

Know the Environment

6-50. Before organizational leaders can effectively communicate, they must assess the environment—people, events, and systems—and tailor their message to the target audience. Organizational leaders constantly communicate by persuading and conveying intent, standards, goals, and priorities at four levels within the Army: their people, their own and higher staffs, their subordinate leaders and commanders, and their superiors. There may also be occasions that require organizational leaders to speak to audiences outside the Army such as the media or community groups. They may have to repeat the message to different audiences and retune it for different echelons, but only leaders can reinforce their true intent.

Know the Boss

6-51. Working to communicate consistently with the boss is especially important for organizational leaders. Organizational leaders have to figure out how to reach the boss. They must assess how the boss communicates and how the boss receives information. For some leaders, direct and personal contact is best; others may be more comfortable with weekly meetings, e-mail, or letters. Knowing the boss's intent, priorities, and thought processes greatly enhances organizational success. An organizational leader who communicates well with the boss minimizes friction between the organization and the higher headquarters and improves the overall organizational climate.

Know the Subordinates

If it's dumb it's not our policy.

Lieutenant General Walter F. Ulmer Jr.
Former Commanding General, III Corps

6-52. The mere presence of an organizational leader somewhere communicates the leader's character and what the leader values. The organizational leader who hurries through a talk about caring for subordinates, then passes up an opportunity to speak face-to-face with some soldiers, does more than negate the message; he undercuts whatever trust his subordinates may have had.

6-53. Because organizational leaders know themselves, they also know that others bring the sum total of their experience to their duties. They analyze interpersonal contact to gather meaning; they look for the message behind the words. In this way, they gain a greater understanding of peers, subordinates, and superiors. Improving communication skills becomes a major self-development challenge for them. By stating their intent openly, organizational leaders give subordinates an open door for feedback on unintended consequences or just bad policy: "Hey sir, did you really mean it when you said, 'If it's dumb it's not our policy?' OK, well what about...?" Leaders must be seen to be heard.

Know the Staff

6-54. Organizational leaders must understand what's going on within their own and the next-higher echelon staff. Networking allows them to improve communication and mission accomplishment by giving them a better understanding of the overall environment. Networking requires leaders to constantly interact and share thoughts, ideas, and priorities. Informed staffs can then turn policies, plans, and programs into realities.

6-55. Organizational leaders must also know the focus of the next-higher staff and commander. Through taking time to interact with the next-higher staff, organizational leaders gain a greater understanding of the boss's priorities and also help set the conditions for their own requirements. Constantly sensing—observing, talking, questioning, and actively listening

to—what's going on helps organizational leaders better identify and solve potential problems and allows them to anticipate decisions and put their outfit in the best possible position to execute.

Know the Best Method

6-56. To disseminate information accurately and rapidly, organizational leaders must also develop an effective communications network. Some of these networks—such as the chain of command, the family support network, the NCO support channel, and staff relationships—simply need to be recognized and exploited. Other informal chains must be developed. Different actions may require different networks.

6-57. The more adept organizational leaders become in recognizing, establishing, and using these networks, the more successful the outcome, especially as they become comfortable using a wider range of communications forums. Memorandums, notes, and e-mail as well as formal and informal meetings, interactions, and publications are tools of an effective communicator. Organizational leaders must know the audiences these methods reach and use them accordingly.

DECISION MAKING

The key is not to make quick decisions, but to make timely decisions.

General Colin Powell
Former Chairman, Joint Chiefs of Staff

6-58. Organizational leaders are far more likely than direct leaders to be required to make decisions with incomplete information. They determine whether they have to decide at all, which decisions to make themselves, and which ones to push down to lower levels. To determine the right course of action, they consider possible second- and third-order effects and think farther into the future—months, or even years, out in the case of some directorates.

6-59. Organizational leaders identify the problem, collect input from all levels, synthesize that input into solutions, and then choose and execute the best solution in time to make a difference. To maximize the use of resources and have the greatest effect on developing an effective organization, organizational leaders move beyond a reacting, problem-*solving* approach to an anticipating, problem-*preemption* method. While there will always be emergencies and unforeseen circumstances, organizational leaders focus on anticipating future events and making decisions about the systems and people necessary to minimize crises. Vision is essential for organizational leaders.

6-60. During operations, the pace and stress of action increase over those of training. Organizational leaders use the MDMP to make tactical decisions; however, they must add their conceptual skill of systems understanding to their knowledge of tactics when considering courses of action. Organizational leaders may be tempted—because of pressure, the threat, fear, or fatigue—to abandon sound decision making by reacting to short-term demands. The same impulses may result in focusing too narrowly on specific events and losing their sense of time and timing. But there's no reason for organizational leaders to abandon proven decision-making processes in crises, although they shouldn't hesitate to modify a process to fit the situation. In combat, success comes from creative, flexible decision making by leaders who quickly analyze a problem, anticipate enemy actions, and rapidly execute their decisions. (Remember GEN Grant's actions at Vicksburg.) Leaders who delay or attempt to avoid a decision may cause unnecessary casualties and even mission failure.

6-61. Effective and timely decision making—both the commander's and subordinates'—is crucial to success. As part of decision making, organizational leaders establish responsibility and accountability among their subordinates. They delegate decision making authority as far as it will go, empowering and encouraging subordinates to make decisions that affect their areas of responsibility or to further delegate that authority to their own subordinates.

6-62. Effective organizational leaders encourage initiative and risk-taking. They remember that they are training leaders and soldiers; the goal is a better-trained team, not some ideal outcome. When necessary, they support subordinates' bad decisions, but only those made attempting to follow the commander's intent. Failing through want of experience or luck is forgivable. Negligence, indecision, or attempts to take an easy route should never be tolerated.

6-63. As GEN Powell's comment makes clear, a decision's timeliness is as important as the speed at which it is made. Just as for direct leaders, a good decision now is better than a perfect one too late. Leaders who are good at handling the decision-making process will perform better when the OPTEMPO speeds up. Leaders who don't deliver timely decisions leave their subordinates scrambling and trying to make up for lost time. Better to launch the operation with a good concept and let empowered subordinates develop subsequent changes to the plan than to court failure by waiting too long for the perfect plan.

6-64. In tough moments, organizational leaders may need the support of key subordinates to close an issue. Consider MG George G. Meade's position at Gettysburg. In command of the Army of the Potomac for only a few days, MG Meade met with his subordinates on the night of 2 July 1863 after two days of tough fighting. Uncertainty hung heavy in the air. MG Meade's decision to stand and fight, made with the support of his corps commanders, influenced the outcome of the battle and became a turning point of the Civil War.

6-65. Coping with uncertainty is normal for all leaders, increasingly so for organizational leaders. Given today's information technology, the dangerous temptation to wait for all available information before making a decision will persist. Even though this same technology may also bring the unwanted attention of a superior, leaders should not allow it to unduly influence their decisions. Organizational leaders are where they are because of their experience, intuition, initiative, and judgment. Events move quickly, and it's more important for decisive organizational leaders to recognize and seize opportunities, thereby creating success, than to wait for all the facts and risk failure.

6-66. In the end, leaders bear ultimate responsibility for their organizations' success or failure. If the mission fails, they can't lay the blame elsewhere but must take full responsibility. If the mission succeeds, good leaders give credit to their subordinates. While organizational leaders can't ensure success by being all-knowing or present everywhere, they can assert themselves throughout the organization by being decisive in times of crisis and quick to seize opportunities. In combat, leaders take advantage of fleeting windows of opportunity: they see challenges rather than obstacles; they seek solutions rather than excuses; they fight through uncertainty to victory.

MOTIVATING

It is not enough to fight. It is the spirit which we bring to the fight that decides the issue. It is morale that wins the victory.

General of the Army George C. Marshall

6-67. Interpersonal skills involved in creating and sustaining ethical and supportive climates are required at the organizational as well as the direct leadership level. As Chapter 3 explains, the organizational, unit, or command climate describes the environment in which subordinates work. Chapter 5 discusses how direct leaders focus their motivational skills on individuals or small groups of subordinates. While direct leaders are responsible for their organizations' climate, their efforts are constrained (or reinforced) by the larger organization's climate. Organizational leaders shape that larger environment. Their primary motivational responsibility is to establish and maintain the climate of their entire organization.

6-68. Disciplined organizations evolve within a positive organizational climate. An organization's climate springs from its leader's attitudes, actions, and priorities. Organizational leaders set the tone most powerfully through a personal example that brings Army values to life. Upon assuming an organizational leadership position, a leader determines the organizational climate by assessing

the organization from the bottom up. Once this assessment is complete, the leader can provide the guidance and focus (purpose, direction, and motivation) required to move the organizational climate to the desired end state.

6-69. A climate that promotes Army values and fosters the warrior ethos encourages learning and promotes creative performance. The foundation for a positive organizational climate is a healthy ethical climate, but that alone is insufficient. Characteristics of successful organizational climates include a clear, widely known intent; well-trained and confident soldiers; disciplined, cohesive teams; and trusted, competent leadership.

6-70. To create such a climate, organizational leaders recognize mistakes as opportunities to learn, create cohesive teams, and reward leaders of character and competence. Organizational leaders value honest feedback and constantly use all available means to maintain a feel for the environment. Staff members who may be good sources for straightforward feedback may include equal opportunity advisors and chaplains. Methods may include town hall meetings, surveys, and councils. And of course, personal observation—getting out and talking to DA civilians, soldiers, and family members—brings organizational leaders face-to-face with the people affected by their decisions and policies. Organizational leaders' consistent, sincere effort to see what's really going on and fix things that are not working right can result in mutual respect throughout their organizations. They must know the intricacies of the job, trust their people, develop trust among them, and support their subordinates.

6-71. Organizational leaders who are positive, fair, and honest in their dealings and who are not afraid of constructive criticism encourage an atmosphere of openness and trust. Their people willingly share ideas and take risks to get the job done well because their leaders strive for more than compliance; they seek to develop subordinates with good judgment.

6-72. Good judgment doesn't mean lockstep thinking. Thinking "outside the box" isn't the same as indiscipline. In fact, a disciplined organization systematically encourages creativity and taking prudent risks. The leader convinces subordinates that anything they break can be fixed, except life or limb. Effective organizational leaders actively listen to, support, and reward subordinates who show disciplined initiative. All these things create opportunities for subordinates to succeed and thereby build their confidence and motivation.

6-73. However, it's not enough that individuals can perform. When people are part of a disciplined and cohesive team, they gain proficiency, are motivated, and willingly subordinate themselves to organizational needs. People who sense they're part of a competent, well-trained team act on what the team needs; they're confident in themselves and feel a part of something important and compelling. These team members know that what they do matters and discipline themselves.

The 505th Parachute Infantry Regiment at Normandy

On 7 June 1944, the day after D-Day, nearly 600 paratroopers of the 505th Parachute Infantry Regiment were in position in the town of Ste. Mère Église in Normandy to block any German counterattack of the Allied invasion force. Although outnumbered by an enemy force of over 6,000 soldiers, the paratroopers attacked the German flank and prevented the enemy's assault. The paratroopers were motivated and well-trained, and they all understood the absolute necessity of preventing the German counterattack. Even in the fog of war, they did what needed to be done to achieve victory. Their feat is especially noteworthy since many landed outside their planned drop zones and had to find their units on their own. They did so quickly and efficiently in the face of the enemy.

The 505th combined shared purpose, a positive and ethical climate, and cohesive, disciplined teams to build the confidence and motivation necessary to fight and win in the face of uncertainty and adversity. Both leaders and soldiers understood that no plan remains intact after a unit crosses the line of departure. The leaders' initiative allowed the disciplined units to execute the mission by following the commander's intent, even when the conditions on the battlefield changed.

OPERATING ACTIONS

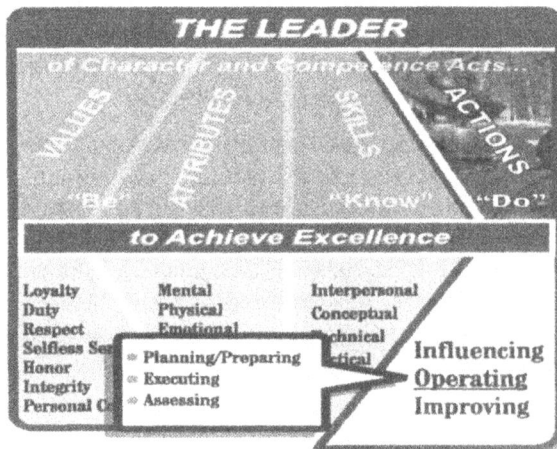

Figure 6-6. Organizational Leader Actions—Operating

6-74. Organizational leaders see, decide, and act as they perform the operating actions shown in Figure 6-6. They emphasize teamwork and cooperation over competition. They provide their intent so subordinates can accomplish the mission, no matter what happens to the original plan. Because organizational leaders primarily work through subordinates, empowerment and delegation are indispensable. As a result of communicating with subordinates, listening to their responses, and obtaining feedback from their assessments, organizational leaders are better equipped to make decisions.

6-75. Organizational-level commanders usually use the MDMP for tactical decision making and planning. However, those who command in the joint environment must use the Joint Operation Planning and Execution System (JOPES) methodology. Both the MDMP and JOPES allow organizational commanders to apply the factors discussed in this chapter.

SYSTEMS PLANNING AND PREPARING

[A] good plan violently executed now *is better than a perfect plan next week.*

General George S. Patton Jr.
War As I Knew It

6-76. GEN Patton wasn't belittling the importance of planning; he was emphasizing the balance necessary for successful operations. Planning, getting ready for the future by laying out how leaders aim to turn their intent into reality, is something leaders do every day and something the Army does very well. However, organizational leaders plan for the systems that support training and operations as well as for the actual training event or operation. Systems planning involves seven steps:

- Establish intent.
- Set goals.
- Determine objectives.
- Determine tasks.
- Establish priorities.
- Follow up.

Establish Intent

6-77. The first step in systems planning is for the organizational leader to have a clear intent for what he wants the organization to be. What will it look like at some future point? Spending extra time visualizing the end state up front is more important than quickly jumping into the mechanics of planning. Obviously, the actual mission is critical in determining this end state. The organizational leader's intent should be announced at the earliest practicable time after it has been formulated so the staff and subordinate commanders can have maximum time to plan. For a division, the intent might be—

- The best infantry division in the world.
- Supported by the finest installation in the Army.
- Trained and ready to deploy anywhere in the world to fight and win.
- But flexible enough to accomplish any other mission the nation asks us to perform.
- A values-based organization that takes care of its soldiers, DA civilians, and families.

6-78. Organizational leaders must determine how this intent affects the various systems for which they are responsible. By their actions and those of their subordinates and by using their presence to be heard, organizational leaders bring meaning for their intent to their people.

Set Goals

6-79. Once they have established their intent, organizational leaders, with the help of their team of subordinate leaders and staffs, set specific goals for their organizations. Goals frame the organizational leader's intent. For instance, the goal "Improve fire control and killing power" could support that part of the intent that states the division will be "trained and ready to deploy anywhere in the world to fight and win." Organizational leaders are personally involved in setting goals and priorities to execute their intent and are aware that unrealistic goals punish subordinates.

Determine Objectives

6-80. In the third step, organizational leaders establish objectives that are specific and measurable. For example, an objective that supports the goal of improving fire control and killing power could be "Fifty percent of the force must fire expert on their personal weapons." Establishing objectives is difficult because the process requires making precise calls from a wide variety of options. Since time and resources are limited, organizational leaders make choices about what can and cannot be accomplished. They check key system nodes to monitor subsystem functions.

Determine Tasks

6-81. The fourth step involves determining the measurable, concrete steps that must be taken on the way to the objective. For example, the commander of a forward-stationed division might ensure family readiness by ordering that any newly arriving soldier with a family may not be deployed without having a vehicle in country and household goods delivered.

Establish Priorities

6-82. The fifth step is to establish a priority for the tasks. This crucial step lets subordinates know how to spend one of their most critical resources: time.

6-83. This system of establishing priorities is important for the organization; organizational leaders must also practice it personally. In fact, a highly developed system of time management may be the only way for organizational leaders to handle all the demands upon them. There's rarely enough time to do everything, yet they must make the time to assess and synthesize information and make timely decisions. Leaders who recognize distractions are better equipped to handle their time well.

Prepare

6-84. Though organizational leaders have more complex missions than direct leaders,

they also have more assets: a staff and additional subordinate leaders, specialists, and equipment allow their preparation to be diverse and complete. Direct leaders prepare by getting individuals moving in the right direction; organizational leaders take a step back and check to make sure the systems necessary to support the mission are in place and functioning correctly.

Follow Up

6-85. The final step in systems planning is to follow up: Does the team understand the tasks? Is the team taking the necessary actions to complete them? Check the chain of command again: does everyone have the word? Organizational leader involvement in this follow up validates the priorities and demonstrates that the leader is serious about seeing the mission completed. Organizational leaders who fail to follow up send a message that the priorities are not really that important and that their orders are not really binding.

The "Paperwork Purge"

The division's new chief of staff was surprised at how much time subordinates spent at meetings; it seemed they had time for little else. After observing the way things worked for two weeks, the chief did away with most of the scheduled meetings, telling the staff, "We'll meet when we need to meet, and not just because it's Friday morning." What's more, the chief required an agenda for each meeting ahead of time: "That way, people can do their homework and see who needs to be there and who doesn't." The chief was always on time for meetings and started at the time specified on the agenda. There were no interruptions of whomever had the floor, and the long, meandering speeches that had marked previous meetings were cut short.

The chief put a one-page limit on briefing papers for the boss. This meant subordinates learned to write concisely. Each staff section did a top-to-bottom review of procedures that had been in place as long as anyone could remember. Anything that couldn't be justified was thrown out. The chief handled most of the correspondence that came across his desk with a quick note written on the original and told the staff to do the same.

The chief made the staff justify requirements they sent to subordinate organizations, with the comment, "If you can't tell them why it's important, then maybe it's not important." The explanation also helped subordinate elements determine their own priorities: "You can't keep sending stuff down saying, 'This is critical!' It gets to be like the boy who cried wolf."

The "Paperwork Purge" (continued)

Of course, the staff didn't take the new chief quite seriously at first, and after a week of reviewing old policy letters, some staff sections let the requirement slide. Then the chief showed up one day and had them give him a rundown on all the policies left after what everyone was calling "the big paperwork purge." A few more outdated requirements fell by the wayside that afternoon. More important, the staff got the message that the chief followed through on decisions.

Finally, and most startling, the chief told staff members that now and then they should sit quietly and stare into space: "You're getting paid to think, and every once in a while you've got to stop moving to do that well."

6-86. Keeping their intent in mind, organizational leaders fight distracters, make time to reflect, and seek to work more efficiently. Despite the pressure of too much to do in too little time, they keep their sense of humor and help those around them do the same.

THE CREATIVE STAFF PROCESS

None of us is as smart as all of us.

A Former Brigade Commander

6-87. The size and complexity of the organizations led by organizational leaders requires well-trained, competent staffs. Training these staffs is a major responsibility of organizational leaders. The chief of staff or executive officer is the organizational leader's right hand in that effort.

6-88. In the 100 days leading to the Battle of Waterloo, Napoleon had to campaign without his intensely loyal and untiring chief of staff, Berthier. In all his other campaigns, Berthier had transformed Napoleon's orders into instructions to the marshals, usually in quadruplicate with different riders carrying four copies to the same marshal over different routes. Berthier's genius for translating Napoleon's intent into tasks for each corps underlay the French Army's versatile, fluid maneuver style. Without Berthier and with an increasingly rigid Napoleon disdaining advice from any source, Napoleon's formations lost a good deal of their flexibility and speed.

6-89. Great staffs work in concert with the leader to turn intent into reality. A single leader in isolation has no doubt done great things and made good decisions. However, the organizational leader alone can't consistently make the right decisions in an environment where operational momentum never stops.

6-90. Building a creative, thinking staff requires the commander's time, maturity, wisdom, and patience. Although managing information is important, the organizational leader needs to invest in both quality people and in training them to think rather than just process information. Several factors contribute to building a creative, thinking staff.

The Right People

6-91. A high-performing staff starts with putting the right people in the right places. Organizational leaders are limited to their organization's resources, but have many choices about how to use them. They assemble, from throughout their organizations, people who think creatively, possess a vast array of technical skills, are trained to solve problems, and can work together. They take the time to evaluate the staff and implement a training program to improve it as a whole. They avoid micromanaging the staff, instead trusting and empowering it to think creatively and provide answers.

The Chief of Staff

6-92. The staff needs its own leader to take charge—someone who can focus it, work with it, inspire it, and move it to decisive action in the absence of the commander. The sections of the staff work as equals, yet without superb leadership they won't perform exceptionally.

To make a staff a true team, an empowered deputy must be worthy of the staff and have its respect. The chief of staff must have the courage to anticipate and demand the best possible quality. On the other hand, the chief must take care of the hardworking people who make up the staff and create an environment that fosters individual initiative and develops potential. (FM 100-34 discusses the role of the chief of staff.)

Challenging Problems

6-93. A staff constantly needs challenging problems to solve if it's to build the attitude that it can overcome any obstacle. Tackling problems with restricted time and resources improves the staff members' confidence and proficiency, as long as they get an opportunity to celebrate successes and to recharge their batteries. Great confidence comes from training under conditions more strenuous than they would likely face otherwise.

Clear Guidance

6-94. The commander constantly shares thoughts and guidance with the staff. Well-trained staffs can then synthesize data according to those guidelines. Computers, because of their ability to handle large amounts of data, are useful analytical tools, but they can do only limited, low-order synthesis. There's no substitute for a clear commander's intent, clearly understood by every member of the staff.

EXECUTING

The American soldier demonstrated that, properly equipped, trained, and led, he has no superior among all the armies of the world.

General Lucian K. Truscott
Former Commanding General, 5th Army

6-95. Planning and preparation for branches and sequels of a plan and contingencies for future operations may continue, even during execution. However, execution is the purpose for which the other operating actions occur; at some point, the organizational leader commits to action, spurs his organization forward, and sees the job through to the end. (FM 100-34 and FM 101-5 discuss branches and sequels.)

6-96. In combat, organizational leaders integrate and synchronize all available elements of the combined arms team, empower subordinates, and assign tasks to accomplish the mission. But the essence of warfighting for organizational leaders is their will. They must persevere despite limitations, setbacks, physical exhaustion, and declining mental and emotional reserves. They then directly and indirectly energize their units—commanders and soldiers—to push through confusion and hardship to victory.

6-97. Whether they're officers, NCOs, or DA civilians, the ultimate responsibility of organizational-level leaders is to accomplish the mission. To this end, they must mass the effects of available forces on the battlefield, to include supporting assets from other services. The process starts before the fight as leaders align forces, resources, training, and other supporting systems.

Combined Arms and Joint Warfighting

6-98. Brigades and battalions usually conduct single-service operations supported by assets from other services. In contrast, the large areas of responsibility in which divisions and corps operate make division and corps fights joint by nature. Joint task forces (JTFs) are also organizational-level formations. Therefore, organizational leaders and their staffs at division-level and higher must understand joint procedures and concerns at least as well as they understand Army procedures and concerns. In addition, it's not unusual for a corps to control forces of another nation; divisions do also, but not as frequently. This means that corps and division headquarters include liaison officers from other nations. In some cases, these staffs may have members of other nations permanently assigned: such a staff is truly multinational.

6-99. Today's operations present all Army leaders—but particularly organizational leaders—with a nonlinear, dynamic environment ranging the full spectrum of continuous operations. These dispersed conditions create an information-intense environment that challenges leaders to synchronize their efforts with nonmilitary and often nongovernmental agencies.

Empowering

Never tell people how *to do things. Tell them* what *to do and they will surprise you with their ingenuity.*

General George S. Patton Jr.
War As I Knew It

6-100. To increase the effects of their will, organizational leaders must encourage initiative in their subordinates. Although unity of command is a principle of war, at some level a single leader alone can no longer control all elements of an organization and personally direct the accomplishment of every aspect of its mission. As leaders approach the brigade or directorate level, hard work and force of personality alone cannot carry the organization. Effective organizational leaders delegate authority and support their subordinates' decisions, while holding subordinates accountable for their actions.

6-101. Delegating successfully involves convincing subordinates that they're empowered, that they indeed have the freedom to act independently. Empowered subordinates have, and know they have, more than the responsibility to get the job done. They have the authority to operate in the way they see fit and are limited only by the leader's intent.

6-102. To do that, the organizational leader gives subordinates the mission, motivates them, and lets them go. Subordinates know that the boss trusts them to make the right things happen; this security motivates them, in turn, to lead their people with determination. They know the boss will underwrite honest mistakes, well-intentioned mistakes—not stupid, careless, or repeated ones. So for the boss, empowering subordinates means building the systems and establishing the climate that gives subordinates the rein to do the job within the bounds of acceptable risk. It means setting organizational objectives and delegating tasks to ensure parallel, synchronized progress.

6-103. Delegation is a critical task: Which subordinates can be trusted with independent action? Which need a short rein? In fluid situations—especially in combat, where circumstances can change rapidly or where leaders may be out of touch or become casualties—empowered subordinates will pursue the commander's intent as the situation develops and react correctly to changes that previous orders failed to anticipate. However, as important as delegation is to the success of organizations, it does not imply in any way a reduction of the commander's responsibility for the outcome. Only the commander is accountable for the overall outcome, the success or failure, of the mission.

ASSESSING

6-104. The ability to assess a situation accurately and reliably—a critical tool in the leader's arsenal—requires instinct and intuition based on experience and learning. It also requires a feel for the reliability and validity of information and its sources. Organizational assessment is necessary to determine organizational weaknesses and preempt mishaps. Accurately determining causes is essential to training management, developing subordinate leadership, and process improvement.

6-105. There are several different ways to gather information: asking subordinates questions to find out if the word is getting to them, meeting people, and checking for synchronized plans are a few. Assessing may also involve delving into the electronic databases upon which situational understanding depends. Assessment techniques are more than measurement tools; in fact, the way a leader assesses something can influence the process being assessed. The techniques used may produce high quality, useful feedback; however, in a dysfunctional command climate, they can backfire and send the wrong message about priorities.

6-106. Staff and subordinates manage and process information for a leader, but this doesn't relieve the leader from the responsibility of analyzing information as part of the decision-making process. Leaders obtain information from various sources so they can compare and make judgments about the accuracy of sources.

6-107. As Third Army commander during World War II, GEN George Patton did this continuously. Third Army staff officers visited front-line units daily to gather the latest available information. In addition, the 6th Cavalry Group, the so-called "Household Cavalry" monitored subordinate unit reconnaissance nets and sent liaison patrols to visit command and observation posts of units in contact. These liaison patrols would exchange information with subordinate unit G2s and G3s and report tactical and operational information directly to the Third Army forward headquarters (after clearing it with the operations section of the unit they were visiting).

6-108. In addition to providing timely combat information, the Household Cavalry and staff visits reduced the number of reports Third Army headquarters required and created a sense of cohesiveness and understanding not found in other field armies. Other organizational leaders have accomplished the same thing using liaison officers grounded in their commander's intent. Whatever the method they choose, organizational leaders must be aware of the second- and third-order effects of having "another set of eyes."

6-109. In the world of digital command and control, commanders may set screens on various command and control systems to monitor the status of key units, selected enemy parameters, and critical planning and execution timelines. They may establish prompts in the command and control terminal that warn of imminent selected events, such as low fuel levels in maneuver units, tight fighter management timelines among aviation crews, or massing enemy artillery.

6-110. A leader's preconceived notions and opinions (such as "technology undermines basic skills" or "technology is the answer") can interfere with objective analysis. It's also possible to be too analytical, especially with limited amounts of information and time. Therefore, when analyzing information, organizational leaders guard against dogmatism, impatience, or overconfidence that may bias their analysis.

6-111. The first step in designing an assessment system is to determine the purpose of the assessment. While purposes vary, most fall into one of the following categories:

- Evaluate progress toward organizational goals (using an emergency deployment readiness exercise to check unit readiness or monitoring progress of units through stages of reception, staging, onward movement, and integration).
- Evaluate the efficiency of a system, that is, the ratio of the resources expended to the results gained (comparing the amount of time spent performing maintenance to the organization's readiness rate).
- Evaluate the effectiveness of a system, that is, the quality of the results it produces (analyzing the variation in Bradley gunnery scores).
- Compare the relative efficiency or effectiveness against standards.
- Compare the behavior of individuals in a group with the prescribed standards (APFT or gunnery scores).
- Evaluate systems supporting the organization (following up "no pay dues" to see what the NCO support channel did about them).

6-112. Organizational leaders consider the direct and indirect costs of assessing. Objective costs include the manpower required to design and administer the system and to collect, analyze, and report the data. Costs may also include machine-processing times and expenses related to communicating the data. Subjective costs include possible confusion regarding organizational priorities and philosophies, misperceptions regarding trust and decentralization, fears over unfair use of collected data, and the energy expended to collect and refine the data.

6-113. Organizational leaders ask themselves these questions: What's the standard? Does the standard make sense to all concerned? Did we meet it? What system measures it? Who's responsible for the system? How do we reinforce or correct our findings? One of the greatest contributions organizational leaders can make to their organizations is to assess their own leadership actions: Are you doing things the way you would to support the nation at war? Will your current systems serve equally well under the stress and strain of continuous fighting? If not, why not?

6-114. It follows that organizational leaders who make those evaluations every day will also hold their organizations to the highest standards. When asked, their closest subordinates will give them informal AARs of their leadership behaviors in the critical situations. When they arrange to be part of official AARs, they can invite subordinates to comment on how they could have made things go better. Organizational leader errors are very visible; their results are probably observed and felt by many subordinates. Thus, there's no sense in not admitting, analyzing. and learning from these errors. A bit of reflection in peacetime may lead to greater effectiveness in war.

6-115. The 1991 ground war in the Iraqi desert lasted only 100 hours, but it was won through hard work over a period of years, in countless field exercises on ranges and at the combat training centers. The continual assessment process allowed organizational leaders to trade long hours of hard work in peacetime for operations in war.

6-116. Organizational leaders are personally dedicated to providing tough, battle-focused training so that the scrimmage is always harder than the game. They must ensure that in training, to the extent that resources and risks allow, nothing is simulated. Constant assessments refine training challenges, forge confidence, and foster the quiet, calculating, and deadly warrior ethos that wins battles and campaigns. (FM 25-100 and FM 25-101 discuss battle focus and training assessment.)

IMPROVING ACTIONS

The creative leader is one who will rewrite doctrine, employ new weapons systems, develop new tactics and who pushes the state of the art.

John O. Marsh Jr.
Former Secretary of the Army

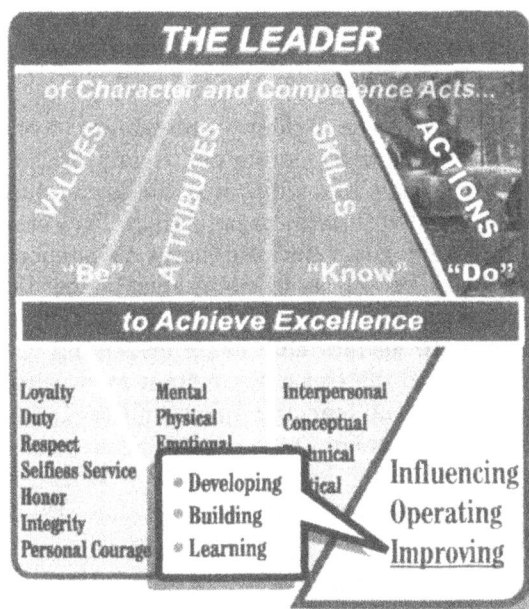

Figure 6-7. Organizational Leader Actions—Improving

6-117. Improving actions are what all leaders do today to make their organization and subordinates better tomorrow, next month, next year, five years from now. The responsibility for how the Army fights the next war lies with today's leaders; the work to improve the organization in the long term never ends. Leaders teaching subordinates to do the leader's job in combat is the hallmark of the profession of arms.

6-118. The payoff for improving actions might not be evident for years. In fact, leaders at all levels may never see the benefit of developing subordinates, since those subordinates go on to work for someone else. But this doesn't stop them from taking pride in their subordinates' development and performance; a subordinate's success is a great measure of a leader's success. Further, it's often difficult to draw a cause-and-effect line from what leaders do today to how it pays off tomorrow. Precisely because of these difficulties, organizational leaders ensure the goals they establish include improving people and organizations. They also make sure they communicate this to their subordinates.

6-119. The developing, building, and learning actions may be more difficult at the organizational level because the leaders themselves must rely more on indirect leadership methods. The challenge is greater because of the size of the organization, but the rewards increase as well: organizational leaders can influence large numbers of people and improve large segments of the Army.

DEVELOPING

Let us set for ourselves a standard so high that it will be a glory to live up to it, and then let us live up to it and add a new laurel to the crown of America.

Woodrow Wilson
28th President of the United States

6-120. Just as leadership begins at the top, so does developing. Organizational leaders keep a focus on where the organization needs to go and what all leaders must be capable of accomplishing. They continually develop themselves and mentor their subordinate leaders. As discussed in Chapter 3, leaders search for and take advantage of opportunities to mentor their subordinates. At the organizational level, commanders ensure that systems and conditions are in place for the mentoring of all organizational members.

6-121. Effective organizational leaders grow leaders at all levels of their organization. Just as they prepare their units for in-stride breaches, for example, they combine existing opportunities into a coherent plan for leadership development. Leaders get much of their development when they practice what they've learned and receive straightforward feedback in rigorously honest AARs. Feedback also comes from self-assessments as well as from peers, subordinates, and supervisors.

6-122. Organizational leaders design and integrate leader development programs into everyday training. They aim to capture learning in common duties, ensure timely feedback, and allow reflection and analysis. As Frederick the Great said, "What good is experience if you do not reflect?" Simply scheduling officer and NCO professional development sessions isn't enough for genuine, lasting leader development. Letting "operating" overwhelm "improving" threatens the future.

6-123. Leadership development is purposeful, not accidental. Everyday mission requirements are opportunities to grow leaders. Based on assessment of their subordinate leaders, organizational leaders describe how they intend to deliberately influence leader development through a comprehensive leadership development program that captures and harnesses what's already occurring in the organization. A leadership development program must provide for learning skills, practicing actions, and receiving feedback.

6-124. Organizational leaders assess their organizations to determine organization-specific developmental needs. They analyze their mission, equipment, and long-term schedule as well as the experience and competence of their subordinate leaders to determine leadership requirements. In addition to preparing their immediate subordinates to take their place, organizational leaders must also prepare subordinate leaders selected for specific duties to actually execute them.

6-125. Based on their assessment, organizational leaders define and clearly articulate their goals and objectives for leadership development within the organization. They create program goals and objectives to support their focus as well as to communicate specific responsibilities for subordinate leaders. These subordinate leaders help bring leadership development to life through constant mentoring and experiential learning opportunities. Leadership development is an important responsibility shared by leaders at every level. It becomes their greatest contribution—their legacy.

6-126. The development technique used depends on the leaders involved. Learning by making mistakes is possible, but having subordinates develop habits of succeeding is better for instilling self-confidence and initiative. Newly assigned assistant operations officers may need time to visit remote sites over which they have day-to-day control. They may need time to visit higher headquarters to establish rapport with those action officers they will have to deal with under the pressure of a tense operational situation. They may have to see the tactical operations center set up in the field and get a chance to review its SOPs before a big exercise. These activities can only happen if the organizational leader supports the leader development program and demands it take place in spite of the pressures of daily (routine) business.

6-127. There are many ways to tackle leader development. For example, instead of pursuing long-term training programs for his career civilians, one DA civilian leader enrolled both his DA civilian managers and military officers in graduate programs while they continued to work full-time. This approach allowed the directorate to provide development opportunities to four times as many personnel at one-third the cost of long-term training. In addition, the students were able to apply what they were learning directly to their jobs, thus providing immediate benefit to the organization.

6-128. In addition to educational programs, innovative interagency exchange assignments can cross-level the knowledge, skills, and experience of DA civilian leaders. Whether taking on new interns or expanding the perspectives of seasoned managers, the DA civilian component mirrors the uniformed components in its approach to broad-based leadership opportunities.

6-129. Often developmental programs involve historical events similar to current operational challenges. Such situations allow all to share a sense of what works and what does not from what worked before and what did not. This analysis can also be applied to recent organizational experiences. For example, in preparation for a CTC rotation, leaders review their own as well as others' experiences to determine valuable lessons learned. They master the individual and collective tasks through a training program that sets up soldiers and leaders for success. Based on internal AARs, they continue to learn, practice, and assess. CTCs also provide individual leaders with invaluable experience in operating under harsh conditions. Organizations execute missions, receive candid feedback and coaching to facilitate lessons learned, then execute again.

6-130. Commanders must take the time to ensure they do developmental counseling. Nothing can replace the face-to-face contribution made by a commander mentoring a subordinate. Developing the most talented and (often) those with the greatest challenges requires a great amount of time and energy, but it's an essential investment for the future.

BUILDING

Building Combat Power

6-131. Emphasis on winning can't waver during training, deploying, and fighting. By developing the right systems and formulating appropriate contingency plans, organizational leaders ensure that the organization is prepared for a variety of conditions and uncertainties. In wartime, building combat power derives from task organization, resourcing, and preparing for execution while still meeting the human needs of the organization. Commanders must preserve and recycle organizational energy throughout the campaign. In peacetime, the main component of potential combat power is embedded collective skill and organizational readiness stemming from hard, continuous, and challenging training to standard.

Building Teams

All United States military doctrine is based upon reliance on the ingenuity of the individual working on his own initiative as a member of a team and using the most modern weapons and equipment which can be provided him.

General Manton S. Eddy
Commanding General, XII Corps, World War II

6-132. Organizational leaders rely on others to follow and execute their intent. Turning a battlefield vision or training goals into reality takes the combined efforts of many teams inside and outside of the leader's organization.

Organizational leaders build solid, effective teams by developing and training them and sustain those teams by creating healthy organizational climates.

6-133. Organizational leaders work consistently to create individual and team ownership of organizational goals. By knowing their subordinates—their aspirations, fears, and concerns—organizational leaders can ensure their subordinate organizations and leaders work together. Taking time to allow subordinates to develop ways to meet organizational missions fosters ownership of a plan. The FM 25-100 training management process, in which subordinate organizations define supporting tasks and suggest the training required to gain and maintain proficiency, is an example of a process that encourages collective investment in training. That investment leads to a commitment that not only supports execution but also reduces the chances of internal conflict.

6-134. Subordinates work hard and fight tenaciously when they're well-trained and feel they're part of a good team. Collective confidence comes from winning under challenging and stressful conditions. People's sense of belonging comes from technical and tactical proficiency—as individuals and then collectively as a team—and the confidence they have in their peers and their leaders. As cohesive teams combine into a network, a team of teams, organizations work in harness with those on the left and right to fight as a whole. The balance among three good battalions is more important than having a single outstanding one. Following that philosophy necessarily affects resource allocation and task assignment.

6-135. Organizational leaders build cohesive organizations. They overcome, and even capitalize on, diversity of background and experience to create the energy necessary to achieve organizational goals. They resolve conflicts among subordinate leaders as well as any conflicts between their own organization and others.

6-136. For example, subordinate leaders may compete for limited resources while pursuing their individual organization's goals. Two battalion commanders may both want and need a certain maneuver training area to prepare for deployment, so they both present the issue professionally and creatively to their commander. The brigade commander must then weigh and decide between the different unit requirements, balancing their competing demands with the greater good of the entire organization and the Army. An even better situation would be if the organizational climate facilitates teamwork and cooperation that results in the subordinate commanders themselves producing a satisfactory solution.

6-137. Similarly, the brigade commander's own interests may at times conflict with that of other organizations. He must maintain a broad perspective and develop sensible solutions for positive resolution with his contemporaries. In both these cases, subordinates observe the actions of leaders and pattern their attitudes and actions after them. Everyone, even experienced leaders, looks up the chain of command for the example of "how we do it here," how to do it right. Organizational leaders empower their subordinates with a powerful personal example.

6-138. Like direct leaders, organizational leaders build teams by keeping team members informed. They share information, solicit input, and consider suggestions. This give-and-take also allows subordinates a glimpse into the mind of their leaders, which helps them prepare for the day when they fill that job. The leader who sends these messages—"I value your opinion; I'm preparing you for greater responsibilities; you're part of the team"—strengthens the bonds that hold the team together.

6-139. Team building produces trust. Trust begins with action, when leaders demonstrate discipline and competence. Over time, subordinates learn that leaders do what they say they'll do. Likewise leaders learn to trust their subordinates. That connection, that mutual assurance, is the link that helps organizations accomplish the most difficult tasks. (FM 100-34 discusses the importance of building trust for command and control.)

LEARNING

6-140. Organizational leaders create an environment that supports people within their organizations learning from their own experiences and the experiences of others. How leaders react to failure and encourage success now is critical to reaching excellence in the future. Subordinates who feel they need to hide mistakes deprive others of valuable lessons. Organizational leaders set the tone for this honest sharing of experiences by acknowledging that not all experiences (even their own) are successful. They encourage subordinates to examine their experiences, and make it easy for them to share what they learn.

6-141. Learning is continuous and occurs throughout an organization: someone is always experiencing something from which a lesson can be drawn. For this reason, organizational leaders ensure continual teaching at all levels; the organization as a whole shares knowledge and applies relevant lessons. They have systems in place to collect and disseminate those lessons so that individual mistakes become organizational tools. This commitment improves organizational programs, processes, and performances.

SECTION III

A HISTORICAL PERSPECTIVE OF ORGANIZATIONAL LEADERSHIP—GENERAL RIDGWAY IN KOREA

6-142. Few leaders have better exemplified effective organizational leadership in combat than GEN Matthew B. Ridgway. GEN Ridgway successfully led the 82d Airborne Division and XVIII Airborne Corps in the ETO during World War II and Eighth (US) Army during the Korean War. His actions during four months in command of Eighth Army prior to his appointment as UN Supreme Commander bring to life the skills and actions described throughout this chapter.

6-143. At the outbreak of the Korean War in June 1950, GEN Ridgway was assigned as the Army Deputy Chief of Staff, Operations. In an agreement between the Army Chief of Staff, GEN J. Lawton Collins, and the UN Supreme Commander, GA Douglas MacArthur, GEN Ridgway was identified early as the replacement for the Eighth Army commander, GEN Walton H. Walker, in the event GEN Walker was killed in combat.

6-144. That year, on 23 December, GEN Walker died in a jeep accident. Following approval by Secretary of Defense George C. Marshall and President Truman, GEN Ridgway was ordered to take command of Eighth Army. At that time, Eighth Army was defending near the 38th parallel, having completed a 300-mile retreat after the Chinese intervention and stunning victory on the Chongchin River.

6-145. The UN defeat had left its forces in serious disarray. One of Eighth Army's four American divisions, the 2d, needed extensive replacements and reorganization. Two other divisions, the 25th and 1st Cavalry, were seriously battered. Of the Republic of Korea divisions, only the 1st was in good fighting shape. A British brigade was combat ready, but it too had suffered substantial losses in helping cover the retreat.

6-146. Within 24 hours of GEN Walker's death, GEN Ridgway was bound for Korea. During the long flight from Washington, DC, to GA MacArthur's headquarters in Japan, GEN Ridgway had an opportunity to reflect on what lay ahead. He felt this problem was like so many others he had experienced: "Here's the situation—what's your solution?" He began to formulate his plan of action. He determined each step based on his assessment of the enemy's strengths and capabilities as well as his own command's strengths and capabilities.

6-147. The necessary steps seemed clear: gain an appreciation for the immediate situation from GA MacArthur's staff, establish his presence as Eighth Army commander by sending a statement of his confidence in them, and then meet with his own staff to establish his priorities. His first message to his new command was straight to the point: "You will have my utmost. I shall expect yours."

6-148. During the flight from Japan to his forward command post, GEN Ridgway carefully looked at the terrain upon which he was to fight. The battered Eighth Army had to cover a rugged, 100-mile-long front that restricted both maneuver and resupply. Poor morale presented a further problem. Many military observers felt that Eighth Army lacked spirit and possessed little stomach for continuing the bruising battle with the Chinese.

6-149. For three days GEN Ridgway traveled the army area by jeep, talking with commanders who had faced the enemy beyond the Han River. GEN Ridgway wrote later,

I held to the old-fashioned idea that it helped the spirits of the men to see the Old Man up there, in the snow and the sleet and the mud, sharing the same cold, miserable existence they had to endure.

6-150. GEN Ridgway believed a commander should publicly show a personal interest in the well-being of his soldiers. He needed to do something to attract notice and display his concern for the front-line fighters. Finding that one of his units was still short of some winter equipment, GEN Ridgway dramatically ordered that the equipment be delivered within 24 hours. In response, the logistical command made a massive effort to comply, flying equipment from Pusan to the front lines. Everyone noticed. He also ordered—and made sure the order was known—that the troops be served hot meals, with any failures to comply reported directly to him.

6-151. GEN Ridgway was candid, criticizing the spirit of both the commanders and soldiers of Eighth Army. He talked with riflemen and generals, from front-line foxholes to corps command posts. He was appalled at American infantrymen who didn't patrol, who had no knowledge of the terrain in which they fought, and who failed to know the whereabouts of their enemy. Moreover, this army was road-bound and failed to occupy commanding terrain overlooking its positions and supply lines. GEN Ridgway also sensed that Eighth Army—particularly the commanders and their staffs—kept looking over their shoulders for the best route to the rear and planned only for retreat. In short, he found his army immobilized and demoralized.

6-152. An important part of GEN Ridgway's effort to instill fighting spirit in Eighth Army was to order units to close up their flanks and tie in with other units. He said he wanted no units cut off and abandoned, as had happened to the 3d Battalion, 8th Cavalry at Unsan, Task Force Faith at Chosin Reservoir, and the 2d Division at Kuni-ri. GEN Ridgway felt that it was essential for soldiers to know they would not be left to fend for themselves if cut off. He believed that soldiers would be persuaded to stand and fight only if they realized help would come. Without that confidence in the command and their fellow soldiers, they would pull out, fearing to be left behind.

6-153. As he visited their headquarters, GEN Ridgway spoke to commanders and their staffs. These talks contained many of his ideas about proper combat leadership. He told his commanders to get out of their command posts and up to the front. When commanders reported on terrain, GEN Ridgway demanded that they base their information on personal knowledge and that it be correct.

6-154. Furthermore, he urged commanders to conduct intensive training in night fighting and make full use of their firepower. He also required commanders to personally check that their men had adequate winter clothing, warming tents, and writing materials. In addition, he encouraged commanders to locate wounded who had been evacuated and make every effort to return them to their old units. Finally, the army commander ordered his officers to stop wasting resources, calling for punishment of those who lost government equipment.

6-155. During its first battle under GEN Ridgway's command in early January 1951, Eighth Army fell back another 70 miles and lost Seoul, South Korea's capital. Major commanders didn't carry out orders to fall back in an orderly fashion, use field artillery to inflict the heaviest possible enemy casualties, and counterattack in force during daylight hours. Eighth Army's morale and sense of purpose reached their lowest point ever.

6-156. Eighth Army had only two choices: substantially improve its fighting spirit or get out of Korea. GEN Ridgway began to restore his men's fighting spirit by ordering aggressive patrolling into areas just lost. When patrols found the enemy few in number and not aggressive, the army commander increased the number and size of patrols. His army discovered it could drive back the Chinese without suffering overwhelming casualties. Buoyed by these successes, GEN Ridgway ordered a general advance along Korea's west coast, where the terrain was more open and his forces could take advantage of its tanks, artillery, and aircraft.

6-157. During this advance, GEN Ridgway also attempted to tell the men of Eighth Army why they were fighting in Korea. He sought to build a fighting spirit in his men based on unit and soldier pride. In addition, he called on them to defend Western Civilization from Communist degradation, saying:

In the final analysis, the issue now joined right here in Korea is whether Communism or individual freedom shall prevail; whether the flight of the fear-driven people we have witnessed here shall be checked, or shall at some future time, however distant, engulf our own loved ones in all its misery and despair.

6-158. In mid-February of 1951, the Chinese and North Koreans launched yet another offensive in the central area of Korea, where US tanks could not maneuver as readily and artillery could be trapped on narrow roads in mountainous terrain. In heavy fights at Chipyon-ni and Wonju, Eighth Army, for the first time, repulsed the Communist attacks. Eighth Army's offensive spirit soared as GEN Ridgway quickly followed up with a renewed attack that took Seoul and regained roughly the same positions Eight Army had held when he first took command. In late March, Eighth Army pushed the Communist forces north of the 38th parallel.

6-159. GEN Ridgway's actions superbly exemplify those expected of organizational leaders. His knowledge of American soldiers, units, and the Korean situation led him to certain expectations. Those expectations gave him a baseline from which to assess his command once he arrived. He continually visited units throughout the army area, talked with soldiers and their commanders, assessed command climate, and took action to mold attitudes with clear intent, supreme confidence, and unyielding tactical discipline.

6-160. He sought to develop subordinate commanders and their staffs by sharing his thoughts and expectations of combat leadership. He felt the pulse of the men on the front, shared their hardships, and demanded they be taken care of. He pushed the logistical systems to provide creature comforts as well as the supplies of war. He eliminated the skepticism of purpose, gave soldiers cause to fight, and helped them gain confidence by winning small victories. Most of all, he led by example.

6-161. In April GEN Ridgway turned Eighth Army over to GEN James A. Van Fleet. In under four months, a dynamic, aggressive commander had revitalized and transformed a traumatized and desperate army into a proud, determined fighting force. GA Omar N. Bradley, Chairman of the Joint Chiefs of Staff, summed up GEN Ridgway's contributions:

It is not often that a single battlefield commander can make a decisive difference. But in Korea Ridgway would prove to be that exception. His brilliant, driving, uncompromising leadership would turn the tide of battle like no other general's in our military history.

SUMMARY

6-162. This chapter has covered how organizational leaders train and lead staffs, subordinate leaders, and entire organizations. The influence of organizational leaders is primarily indirect: they communicate and motivate through staffs and subordinate commanders. Because their leadership is much more indirect, the eventual outcomes of their actions are often difficult to foresee. Nor do organizational leaders receive the immediate feedback that direct leaders do.

6-163. Still, as demonstrated by GEN Ridgway in Korea, the presence of commanders at the critical time and place boosts confidence and performance. Regardless of the type of organization they head, organizational leaders direct operations by setting the example, empowering their subordinates and organizations and supervising them appropriately. Organizational

leaders concern themselves with combat power—how to build, maintain, and recover it. That includes developing systems that will provide the organization and the Army with its next generation of leaders. They also improve conditions by sustaining an ethical and supportive climate, building strong cohesive teams and organizations, and improving the processes that work within the organization.

6-164. Strategic leaders provide leadership at the highest levels of the Army. Their influence is even more indirect and the consequences of their actions more delayed than those of organizational leaders. Because of this, strategic leaders must develop additional skills based on those they've mastered as direct and organizational leaders. Chapter 7 discusses these and other aspects of strategic leadership.

Chapter 7

Strategic Leadership

It became clear to me that at the age of 58 I would have to learn new tricks that were not taught in the military manuals or on the battlefield. In this position I am a political soldier and will have to put my training in rapping-out orders and making snap decisions on the back burner, and have to learn the arts of persuasion and guile. I must become an expert in a whole new set of skills.

General of the Army George C. Marshall

7-1. Strategic leaders are the Army's highest-level thinkers, warfighters, and political-military experts. Some work in an institutional setting within the United States; others work in strategic regions around the world. They simultaneously sustain the Army's culture, envision the future, convey that vision to a wide audience, and personally lead change. Strategic leaders look at the environment outside the Army today to understand the context for the institution's future role. They also use their knowledge of the current force to anchor their vision in reality. This chapter outlines strategic leadership for audiences other than the general officers and Senior Executive Service DA civilians who actually lead there. Those who support strategic leaders need to understand the distinct environment in which these leaders work and the special considerations it requires.

7-2. Strategic leadership requires significantly different techniques in both scope and skill from direct and organizational leadership. In an environment of extreme uncertainty, complexity, ambiguity, and volatility, strategic leaders think in multiple time domains and operate flexibly to manage change. Moreover, strategic leaders often interact with other leaders over whom they have minimal authority.

7-3. Strategic leaders are not only experts in their own domain—warfighting and leading large military organizations—but also are astute in the departmental and political environments of the nation's decision-making process. They're expected to deal competently with the public sector, the executive branch, and the legislature. The complex national security

environment requires an in-depth knowledge of the political, economic, informational, and military elements of national power as well as the interrelationship among them. In short, strategic leaders not only know themselves and their own organizations but also understand a host of different players, rules, and conditions.

7-4. Because strategic leaders implement the National Military Strategy, they deal with the elements that shape that strategy. The most important of these are Presidential Decision Memorandums, Department of State Policies, the will of the American people, US national security interests, and the collective strategies—theater and functional—of the combatant commanders (CINCs). Strategic leaders operate in intricate networks of competing constituencies and cooperate in endeavors extending beyond their establishments. As institutional leaders, they represent their organizations to soldiers, DA civilians, citizens, statesmen, and the media, as well as to other services and nations. Communicating effectively with these different audiences is vital to the organization's success.

7-5. Strategic leaders are keenly aware of the complexities of the national security environment. Their decisions take into account factors such as congressional hearings, Army budget constraints, reserve component issues, new systems acquisition, DA civilian programs, research, development, and interservice cooperation. Strategic leaders process information from these areas quickly, assess alternatives based on incomplete data, make decisions, and garner support. Often, highly developed interpersonal skills are essential to building consensus among civilian and military policy makers. Limited interpersonal skills can limit the effect of other skills.

7-6. While direct and organizational leaders have a short-term focus, strategic leaders have a "future focus." Strategic leaders spend much of their time looking toward the mid-term and positioning their establishments for long-term success, even as they contend with immediate issues. With that perspective, strategic leaders seldom see the whole life span of their ideas; initiatives at this level may take years to come to fruition. Strategic leaders think, therefore, in terms of strategic systems that will operate over extended time periods. They ensure these systems are built in accord with the six imperatives mentioned in Chapter 6—quality people, training, force mix, doctrine, modern equipment, and leader development—and they ensure that programs and resources are in place to sustain them. This systems approach sharpens strategic leaders' "future focus" and helps align separate actions, reduce conflict, and improve cooperation.

SECTION I
STRATEGIC LEADERSHIP SKILLS

7-7. The values and attributes demanded of Army leaders are the same at all leadership levels. Strategic leaders live by Army values and set the example just as much as direct and organizational leaders, but they face additional challenges. Strategic leaders affect the culture of the entire Army and may find themselves involved in political decision making at the highest national or even global levels. Therefore, nearly any task strategic leaders set out to accomplish requires more coordination, takes longer, has a wider impact, and produces longer-term effects than a similar organizational-level task.

7-8. Strategic leaders understand, embody, and execute values-based leadership. The political and long-term nature of their decisions doesn't release strategic leaders from the current demands of training, readiness, and unforeseen crises; they are responsible to continue to work toward the ultimate goals of the force, despite the burden of those events. Army values provide the constant reference for actions in the stressful environment of strategic leaders. Strategic leaders understand, embody, and execute leadership based on Army values.

INTERPERSONAL SKILLS

7-9. Strategic leaders continue to use interpersonal skills developed as direct and organizational leaders, but the scope, responsibilities, and authority of strategic positions require leaders with unusually sophisticated interpersonal skills. Internally, there are more levels of

people to deal with; externally, there are more interactions with outside agencies, with the media, even with foreign governments. Knowing the Army's needs and goals, strategic leaders patiently but tenaciously labor to convince the proper people about what the Army must have and become. Figure 7-1 lists strategic leader interpersonal skills.

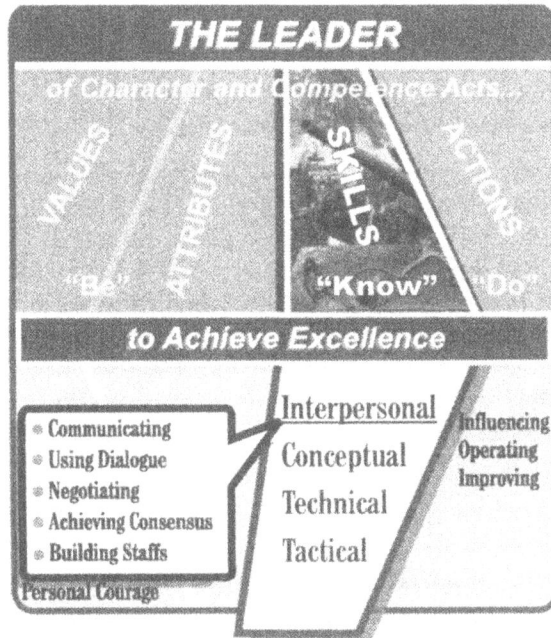

Figure 7-1. Strategic Leader Skills—Interpersonal

7-10. Strategic leaders and their staffs develop networks of knowledgeable individuals in organizations and agencies that influence their own organizations. Through penetrating assessments, these leaders seek to understand the personal strengths and weaknesses of all the main players on a particular issue. Strategic leaders are adept at reading other people, and they work to completely control their own actions and reactions. Armed with improved knowledge of others, self-control, and established networks, strategic leaders influence external events by providing leadership, timely and relevant information, and access to the right people and agencies.

COMMUNICATING

7-11. Communication at the strategic level is complicated by the wide array of staff, functional, and operational components interacting with each other and with external agencies. These complex relationships require strategic leaders to employ comprehensive communications skills as they represent their organizations. One of the most prominent differences between strategic leaders and leaders at other levels is the greater importance of symbolic communication. The example strategic leaders set, their decisions, and their actions have meaning beyond their immediate consequences to a much greater extent than those of direct and organizational leaders.

7-12. Thus, strategic leaders identify those actions that send messages. Then they use their positions to send the desired messages to their organizations and ensure that the right audiences hear them. The messages strategic leaders send set the example in the largest sense. For instance, messages that support traditions, Army values, or a particular program indicate the strategic leader's priorities.

7-13. Thus, strategic leaders communicate not only to the organization but also to a large external audience that includes the political leadership, media, and the American people. To influence those audiences, strategic leaders seek to convey integrity and win trust. As GA Marshall noted, they become expert in "the art of persuasion."

7-14. Strategic leaders commit to a few common, powerful, and consistent messages and repeat them over and over in different forms and settings. They devise a communications campaign plan, written or conceptual, that outlines how to deal with each target group. When preparing to address a specific audience, they determine its composition and agenda so they know how best to reach its members. Finding some apparent success with the medium, frequency, and words of the message, strategic leaders determine the best way to measure the message's effectiveness and continually scan and assess the environment to make sure that the message is going to all the right groups.

USING DIALOGUE

7-15. One of the forms of communication that strategic leaders use to persuade individuals, rather than groups, is dialogue. Dialogue is a conversation between two or more people. It requires not only active listening, but carefully considering what's said (and not said), logically assessing it without personal bias, and specifying issues that are not understood or don't make sense within the strategic leader's frame of reference. By using dialogue to thoroughly exchange points of view, assumptions, and concepts, strategic leaders gather information, clarify issues, and enlist support of subordinates and peers.

NEGOTIATING

7-16. Many relationships between strategic-level organizations are lateral and without clear subordination. Often, strategic leaders rely heavily on negotiating skills to obtain the cooperation and support necessary to accomplish a mission or meet the command's needs. For example, commanders of the national contingents that made up the North Atlantic Treaty Organization (NATO) implementation force (IFOR) sent to Bosnia to support the 1995 Dayton peace accords all had limitations imposed on the extent of their participation. In addition, they all had direct lines to their home governments, which they used when they believed IFOR commanders exceeded those limits. NATO strategic leaders had to negotiate some actions that ordinarily would have required only issuing orders. They often had to interpret a requirement to the satisfaction of one or more foreign governments.

7-17. Successful negotiation requires a range of interpersonal skills. Good negotiators are also able to visualize several possible end states while maintaining a clear idea of the best end state from the command's perspective. One of the most important skills is the ability to stand firm on nonnegotiable points while simultaneously communicating respect for other participants and their negotiating limits. In international forums, firmness and respect demonstrate that the negotiator knows and understands US interests. That understanding can help the negotiator persuade others of the validity of US interests and convince others that the United States understands and respects the interests of other states.

7-18. A good negotiator is particularly skilled in active listening. Other essential personal characteristics include perceptiveness and objectivity. Negotiators must be able to diagnose

unspoken agendas and detach themselves from the negotiation process. Successful negotiating involves communicating a clear position on all issues while still conveying willingness to bargain on negotiable issues, recognizing what's acceptable to all concerned, and achieving a compromise that meets the needs of all participants to the greatest extent possible.

7-19. Sometimes strategic leaders to put out a proposal early so the interchange and ultimate solution revolve around factors important to the Army. However, they are confident enough to resist the impulse to leave their thumbprints on final products. Strategic leaders don't have to claim every good idea because they know they will have more. Their understanding of selfless service allows them to subordinate personal recognition to negotiated settlements that produce the greatest good for their establishment, the Army, and the nation or coalition.

ACHIEVING CONSENSUS

7-20. Strategic leaders are skilled at reaching consensus and building and sustaining coalitions. They may apply these skills to tasks as diverse as designing combatant commands, JTFs, and policy working groups or determining the direction of a major command or the Army as an institution. Strategic leaders routinely weld people together for missions lasting from months to years. Using peer leadership rather than strict positional authority, strategic leaders oversee progress toward their visualized end state and monitor the health of the relationships necessary to achieve it. Interpersonal contact sets the tone for professional relations: strategic leaders are tactful and discreet.

7-21. GA Eisenhower's creation of SHAEF during World War II (which was mentioned in Chapter 2) is an outstanding example of coalition building and sustainment. GA Eisenhower insisted on unity of command over the forces assigned to him. He received this authority from both the British and US governments but exercised it through an integrated command and staff structure that related influence roughly to the contribution of the nations involved. The sections within SHAEF all had chiefs of one nationality and deputies of another.

7-22. GA Eisenhower also insisted that military, rather than political, criteria would predominate in his operational and strategic decisions as Supreme Allied Commander. His most controversial decisions, adoption of the so-called broad-front strategy and the refusal to race the Soviet forces to Berlin, rested on his belief that maintaining the Anglo-American alliance was a national interest and his personal responsibility. Many historians argue that this feat of getting the Allies to work together was his most important contribution to the war.

Allied Command During the Battle of the Bulge

A pivotal moment in the history of the Western Alliance arrived on 16 December 1944, when the German Army launched a massive offensive in a lightly held-sector of the American line in the Ardennes Forest. This offensive, which became known as the Battle of the Bulge, split GEN Omar Bradley's Twelfth Army Group. North of the salient, British GEN Bernard Montgomery commanded most of the Allied forces, so GA Eisenhower shifted command of the US forces there to GEN Montgomery rather than have one US command straddle the gap. GEN Bradley, the Supreme Allied Commander reasoned, could not effectively control forces both north and south of the penetration. It made more sense for GEN Montgomery to command all Allied forces on the northern shoulder and GEN Bradley all those on its southern shoulder. GA Eisenhower personally telephoned GEN Bradley to tell his old comrade of the decision. With the SHAEF staff still present, GA Eisenhower passed the order to his reluctant subordinate, listened to GEN Bradley's protests, and then said sharply, "Well, Brad, those are my orders."

According to historian J.D. Morelock, this short conversation, more than any other action taken by GA Eisenhower and the SHAEF staff during the battle, "discredited the German assumption that nationalistic fears and rivalries would inhibit prompt and effective steps to meet the German challenge." It demonstrated GA Eisenhower's "firm grasp of the true nature of an allied command" and it meant that Hitler's gamble to win the war had failed.

7-23. Across the Atlantic Ocean, GA George C. Marshall, the Army Chief of Staff, also had to seek consensus with demanding peers, none more so than ADM Ernest J. King, Commander in Chief, US Fleet and Chief of Naval Operations. GA Marshall expended great personal energy to ensure that interservice feuding at the top didn't mar the US war effort. ADM King, a forceful leader with strong and often differing views, responded in kind. Because of the ability of these two strategic leaders to work in harmony, President Franklin D. Roosevelt had few issues of major consequence to resolve once he had issued a decision and guidance.

7-24. Opportunities for strategic leadership may come at surprising moments. For instance, Joshua Chamberlain's greatest contribution to our nation may have been not at Gettysburg or Petersburg, but at Appomattox. By that time a major general, Chamberlain was chosen to command the parade at which GEN Lee's Army of Northern Virginia laid down its arms and colors. GEN Grant had directed a simple ceremony that recognized the Union victory without humiliating the Confederates.

7-25. However, MG Chamberlain sensed the need for something even greater. Instead of gloating as the vanquished army passed, he directed his bugler to sound the commands for attention and present arms. His units came to attention and rendered a salute, following his order out of respect for their commander, certainly not out of sudden warmth for recent enemies. That act set the tone for reconciliation and reconstruction and marks a brilliant leader, brave in battle and respectful in peace, who knew when, where, and how to lead.

BUILDING STAFFS

The best executive is the one who has sense enough to pick good men to do what he wants done, and self-restraint enough to keep from meddling with them while they do it.

Theodore Roosevelt
26th President of the United States

7-26. Until Army leaders reach the highest levels, they cannot staff positions and projects as they prefer. Strategic leaders have not only the authority but also the responsibility to pick the best people for their staffs. They seek to put the right people in the right places, balancing strengths and weaknesses for the good of the nation. They mold staffs able to package concise, unbiased information and build networks across organizational lines. Strategic leaders make so many wide-ranging, interrelated decisions that they must have imaginative staff members who know the environment, foresee consequences of various courses of action, and identify crucial information accordingly.

7-27. With their understanding of the strategic environment and vision for the future, strategic leaders seek to build staffs that compensate for their weaknesses, reinforce their vision, and ensure institutional success. Strategic leaders can't afford to be surrounded by staffs that blindly agree with everything they say. Not only do they avoid surrounding themselves with "yes-men," they also reward staff members for speaking the truth. Strategic leaders encourage their staffs to participate in dialogue with them, discuss alternative points of view, and explore all facts, assumptions, and implications. Such dialogue assists strategic leaders to fully assess all aspects of an issue and helps clarify their intent and guidance.

7-28. As strategic leaders build and use their staffs, they continually seek honesty and competence. Strategic-level staffs must be able to discern what the "truth" is. During World War II, GA Marshall's ability to fill his staff and commands with excellent officers made a difference in how quickly the Army could create a wartime force able to mobilize, deploy, fight, and win. Today's strategic leaders face an environment more complex than the one GA Marshall faced. They often have less time than GA Marshall had to assess situations, make plans, prepare an appropriate response, and execute. The importance of building courageous, honest, and competent staffs has correspondingly increased.

CONCEPTUAL SKILLS

From an intellectual standpoint, Princeton was a world-shaking experience. It fundamentally changed my approach to life. The basic thrust of the curriculum was to give students an appreciation of how complex and diverse various political systems and issues are....The bottom line was that answers had to be sought in terms of the shifting relationships of groups and individuals, that politics pervades all human activity, a truth not to be condemned but appreciated and put to use.

Admiral William Crowe
Former Chairman, Joint Chiefs of Staff

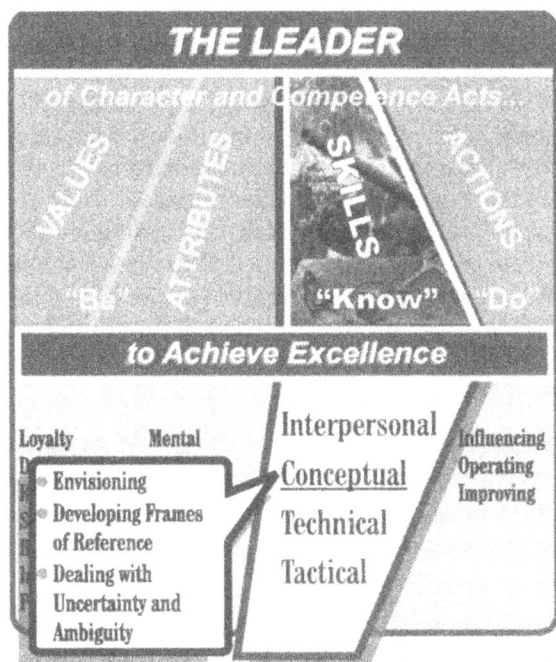

Figure 7-2. Strategic Leader Skills—Conceptual

7-29. Strategic leaders, more than direct and organizational leaders, draw on their conceptual skills to comprehend national, national security, and theater strategies, operate in the strategic and theater contexts, and improve their vast, complex organizations. The variety and scope of their concerns demand the application of more sophisticated concepts.

7-30. Strategic leaders need wisdom—and wisdom isn't just knowledge. They routinely deal with diversity, complexity, ambiguity, change, uncertainty, and conflicting policies. They are responsible for developing well-reasoned positions and providing their views and advice to our nation's highest leaders. For the good of the Army and the nation, strategic leaders seek to determine what's important now and what will be important in the future. They develop the necessary wisdom by freeing themselves to stay in touch with the force and spending time thinking, simply thinking.

ENVISIONING

It is in the minds of the commanders that the issue of battle is really decided.

Sir Basil H. Liddell Hart

7-31. Strategic leaders design compelling visions for their organizations and inspire a collaborative effort to articulate the vision in detail. They then communicate that vision clearly and use it to create a plan, gain support, and focus subordinates' work. Strategic leaders have the further responsibility of defining for their diverse organizations what counts as success in achieving the vision. They monitor their progress by drawing on personal observations, review and analysis, strategic management plans, and informal discussions with soldiers and DA civilians.

7-32. Strategic leaders look realistically at what the future may hold. They consider things they know and things they can anticipate. They incorporate new ideas, new technologies, and new capabilities. The National Security Strategy and National Military Strategy guide strategic leaders as they develop visions for their organizations. From a complicated mixture of ideas, facts, conjecture, and personal experience they create an image of what their organizations need to be.

7-33. Once strategic leaders have developed a vision, they create a plan to reach that end state. They consider objectives, courses of action to take the organization there, and resources needed to do the job. The word "vision" implies that strategic leaders create a conceptual model of what they want. Subordinates will be more involved in moving the organization forward if they can "see" what the leader has in mind. And because moving a large organization is often a long haul, subordinates need some sign that they're making progress. Strategic leaders therefore provide intermediate objectives that act as milestones for their subordinates in checking their direction and measuring their progress.

7-34. The strategic leader's vision provides the ultimate sense of purpose, direction, and motivation for everyone in the organization. It is at once the starting point for developing specific goals and plans, a yardstick for measuring what the organization accomplishes, and a check on organizational values. Ordinarily, a strategic leader's vision for the organization may have a time horizon of years, or even decades. In combat, the horizon is much closer, but strategic leaders still focus far beyond the immediate actions.

7-35. The strategic leader's vision is a goal, something the organization strives for (even though some goals may always be just out of reach). When members understand the vision, they can see it as clearly as the strategic leader can. When they see it as worthwhile and accept it, the vision creates energy, inspiration, commitment, and a sense of belonging.

7-36. Strategic leaders set the vision for their entire organization. They seek to keep the vision consistent with the external environment, alliance or coalition goals, the National Security Strategy, and the National Military Strategy. Subordinate leaders align their visions and intent with their strategic leader's vision. A strategic leader's vision may be expressed in everything from small acts to formal, written policy statements.

7-37. Joint Vision 2010 and Army Vision 2010, which is derived from it, are not based on formal organizations; rather they array future technologies and force structure against emerging threats. While no one can yet see exactly what that force will look like, the concepts themselves provide an azimuth and a point on the horizon. Achieving well-publicized milepost initiatives shows that the Army as an institution is progressing toward the end state visualized by its strategic leaders.

DEVELOPING FRAMES OF REFERENCE

7-38. All Army leaders build a personal frame of reference from schooling, experience, self-study, and reflection on current events and history. Strategic leaders create a comprehensive frame of reference that encompasses their organization and places it in the strategic environment. To construct a useful frame, strategic leaders are open to new experiences and to comments from others, including subordinates. Strategic leaders are reflective, thoughtful, and unafraid to rethink past experiences and to learn from them. They are comfortable with the abstractions and concepts common in the strategic environment. Moreover, they understand the circumstances surrounding them, their organization, and the nation.

7-39. Much like intelligence analysts, strategic leaders look at events and see patterns that others often miss. These leaders are likely to identify and understand a strategic situation and, more important, infer the outcome of interventions or the absence of interventions. A strategic leader's frame of reference helps identify the information most relevant to a strategic situation so that the leader can go to the heart of a matter without being distracted. In the new information environment, that talent is more important than ever. Cosmopolitan strategic leaders, those with comprehensive frames of reference and the wisdom that comes from thought and reflection, are well equipped to deal with events having complex causes and to envision creative solutions.

7-40. A well-developed frame of reference also gives strategic leaders a thorough understanding of organizational subsystems and their interacting processes. Cognizant of the relationships among systems, strategic leaders foresee the possible effects on one system of actions in others. Their vision helps them anticipate and avoid problems.

DEALING WITH UNCERTAINTY AND AMBIGUITY

True genius resides in the capacity for evaluation of uncertain, hazardous, and conflicting information.

Sir Winston Churchill
Prime Minister of Great Britain, World War II

7-41. Strategic leaders operate in an environment of increased volatility, uncertainty, complexity, and ambiguity. Change at this level may arrive suddenly and unannounced. As they plan for contingencies, strategic leaders prepare intellectually for a range of uncertain threats and scenarios. Since even great planning and foresight can't predict or influence all future events, strategic leaders work to shape the future on terms they can control, using diplomatic, informational, military, and economic instruments of national power.

7-42. Strategic leaders fight complexity by encompassing it. They must be more complex than the situations they face. This means they're able to expand their frame of reference to fit a situation rather than reducing a situation to fit their preconceptions. They don't lose sight of Army values and force capabilities as they focus on national policy. Because of their maturity and wisdom, they tolerate ambiguity, knowing they will never have all the information they want. Instead, they carefully analyze events and decide when to make a decision, realizing that they must innovate and accept some risk. Once they make decisions, strategic leaders then explain them to the Army and the nation, in the process imposing order on the uncertainty and ambiguity of the situation. Strategic leaders not only understand the environment themselves; they also translate their understanding to others.

Strategic Flexibility in Haiti

Operation Uphold Democracy, the 1994 US intervention in Haiti conducted under UN auspices, provides an example of strategic leaders achieving success in spite of extreme uncertainty and ambiguity. Prior to the order to enter Haiti, strategic leaders didn't know either D-day or the available forces. Neither did they know whether the operation would be an invitation (permissive entry), an invasion (forced entry), or something in between. To complicate the actual military execution, former President Jimmy Carter, retired GEN Colin Powell, and Senator Sam Nunn were negotiating with LTG Raoul Cedras, commander in chief of the Haitian armed forces, in the Haitian capital even as paratroopers, ready for a combat jump, were inbound.

When LTG Cedras agreed to hand over power, the mission of the inbound JTF changed from a forced to a permissive entry. The basis for the operation wound up being an operation plan based on an "in-between" course of action inferred by the JTF staff during planning. The ability of the strategic leaders involved to change their focus so dramatically and quickly provides an outstanding example of strategic flexibility during a crisis. The ability of the soldiers, sailors, airmen, and Marines of the JTF to execute the new mission on short notice is a credit to them and their leaders at all levels.

7-43. In addition to demonstrating the flexibility required to handle competing demands, strategic leaders understand complex cause-and-effect relationships and anticipate the second- and third-order effects of their decisions throughout the organization. The highly volatile nature of the strategic environment may tempt them to concentrate on the short term, but strategic leaders don't allow the crisis of the moment absorb them completely. They remain focused on their responsibility to shape an organization or policies that will perform successfully over the next 10 to 20 years. Some second- and third-order effects are desirable; leaders can design and pursue actions to achieve them. For example, strategic leaders who continually send—through their actions—messages of trust to subordinates inspire trust in themselves. The third-order effect may be to enhance subordinates' initiative.

TECHNICAL SKILLS

The crucial difference (apart from levels of innate ability) between Washington and the commanders who opposed him was they were sure they knew all the answers, while Washington tried every day and every hour to learn.

James Thomas Flexner
George Washington in the American Revolution

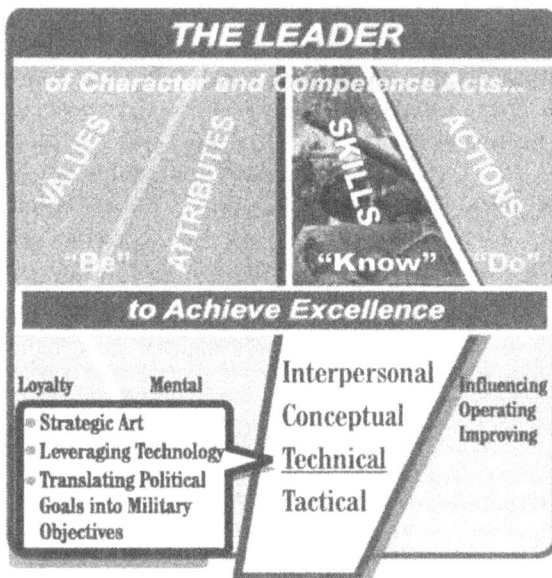

Figure 7-3. Strategic Leader Skills—Technical

7-44. Strategic leaders create their work on a broad canvas that requires broad technical skills of the sort named in Figure 7-3.

STRATEGIC ART

7-45. The strategic art, broadly defined, is the skillful formulation, coordination, and application of ends, ways, and means to promote and defend the national interest. Masters of the strategic art competently integrate the three roles performed by the complete strategist: strategic leader, strategic practitioner, and strategic theorist.

7-46. Using their understanding of the systems within their own organizations, strategic leaders

work through the complexity and uncertainty of the strategic environment and translate abstract concepts into concrete actions. Proficiency in the science of leadership—programs, schedules, and systems, for example—can bring direct or organizational leaders success. For strategic leaders, however, the intangible qualities of leadership draw on their long and varied experience to produce a rare art.

7-47. Strategic leaders do more than imagine and accurately predict the future; they shape it by moving out of the conceptual realm into practical execution. Although strategic leaders never lose touch with soldiers and their technical skills, some practical activities are unique to this level.

7-48. By reconciling political and economic constraints with the Army's needs, strategic leaders navigate to move the force forward using the strategy and budget processes. They spend a great deal of time obtaining and allocating resources and determining conceptual directions, especially those judged critical for future strategic positioning and necessary to prevent readiness shortfalls. They're also charged with overseeing of the Army's responsibilities under Title 10 of the United States Code.

7-49. Strategic leaders focus not so much on internal processes as on how the organization fits into the DOD and the international arena: What are the relationships among external organizations? What are the broad political and social systems in which the organization and the Army must operate? Because of the complex reporting and coordinating relationships, strategic leaders fully understand their roles, the boundaries of these roles, and the expectations of other departments and agencies. Understanding those interdependencies outside the Army helps strategic leaders do the right thing for the programs, systems, and people within the Army as well as for the nation.

7-50. Theater CINCs, with their service component commanders, seek to shape their environments and accomplish long-term national security policy goals within their theaters. They operate through congressional testimony, creative use of assigned and attached military forces, imaginative bilateral and multilateral programs, treaty obligations, person-to-person contacts with regional leaders, and various joint processes. These actions require strategic leaders to apply the strategic art just as much as does designing and employing force packages to achieve military end states.

7-51. GA Douglas MacArthur, a theater CINC during World War II, became military governor of occupied Japan after the Japanese surrender. His former enemies became his responsibility; he had to deal diplomatically with the defeated nation as well as the directives of American civil authorities and the interests of the former Allies. GA MacArthur understood the difference between preliminary (often called military) end state conditions and the broader set of end state conditions that are necessary for the transition from war to peace. Once a war has ended, military force can no longer be the principal means of achieving strategic aims. Thus, a strategic leader's end state vision must include diplomatic, economic, and informational—as well as military—aspects. GA MacArthur's vision, and the actions he took to achieve it, helped establish the framework that preserved peace in the Pacific Ocean and rebuilt a nation that would become a trusted ally.

7-52. A similar institutional example occurred in the summer of 1990. Then, while the Army was in the midst of the most precisely planned force "build-down" in history, Army Chief of Staff Carl Vuono had to halt the process to meet a crisis in the Persian Gulf. GEN Vuono was required to call up, mobilize and deploy forces necessary to meet the immediate crisis while

retaining adequate capabilities in other theaters. The following year he redeployed Third (US) Army, demobilized the activated reserves, and resumed downsizing toward the smallest active force since the 1930s. GEN Vuono demonstrated the technical skill of the strategic art and proved himself a leader of character and competence motivated by Army values.

LEVERAGING TECHNOLOGY

7-53. Leveraging technology—that is, applying technological capabilities to obtain a decisive military advantage—has given strategic leaders advantages in force projection, in command and control, and in the generation of overwhelming combat power. Leveraging technology has also increased the tempo of operations, the speed of maneuver, the precision of firepower, and the pace at which information is processed. Ideally, information technology, in particular, enhances not only communications, but also situational understanding. With all these advantages, of course, comes increasing complexity: it's harder to control large organizations that are moving quickly. Strategic leaders seek to understand emerging military technologies and apply that understanding to resourcing, allocating, and exploiting the many systems under their control.

7-54. Emerging combat, combat support, and combat service support technologies bring more than changes to doctrine. Technological change allows organizations to do the things they do now better and faster, but it also enables them to do things that were not possible before. So a part of leveraging technology is envisioning the future capability that could be exploited by developing a technology. Another aspect is rethinking the form the organization ought to take in order to exploit new processes that previously were not available. This is why strategic leaders take time to think "out of the box."

TRANSLATING POLITICAL GOALS INTO MILITARY OBJECTIVES

7-55. Leveraging technology takes more than understanding; it takes money. Strategic leaders call on their understanding and their knowledge of the budgetary process to determine which combat, combat support, and combat service support technologies will provide the leap-ahead capability commensurate with the cost. Wise Army leaders in the 1970s and 1980s realized that superior night systems and greater standoff ranges could expose fewer Americans to danger yet kill more of the enemy. Those leaders committed money to developing and procuring appropriate weapons systems and equipment. Operation Desert Storm validated these decisions when, for example, M1 tanks destroyed Soviet-style equipment before it could close within its maximum effective range. However strategic leaders are always in the position of balancing budget constraints, technological improvements, and current force readiness against potential threats as they shape the force for the future.

7-56. Strategic leaders identify military conditions necessary to satisfy political ends desired by America's civilian leadership. They must synchronize the efforts of the Army with those of the other services and government agencies to attain those conditions and achieve the end state envisioned by America's political leaders. To operate on the world stage, often in conjunction with allies, strategic leaders call on their international perspective and relationships with policy makers in other countries.

Show of Force in the Philippines

At the end of November 1989, 1,000 rebels seized two Filipino air bases in an attempt to overthrow the government of the Philippines. There had been rumors that someone was plotting a coup to end Philippine President Corazon Aquino's rule. Now rebel aircraft from the captured airfields had bombed and strafed the presidential palace. President Aquino requested that the United States help suppress the coup attempt by destroying the captured airfields. Vice President Dan Quayle and Deputy Secretary of State Lawrence Eagleburger favored US intervention to support the Philippine government. As the principal military advisor to the president, Joint Chiefs of Staff Chairman Colin Powell was asked to recommend a response to President Aquino's request.

GEN Powell applied critical reasoning to this request for US military power in support of a foreign government. He first asked the purpose of the proposed intervention. The State Department and White House answered that the United States needed to demonstrate support for President Aquino and keep her in power. GEN Powell then asked the purpose of bombing the airfields. To prevent aircraft from supporting the coup, was the reply. Once GEN Powell understood the political goal, he recommended a military response to support it.

The chairman recommended to the White House that American jets fly menacing runs over the captured airfields. The goal would be to prevent takeoffs from the airfields by intimidating the rebel pilots rather than destroying rebel aircraft and facilities. This course of action was approved by President George Bush and achieved the desired political goal: it deterred the rebel pilots from supporting the coup attempt. By understanding the political goal and properly defining the military objective, GEN Powell was able to recommend a course of action that applied a measured military response to what was, from the United States' perspective, a diplomatic problem. By electing to conduct a show of force rather than an attack, the United States avoided unnecessary casualties and damage to the Philippine infrastructure.

7-57. Since the end of the Cold War, the international stage has become more confused. Threats to US national security may come from a number of quarters: regional instability, insurgencies, terrorism, and proliferation of weapons of mass destruction to name a few. International drug traffickers and other transnational groups are also potential adversaries. To counter such diverse threats, the nation needs a force flexible enough to execute a wide array of missions, from warfighting to peace operations to humanitarian assistance. And of course, the nation needs strategic leaders with the sound perspective that allows them to understand the nation's political goals in the complex international environment and to shape military objectives appropriate to the various threats.

SECTION II

STRATEGIC LEADERSHIP ACTIONS

Leadership is understanding people and involving them to help you do a job. That takes all of the good characteristics, like integrity, dedication of purpose, selflessness, knowledge, skill, implacability, as well as determination not to accept failure.

Admiral Arleigh A. Burke
Naval Leadership: Voices of Experience

7-58. Operating at the highest levels of the Army, the DOD, and the national security establishment, military and DA civilian strategic leaders face highly complex demands from inside and outside the Army. Constantly changing global conditions challenge their decision-making abilities. Strategic leaders tell the Army story, make long-range decisions, and shape the Army culture to influence the force and its partners inside and outside the United States. They plan for contingencies across the range of military operations and allocate resources to prepare for them, all the while assessing the threat and the force's readiness. Steadily improving the Army, strategic leaders develop their successors, lead changes in the force, and optimize systems and operations. This section addresses the influencing, operating, and improving actions they use.

INFLUENCING ACTIONS

7-59. Strategic leaders act to influence both their organization and its outside environment. Like direct and organizational leaders, strategic leaders influence through personal example as well as by communicating, making decisions, and motivating.

7-60. Because the external environment is diverse and complex, it's sometimes difficult for strategic leaders to identify and influence the origins of factors affecting the organization. This difficulty applies particularly to fast-paced situations like theater campaigns. Strategic leaders meet this challenge by becoming masters of information, influence, and vision.

7-61. Strategic leaders also seek to control the information environment, consistent with US law and Army values. Action in this area can range from psychological operations campaigns to managing media relationships. Strategic leaders who know what's happening with present and future requirements, both inside and

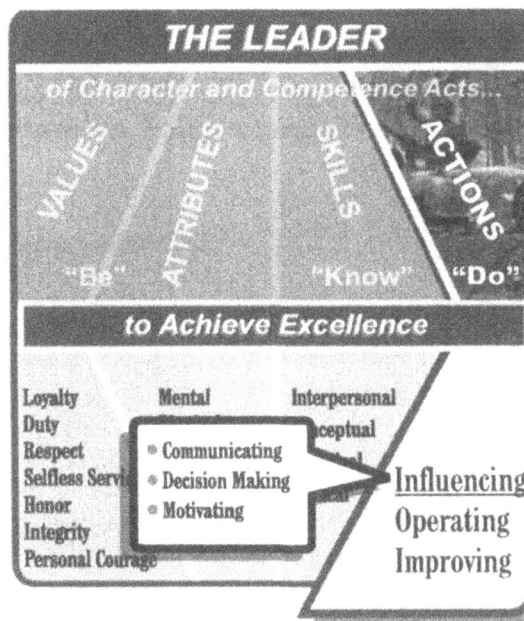

Figure 7-4. Strategic Leader Actions—Influencing

outside the organization, are in a position to influence events, take advantage of opportunities, and move the organization toward its goals.

7-62. As noted earlier, strategic leaders develop the wisdom and frames of reference necessary to identify the information relevant to the situation at hand. In addition, they use interpersonal skills to develop a network of knowledgeable people, especially in those organizations that can influence their own. They encourage staff members to develop similar networks. Through these networks, strategic leaders actively seek information relevant to their organizations and subject matter experts they can call on to assist themselves and their staffs. Strategic leaders can often call on the nation's best minds and information sources and may face situations where nothing less will do.

COMMUNICATING

Moving our Army into the next century is a journey, not a destination; we know where we are going and we are moving out.

General Gordon R. Sullivan
Former Army Chief of Staff

Communicating a Vision

7-63. The skill of envisioning is vital to the strategic leader. But forming a vision is pointless unless the leader shares it with a broad audience, gains widespread support, and uses it as a compass to guide the organization. For the vision to provide purpose, direction, and motivation, the strategic leader must personally commit to it, gain commitment from the organization as a whole, and persistently pursue the goals and objectives that will spread the vision throughout the organization and make it a reality.

7-64. Strategic leaders identify trends, opportunities, and threats that could affect the Army's future and move vigorously to mobilize the talent that will help create strategic vision. In 1991 Army Chief of Staff Gordon R. Sullivan formed a study group of two dozen people to help craft his vision for the Army. In this process, GEN Sullivan considered authorship less important than shared vision:

Once a vision has been articulated and the process of buy-in has begun, the vision must be continually interpreted. In some cases, the vision may be immediately understandable at every level. In other cases, it must be translated—put into more appropriate language—for each part of the organization. In still other cases, it may be possible to find symbols that come to represent the vision.

7-65. Strategic leaders are open to ideas from a variety of sources, not just their own organizations. Some ideas will work; some won't. Some will have few, if any, long-lasting effects; others, like the one in this example, will have effects few will foresee.

Combat Power from a Good Idea

In 1941, as the American military was preparing for war, Congresswoman Edith Nourse Rogers correctly anticipated manpower shortages in industry and in the armed forces as the military grew. To meet this need, she proposed creation of a Women's Army Auxiliary Corps (WAAC) of 25,000 women to fill administrative jobs and free men for service with combat units. After the United States entered the war, when the size of the effort needed became clearer, Congresswoman Rogers introduced another bill for a WAAC of some 150,000 women. Although the bill met stiff opposition in some quarters, a version passed and eventually the Women's Army Corps was born. Congresswoman Rogers' vision of how to best get the job done in the face of vast demands on manpower contributed a great deal to the war effort.

Telling the Army Story

If you have an important point to make, don't try to be subtle or clever. Use a pile-driver. Hit the point once. Then come back and hit it a second time—a tremendous whack!

Sir Winston Churchill
Prime Minister of Great Britain, World War II

7-66. Whether by nuance or overt presentation, strategic leaders vigorously and constantly represent who Army is, what it's doing, and where it's going. The audience is the Army itself as well as the rest of the world. There's an especially powerful responsibility to explain things to the American people, who support their Army with money and lives. Whether working with other branches of government, federal agencies, the media, other militaries, the other services, or their own organizations, strategic leaders rely increasingly on writing and public speaking (conferences and press briefings) to reinforce the Army's central messages. Because so much of this communication is directed at outside agencies, strategic leaders avoid parochial language and remain sensitive to the Army's image.

7-67. Strategic leaders of all times have determined and reinforced the message that speaks to the soul of the nation and unifies the force. In 1973 Army leaders at all levels knew about "The Big Five," the weapons systems that would transform the Army (a new tank, an infantry fighting vehicle, an advanced attack helicopter, a new utility helicopter, and an air defense system). Those programs yielded the M1 Abrams, the M2/M3 Bradley, the AH-64 Apache, the UH-60 Blackhawk, and the Patriot. But those initiatives were more than sales pitches for newer hardware; they were linked to concepts about how to fight and win against a massive Soviet-style force. As a result, fielding the new equipment gave physical form to the new ideas being adopted at the same time. Soldiers could see improvements as well as read about them. The synergism of new equipment, new ideas, and good leadership resulted in the Army of Excellence.

7-68. Today, given the rapid growth of technology, unpredictable threats, and newly emerging roles, Army leaders can't cling to new hardware as the key to the Army's vision. Instead, today's strategic leaders emphasize the Army's core strength: Army values and the timeless character of the American soldier. The Army—trained, ready, and led by leaders of character and competence at all levels—has met and will continue to meet the nation's security needs. That's the message of today's Army to the nation it serves.

7-69. A recent example of successfully telling the Army story occurred during Operation Desert Shield. During the deployment phase, strategic leaders decided to get local reporters to the theater of war to report on mobilized reserve component units from their communities. That decision had several effects. The first-order effect was to get the Army story to the citizens of hometown America. That publicity resulted in an unintended second-order effect: a flood of mail that the nation sent to its deployed soldiers. That mail, in turn, produced a third-order effect felt by American soldiers: a new pride in themselves.

STRATEGIC DECISION MAKING

When I am faced with a decision—picking somebody for a post, or choosing a course of action—I dredge up every scrap of knowledge I can. I call in people. I telephone them. I read whatever I can get my hands on. I use my intellect to inform my instinct. I then use my instinct to test all this data. "Hey, instinct, does this sound right? Does it smell right, feel right, fit right?"

General Colin Powell
Former Chairman, Joint Chiefs of Staff

7-70. Strategic leaders have great conceptual resources; they have a collegial network to share thoughts and plan for the institution's continued success and well being. Even when there's consensual decision making, however, everyone knows who the boss is. Decisions made by strategic leaders—whether CINCs deploying forces or service chiefs initiating budget programs—often result in a major commitment of resources. They're expensive and tough to reverse. Therefore, strategic leaders rely on timely feedback throughout the decision-making process in order to avoid making a decision based on inadequate or faulty information. Their purpose, direction, and motivation flow down; information and recommendations surface from below.

7-71. Strategic leaders use the processes of the DOD, Joint Staff, and Army strategic planning systems to provide purpose and direction to subordinate leaders. These systems include the Joint Strategic Planning System (JSPS), the Joint Operation Planning and Execution System (JOPES), and the Planning, Programming and Budgeting System (PPBS). However, no matter how many systems are involved and no matter how complex they are, providing motivation remains the province of the individual strategic leader.

7-72. Because strategic leaders are constantly involved in this sort of planning and because decisions at this level are so complex and depend on so many variables, there's a temptation to analyze things endlessly. There's always new information; there's always a reason to wait for the next batch of reports or the next dispatch. Strategic leaders' perspective, wisdom, courage, and sense of timing help them know when to decide. In peacetime the products of those decisions may not see completion for 10 to 20 years and may require leaders to constantly adjust them along the way. By contrast, a strategic leader's decision at a critical moment in combat can rapidly alter the course of the war, as did the one in this example.

The D-Day Decision

On 4 June 1944 the largest invasion armada ever assembled was poised to strike the Normandy region of France. Weather delays had already caused a 24-hour postponement and another front of bad weather was heading for the area. If the Allies didn't make the landings on 6 June, they would miss the combination of favorable tides, clear flying weather, and moonlight needed for the assault. In addition to his concerns about the weather, GA Dwight D. Eisenhower, the Supreme Allied Commander, worried about his soldiers. Every hour they spent jammed aboard crowded ships, tossed about and seasick, degraded their fighting ability.

The next possible invasion date was 19 June; however the optimal tide and visibility conditions would not recur until mid-July. GA Eisenhower was ever mindful that the longer he delayed, the greater chance German intelligence had to discover the Allied plan. The Germans would use any additional time to improve the already formidable coastal defenses.

On the evening of 4 June GA Eisenhower and his staff received word that there would be a window of clear weather on the next night, the night of 5-6 June. If the meteorologists were wrong, GA Eisenhower would be sending seasick men ashore with no air cover or accurate naval gunfire. GA Eisenhower was concerned for his soldiers.

"Don't forget," GA Eisenhower said in an interview 20 years later, "some hundreds of thousands of men were down here around Portsmouth, and many of them had already been loaded for some time, particularly those who were going to make the initial assault. Those people in the ships and ready to go were in cages, you might say. You couldn't call them anything else. They were fenced in. They were crowded up, and everybody was unhappy."

GA Eisenhower continued, "Goodness knows, those fellows meant a lot to me. But these are the decisions that have to be made when you're in a war. You say to yourself, I'm going to do something that will be to my country's advantage for the least cost. You can't say without any cost. You know you're going to lose some of them, and it's very difficult."

A failed invasion would delay the end of a war that had already dragged on for nearly five years. GA Eisenhower paced back and forth as a storm rattled the windows. There were no guarantees, but the time had come to act.

He stopped pacing and, facing his subordinates, said quietly but clearly, "OK, let's go."

MOTIVATING

It is the morale of armies, as well as of nations, more than anything else, which makes victories and their results decisive.

Baron Antoine-Henri de Jomini
Precis de l'Art de Guerre, 1838

Shaping Culture

7-73. Strategic leaders inspire great effort. To mold morale and motivate the entire Army, strategic leaders cultivate a challenging, supportive, and respectful environment for soldiers and DA civilians to operate in. An institution with a history has a mature, well-established culture—a shared set of values and assumptions that members hold about it. At the same time, large and complex institutions like the Army are diverse; they have many subcultures, such as those that exist in the civilian and reserve components, heavy and light forces, and special operations forces. Gender, ethnic, religious, occupational, and regional differences also define groups within the force.

Culture and Values

7-74. The challenge for strategic leaders is to ensure that all these subcultures are part of the larger Army culture and that they all share Army values. Strategic leaders do this by working with the best that each subculture has to offer and ensuring that subcultures don't foster unhealthy competition with each other, outside agencies, or the rest of the Army. Rather, these various subcultures must complement each other and the Army's institutional culture. Strategic leaders appreciate the differences that characterize these subcultures and treat all members of all components with dignity and respect. They're responsible for creating an environment that fosters mutual understanding so that soldiers and DA civilians treat one another as they should.

7-75. Army values form the foundation on which the Army's institutional culture stands. Army values also form the basis for Army policies and procedures. But written values are of little use unless they are practiced. Strategic leaders help subordinates adopt these values by making sure that their experience validates

them. In this, strategic leaders support the efforts all Army leaders make to develop the their subordinates' character. This character development effort (discussed in Appendix E) strives to have all soldiers and DA civilians adopt Army values, incorporate them into a personal code, and act according to them.

7-76. Like organizational and direct leaders, strategic leaders model character by their actions. Only experience can validate Army values: subordinates will hear of Army values, then look to see if they are being lived around them. If they are, the Army's institutional culture is strengthened; if they are not, the Army's institutional culture begins to weaken. Strategic leaders ensure Army values remain fundamental to the Army's institutional culture.

7-77. Over time, an institution's culture becomes so embedded in its members that they may not even notice how it affects their attitudes. The institutional culture becomes second nature and influences the way people think, the way they act in relation to each other and outside agencies, and the way they approach the mission. Institutional culture helps define the boundaries of acceptable behavior, ranging from how to wear the uniform to how to interact with foreign nationals. It helps determine how people approach problems, make judgments, determine right from wrong, and establish priorities. Culture shapes Army customs and traditions through doctrine, policies and regulations, and the philosophy that guides the institution. Professional journals, historical works, ceremonies—even the folklore of the organization—all contain evidence of the Army's institutional culture.

Culture and Leadership

7-78. A healthy culture is a powerful leadership tool strategic leaders use to help them guide their large diverse organizations. Strategic leaders seek to shape the culture to support their vision, accomplish the mission, and improve the organization. A cohesive culture molds the organization's morale, reinforcing an ethical climate built on Army values, especially respect. As leaders initiate changes for long-range improvements, soldiers and DA civilians must feel that they're valued as persons, not just as workers or program supporters.

7-79. One way the Army's institutional culture affirms the importance of individuals is through its commitment to leader development: in essence, this commitment declares that people are the Army's future. By committing to broad-based leader development, the Army has redefined what it means to be a soldier. In fact, Army leaders have even changed the appearance of American soldiers and the way they perform. Introducing height and weight standards, raising PT standards, emphasizing training and education, and deglamorizing alcohol have all fundamentally changed the Army's institutional culture.

OPERATING ACTIONS

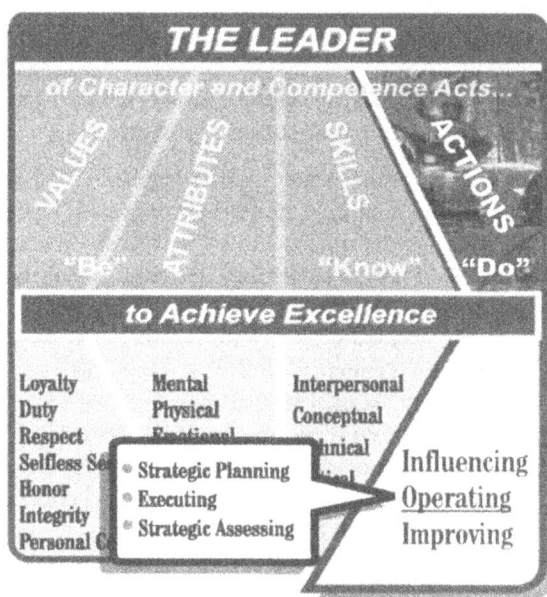

Figure 7-5. Strategic Leader Actions—Operating

7-80. Operating at the strategic level can involve both short-term and long-term actions. The most agile organizations have standing procedures and policies to take the guesswork out of routine actions and allow leaders to concentrate their imagination and energy on the most difficult tasks. Strategic leaders coordinate their organizations' actions to accomplish near-term missions, often without the benefit of direct guidance. Strategic leaders receive general guidance—frequently from several sources, including the national command authority.

7-81. Although they perform many of the same operating actions as organizational and direct leaders, strategic leaders also manage joint, multinational, and interagency relationships. For strategic leaders, planning, preparing, executing, and assessing are nearly continuous, more so than at the other leadership levels, because the larger organizations they lead have continuing missions. In addition, the preparing action takes on a more comprehensive meaning at the strategic leadership level. Leaders at all levels keep one eye on tomorrow. Strategic leaders, to a greater extent than leaders at other levels, must coordinate their organizations' actions, positioning them to accomplish the current mission in a way that will feed seamlessly into the next one. The Army doesn't stop at the end of a field exercise—or even after recovering from a major deployment; there's always another mission about to start and still another one on the drawing board.

STRATEGIC PLANNING

7-82. Strategic-level plans must balance competing demands across the vast structure of the DOD, but the fundamental requirements for strategic-level planning are the same as for direct- and organizational-level planning. At all levels, leaders establish priorities and communicate decisions; however, at the strategic level, the sheer number of players who can influence the organization means that strategic leaders must stay on top of multiple demands. To plan coherently and comprehensively, they look at the mission from other players' points of view. Strategic planning depends heavily on wisely applying interpersonal and conceptual skills. Strategic leaders ask, What will these people want? How will they see things? Have I justified the mission? The interaction among strategic leaders' interpersonal and conceptual skills and their operating actions is highly complex.

7-83. Interpersonal and conceptual understanding helped the Army during Operation Uphold Democracy, the US intervention in Haiti. The success of the plan to collect and disarm former Haitian police and military officials, investigate them, remove them (if required), or retrain them owed much to recognizing the special demands of the Haitian psyche. The population needed a secure and stable environment and a way to know when that condition was in place. The Haitians feared the resurgence of government terror, and any long-term solution had to address their concerns. Strategic planners maintained a focus on the desired end state: US disengagement and a return to a peaceful, self-governing Haiti. In the end, the United States forced Haitian leaders to cooperate, restored the elected president, Jean-Bertrand Aristide, and made provisions for returning control of affairs to the Haitians themselves.

EXECUTING

There are no victories at bargain prices.

General of the Army Dwight D. Eisenhower

Allocating Resources

7-84. Because lives are precious and materiel is scarce, strategic leaders make tough decisions about priorities. Their goal is a capable, prepared, and victorious force. In peacetime, strategic leaders decide which programs get funded and consider the implications of those choices. Allocating resources isn't simply a matter of choosing helicopters, tanks, and missiles for the future Army. Strategic resourcing affects how the Army will operate and fight tomorrow. For example, strategic leaders determine how much equipment can be pre-positioned for contingencies without degrading current operational capabilities.

Managing Joint, Interagency, and Multinational Relationships

7-85. Strategic leaders oversee the relationship between their organizations, as part of the nation's total defense force, and the national policy apparatus. They use their knowledge of how things work at the national and international levels to influence opinion and build consensus for the organization's missions, gathering support of diverse players to achieve their vision. Among their duties, strategic leaders—

- Provide military counsel in national policy forums.

- Interpret national policy guidelines and directions.

- Plan for and maintain the military capability required to implement national policy.

- Present the organization's resource requirements.

Multinational Resource Allocation

Following the breakout and pursuit after the Normandy landings, Allied logistics systems became seriously overstretched. GA Eisenhower, the Supreme Allied Commander, had to make a number of decisions on resource allocation among his three army groups. These decisions had serious implications for the conduct of the war in the ETO. Both GEN George Patton, Commander of the Third (US) Army in GEN Omar Bradley's Twelfth Army Group, and British GEN Bernard Montgomery, the Twenty-first Army Group Commander, argued that sole priority for their single thrusts into the German homeland could win the war. GA Eisenhower, dedicated to preserving the alliance with an Allied success in the West, gave GEN Montgomery only a limited priority for a risky attempt to gain a Rhine bridgehead, and at the same time, slowed GEN Patton's effort to what was logistically feasible under the circumstances. The Supreme Allied Commander's decision was undoubtedly unpopular with his longtime colleague, GEN Patton, but it contributed to alliance solidarity, sent a message to the Soviets, and ensured a final success that did not rely on the still highly uncertain collapse of German defenses.

- Develop strategies to support national objectives.

- Bridge the gap between political decisions made as part of national strategy and the individuals and organizations that must carry out those decisions.

7-86. As part of this last requirement, strategic leaders clarify national policy for subordinates and explain the perspectives that contribute to that national policy. They develop policies reflecting national security objectives and prepare their organizations to respond to missions across the spectrum of military actions.

7-87. Just as direct and organizational leaders consider their sister units and agencies, strategic leaders consider and work with other armed services and government agencies. How important is this joint perspective? Most of the Army's four-star billets are joint or multinational. Almost half of the lieutenant generals hold similar positions on the Joint Staff, with the DOD, or in combatant commands. While the remaining strategic leaders are assigned to organizations that are nominally single service (Forces Command, Training and Doctrine Command, Army Materiel Command), they frequently work outside Army channels. In addition, many DA civilian strategic leaders hold positions that require a joint perspective.

7-88. The complexity of the work created by joint and multinational requirements is twofold. First, communication is more complicated because of the different interests, cultures, and languages of the participants. Even the cultures and jargon of the various US armed services differ dramatically. Second, subordinates may not be subordinate in the same sense as they are in a purely Army organization. Strategic leaders and their forces may fall under international operational control but retain their allegiances and lines of authority to their own national commanders. UN and NATO commands, such as the IFOR, discussed earlier are examples of this kind of arrangement.

7-89. To operate effectively in a joint or multinational environment, strategic leaders exercise a heightened multiservice and international sensitivity developed over their years of experience. A joint perspective results from shared experiences and interactions with leaders of other services, complemented by the leader's habitual introspection. Similar elements in the international arena inform an international perspective. Combing those perspectives with their own Army and national perspectives, strategic leaders—

- Influence the opinions of those outside the Army and help them understand Army needs.

- Interpret the outside environment for people on the inside, especially in the formulation of plans and policies.

Most Army leaders will have several opportunities to serve abroad, sometimes with forces of other nations. Perceptive leaders turn such service into opportunities for self-development and personal broadening.

7-90. Chapter 2 describes building a "third culture," that is, a hybrid culture that bridges the gap between partners in multinational operations. Strategic leaders take the time to learn about their partners' cultures—including political, social and economic aspects—so that they understand how and why the partners think and act as they do. Strategic leaders are also aware that the successful conduct of multinational operations requires a particular sensitivity to the effect that deploying US forces may have on the laws, traditions, and customs of a third country.

7-91. Strategic leaders understand American and Army culture. This allows them to see their own culturally-based actions from the viewpoint of another culture—civilian, military, or foreign. Effective testimony before Congress requires an understanding of how Congress works and how its members think. The same is true concerning dealings with other federal and state agencies, non-governmental organizations, local political leaders, the media, and other people who shape public opinion and national attitudes toward the military. Awareness of the audience helps strategic leaders represent their organizations to outside agencies. Understanding societal values—those values people bring into the Army—helps strategic leaders motivate subordinates to live Army values.

7-92. When the Army's immediate needs conflict with the objectives of other agencies, strategic leaders work to reconcile the differences. Reconciliation begins with a clear understanding of the other agency's position. Understanding the other side's position is the first step in identifying shared interests, which may permit a new outcome better for both parties. There will be times when strategic leaders decide to stick to their course; there will be other times when Army leaders bend to accommodate other organizations. Continued disagreement can impair the Army's ability to serve the nation; therefore, strategic leaders must work to devise Army courses of action that reflect national policy objectives and take into account the interests of other organizations and agencies.

7-93. Joint and multinational task force commanders may be strategic leaders. In certain operations they will work for a CINC but receive guidance directly from the Chairman of the Joint Chiefs of Staff, the Secretary of Defense, the State Department, or the UN. Besides establishing professional relationships within the DOD and US government, such strategic leaders must build personal rapport with officials from other countries and military establishments.

Military Actions Across the Spectrum

7-94. Since the character of the next war has not been clearly defined for them, today's strategic leaders rely on hints in the international environment to provide information on what sort of force to prepare. Questions they consider include these: Where is the next threat? Will we have allies or contend alone? What will our national and military goals be? What will the exit strategy be? Strategic leaders address the technological, leadership, and moral considerations associated with fighting on an asymmetrical battlefield. They're at the center of the tension between traditional warfare and the newer kinds of multiparty conflict emerging outside the

industrialized world. Recent actions like those in Bosnia, Somalia, Haiti, Grenada, and the Persian Gulf suggest the range of possible military contingencies. Strategic leaders struggle with the ramifications of switching repeatedly among the different types of military actions required under a strategy of engagement.

7-95. The variety of potential missions calls for the ability to quickly build temporary organizations able to perform specific tasks. As they design future joint organizations, strategic leaders must also determine how to engineer both cohesion and proficiency in modular units that are constantly forming and reforming.

STRATEGIC ASSESSING

7-96. There are many elements of their environment that strategic leaders must assess. Like leaders at other levels, they must first assess themselves: their leadership style, strengths and weaknesses, and their fields of excellence. They must also understand the present operational environment—to include the will of the American people, expressed in part through law, policy, and their leaders. Finally, strategic leaders must survey the political landscape and the international environment, for these affect the organization and shape the future strategic requirements.

7-97. Strategic leaders also cast a wide net to assess their own organizations. They develop performance indicators to signal how well they're communicating to all levels of their commands and how well established systems and processes are balancing the six imperatives. Assessment starts early in each mission and continues through its end. It may include monitoring such diverse areas as resource use, development of subordinates, efficiency, effects of stress and fatigue, morale, and mission accomplishment. Such assessments generate huge amounts of data; strategic leaders must make clear what they're looking for so their staffs can filter information for them. They must also guard against misuse of assessment data.

World War II Strategic Assessment

Pursuing a "Germany first" strategy in World War II was a deliberate decision based on a strategic and political assessment of the global situation. Military planners, particularly Army Chief of Staff George C. Marshall, worried that US troops might be dispersed and used piecemeal. Strategic leaders heeded Frederick the Great's adage, "He who defends everywhere defends nowhere." The greatest threat to US interests was a total German success in Europe: a defeated Russia and neutralized Great Britain. Still, the Japanese attacks on Pearl Harbor and the Philippines and the threat to the line of communications with Australia tugged forces toward the opposite hemisphere. Indeed, throughout the first months of 1942, more forces headed for the Pacific Theater than across the Atlantic.

However, before the US was at war with anyone, President Franklin Roosevelt had agreed with British Prime Minister Winston Churchill to a "Germany first" strategy. The 1942 invasion of North Africa restored this focus. While the US military buildup took hold and forces flowed into Great Britain for the Normandy invasion in 1944, operations in secondary theaters could and did continue. However, they were resourced only after measuring their impact on the planned cross-channel attack.

IMPROVING ACTIONS

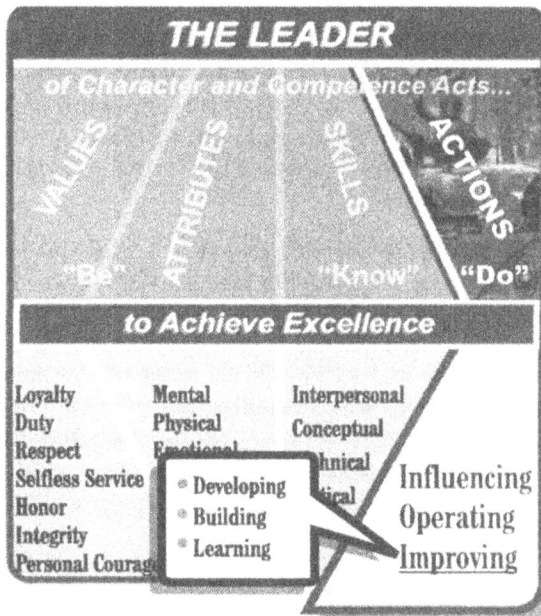

Figure 7-6. Strategic Leader Actions—
Improving

7-98. Improving is institutional investment for the long haul, refining the things we do today for a better organization tomorrow. A fundamental goal of strategic leaders is to leave the Army better than they found it. Improving at this level calls for experimentation and innovation; however, because strategic-level organizations are so complex, quantifying the results of changes may be difficult.

7-99. Improving the institution and organizations involves an ongoing tradeoff between today and tomorrow. Wisdom and a refined frame of reference are tools to understand what improvement is and what change is needed. Knowing when and what to change is a constant challenge: what traditions should remain stable, and which long-standing methods need to evolve? Strategic leaders set the conditions for long-term success of the organization by developing subordinates, leading change, building the culture and teams, and creating a learning environment.

7-100. One technique for the Army as a learning institution is to decentralize the learning and other improving actions to some extent. That technique raises the questions of how to share good ideas across the entire institution and how to incorporate the best ideas into doctrine (thus establishing an Army-wide standard) without discouraging the decentralized learning process that generated the ideas in the first place. Those and other questions face the strategic leaders of the learning organization the Army seeks to become.

DEVELOPING

George C. Marshall learned leadership from John J. Pershing, and Marshall's followers became great captains themselves: Dwight D. Eisenhower, Omar N. Bradley...among them. Pershing and Marshall each taught their subordinates their profession; and, more importantly, they gave them room to grow.

General Gordon R. Sullivan
Former Army Chief of Staff

Mentoring

7-101. Strategic leaders develop subordinates by sharing the benefit of their perspective and experience. People arriving at the Pentagon know how the Army works in the field, but regardless of what they may have read, they don't really know how the institutional Army works. Strategic leaders act as a kind of sponsor by introducing them to the important players and pointing out the important places and activities. But strategic leaders actually become mentors as they, in effect, underwrite the learning, efforts, projects, and ideas of rising leaders. The moral responsibility associated with mentoring is compelling for all leaders; for strategic leaders, the potential significance is enormous.

7-102. More than a matter of required forms and sessions, mentoring by strategic leaders means giving the right people an intellectual boost so that they make the leap to operations and thinking at the highest levels. Because those being groomed for strategic leadership positions are among the most talented Army leaders, the manner in which leaders and subordinates interact also changes. Strategic leaders aim not only to pass on knowledge but also to grow wisdom in those they mentor.

7-103. Since few formal leader development programs exist beyond the senior service colleges, strategic leaders pay special attention to their subordinates' self-development, showing them what to study, where to focus, whom to watch, and how to proceed. They speak to audiences at service schools about what goes on "at

the top" and spend time sharing their perspectives with those who haven't yet reached the highest levels of Army leadership. Today's subordinates will, after all, become the next generation of strategic leaders. Strategic leaders are continually concerned that the Army institutional culture and the climates in subordinate organizations encourage mentoring by others so that growth opportunities are available from the earliest days of a soldier or DA civilian's careers.

Developing Intellectual Capital

7-104. What strategic leaders do for individuals they personally mentor, they also seek to provide to the force at large. They invest in the future of the force in several ways. Committing money to programs and projects and investing more time and resources in some actions than others are obvious ways strategic leaders choose what's important. They also value people and ideas as investments in the future. The concepts that shape the thinking of strategic leaders become the intellectual currency of the coming era; the soldiers and DA civilians who develop those ideas become trusted assets themselves. Strategic leaders must choose wisely the ideas that bridge the gap between today and tomorrow and skillfully determine how best to resource important ideas and people.

7-105. Strategic leaders make difficult decisions about how much institutional development is enough. They calculate how much time it will take to plant and grow the seeds required for the Army's great leaders and ideas in the future. They balance today's operational requirements with tomorrow's leadership needs to produce programs that develop a core of Army leaders with the required skills and knowledge.

7-106. Programs like training with industry, advanced civil schooling, and foreign area officer education complement the training and education available in Army schools and contribute to shaping the people who will shape the Army's future. Strategic leaders develop the institution using Army resources when they are available and those of other services or the public sector when they are not.

7-107. After Vietnam the Army's leadership thought investing in officer development so important that new courses were instituted to revitalize professional education for the force. The establishment of the Training and Doctrine Command revived Army doctrine as a central intellectual pillar of the entire service. The Goldwater-Nichols Act of 1986 provided similar attention and invigoration to professional joint education and joint doctrine.

7-108. Likewise, there has been a huge investment in and payoff from developing the NCO corps. The Army has the world's finest noncommissioned officers, in part because they get the world's best professional development. The strategic decision to resource a robust NCO education system signaled the Army's investment in developing the whole person—not just the technical skills—of its first-line leaders.

7-109. The Army Civilian Training and Education Development System is the Army's program for developing DA civilian leaders. Like the NCO education system, it continues throughout an individual's career. The first course integrates interns into the Army by explaining Army values, culture, customs, and policies. The Leadership Education and Development Course helps prepare leaders for supervisory demands with training in communication, counseling, team building, problem solving, and group development. For organizational managers, the Organizational Leadership for Executives course adds higher-order study on topics such as strategic planning, change management, climate, and culture. DA civilians in the Senior Executive Service have a variety of leadership education options that deal with leadership in both the military and civilian contexts. Together, these programs highlight ways that leadership development of DA civilians parallels that of soldiers.

BUILDING

The higher up the chain of command, the greater is the need for boldness to be supported by a reflective mind, so that boldness does not degenerate into purposeless bursts of blind passion.

Carl von Clausewitz

Building Amid Change

7-110. The Army has no choice but to face change. It's in a nearly constant state of flux, with new people, new missions, new technologies, new equipment, and new information. At the same time, the Army, inspired by strategic leaders, must innovate and create change. The Army's customs, procedures, hierarchical structure, and sheer size make change especially daunting and stressful. Nonetheless, the Army must be flexible enough to produce and respond to change, even as it preserves the core of traditions that tie it to the nation, its heritage and its values.

7-111. Strategic leaders deal with change by being proactive, not reactive. They anticipate change even as they shield their organizations from unimportant and bothersome influences; they use the "change-drivers" of technology, education, doctrine, equipment, and organization to control the direction and pace of change. Many agencies and corporations have "futures" groups charged with thinking about tomorrow; strategic leaders and their advisory teams are the Army's "futures people."

Leading Change

7-112. Strategic leaders lead change by—

- Identifying the force capabilities necessary to accomplish the National Military Strategy.
- Assigning strategic and operational missions, including priorities for allocating resources.
- Preparing plans for using military forces across the spectrum of operations.
- Creating, resourcing, and sustaining organizational systems, including—
  - Force modernization programs.
  - Requisite personnel and equipment.
  - Essential command, control, communications, computers, and intelligence systems.
- Developing and improving doctrine and the training methods to support it.

- Planning for the second- and third-order effects of change.

- Maintaining an effective leader development program and other human resource initiatives.

Change After Vietnam

The history of the post-Vietnam Army provides an example of how strategic leaders' commitment can shape the environment and harness change to improve the institution while continuing to operate.

The Army began seeking only volunteers in the early 1970s. With the all-volunteer force came a tremendous emphasis on doctrinal, personnel, and training initiatives that took years to mature. The Army tackled problems in drug abuse, racial tensions, and education with ambitious, long-range plans and aggressive leader actions. Strategic leaders overhauled doctrine and created an environment that improved training at all levels; the CTC program provided a uniform, rock-solid foundation of a single, well-understood warfighting doctrine upon which to build a trained and ready Army. Simultaneously, new equipment, weapons, vehicles, and uniforms were introduced. The result was the Army of Desert Storm, which differed greatly from the force of 15 years earlier.

None of these changes happened by chance or through evolution. Change depended on the hard work of direct and organizational leaders who developed systematically in an environment directed, engineered, and led by strategic leaders.

7-113. Strategic leaders must guide their organizations through eight stages if their initiatives for change are to make lasting progress. Skipping a step or moving forward prematurely subverts the process and compromises success. Strategic leaders (1) demonstrate a sense of urgency by showing not only the benefits of but the necessity for change. They (2) form guiding coalitions to work the process all the way from concept through implementation. With those groups they (3) develop a vision of the future and strategy for achieving it. Because change is most effective when members embrace it, strategic leaders (4) communicate the vision throughout the institution or organization, and then (5) empower subordinates at all levels for widespread, parallel efforts. They (6) plan for short-term successes to validate the programs and keep the vision credible and (7) consolidate those wins and produce further change. Finally, the leader (8) preserves the change culturally. The result is an institution that constantly prepares for and even shapes the future environment. Strategic leaders seek to sustain the Army as that kind of institution.

LEARNING

A good soldier, whether he leads a platoon or an army, is expected to look backward as well as forward; but he must think only forward.

General of the Army Douglas MacArthur

7-114. The nation expects military professionals as individuals and the Army as an institution to learn from the experience of others and apply that learning to understanding the present and preparing for the future. Such learning requires both individual and institutional commitments. Each military professional must be committed to self-development, part of which is studying military history and other disciplines related to military operations. The Army as an institution must be committed to conducting technical research, monitoring emerging threats, and developing leaders for the next generation. Strategic leaders, by their example and resourcing decisions, sustain the culture and policies that encourage both the individual and the Army to learn.

7-115. Strategic leaders promote learning by emplacing systems for studying the force and the future. Strategic leaders must resource a structure that constantly reflects on how the Army fights and what victory may cost. All that means constantly assessing the culture and deliberately encouraging creativity and learning.

7-116. The notion of the Army as a "learning organization" is epitomized by the AAR concept, which was developed as part of the REALTRAIN project, the first version of engagement simulation. Since then, it has been part of a cultural change, in which realistic "hot washes," such as following tough engagements at CTCs, are now embedded in all training. Twenty years ago, anything like today's AARs would have been rare.

7-117. Efficient and effective operations require aligning various initiatives so that different factions are not working at cross-purposes. Strategic leaders focus research and development efforts on achieving combined arms success. They deal with questions such as: Can these new systems from various sources communicate with one another? What happens during digitization lapses—what's our residual combat capability? Strategic leaders coordinate time lines and budgets so that compatible systems are fielded together. However, they are also concerned that the force have optimal capability across time; therefore, they prepare plans that integrate new equipment and concepts into the force as they're developed, rather than waiting for all elements of a system to be ready before fielding it. Finally, learning what the force should be means developing the structure, training, and leaders those future systems will support and studying the variety of threats they may face.

7-118. The Louisiana Maneuvers in 1941 taught the Army what mechanized warfare would look like and what was needed to prepare for it. The study of the same name 50 years later helped produce the conceptual Force XXI and the first digitized division. Strategic leaders commissioned these projects because the Army is dedicated to learning about operations in new environments, against different threats. The projects were strategic counterparts to the rehearsals that direct and organizational leaders conduct to prepare for upcoming missions.

SECTION III

A HISTORICAL PERSPECTIVE OF STRATEGIC LEADERSHIP— GENERAL OF THE ARMY MARSHALL DURING WORLD WAR II

7-119. GA George C. Marshall was one of the greatest strategic leaders of World War II, of this century, of our nation's history. His example over many years demonstrates the skills and actions this chapter has identified as the hallmarks of strategic leadership.

7-120. Chosen over 34 officers senior to him, GA Marshall became Army Chief of Staff in 1939, a time of great uncertainty about the future of the free world. Part of his appeal for President Roosevelt was his strength of character and personal integrity. The honesty and candor that GA Marshall displayed early in their relationship were qualities the president knew he and the nation would need in the difficult times ahead.

7-121. The new Army Chief of Staff knew he had to wake the Army from its interwar slumber and grow it beyond its 174,000 soldiers—a size that ranked it seventeenth internationally, behind Bulgaria and Portugal. By 1941 he had begun to move the Army toward his vision of what it needed to become: a world-changing force of 8,795,000 soldiers and airmen. His vision was remarkably accurate: by the end of the war, 89 divisions and over 8,200,000 soldiers in Army uniforms had made history.

7-122. GA Marshall reached deep within the Army for leaders capable of the conceptual leaps necessary to fight the impending war. He demanded leaders ready for the huge tasks ahead, and he accepted no excuses. As he found colonels, lieutenant colonels, and even majors who seemed ready for the biggest challenge of their lives, he promoted them ahead of those more senior but less capable and made many of them generals. He knew firsthand that such jumps could be productive. As a lieutenant in the Philippines, he had commanded 5,000 soldiers during an exercise. For generals who could not adjust to the sweeping changes in the Army, he made career shifts as well: he retired them. His loyalty to the institution and the nation came before any personal relationships.

7-123. Merely assembling the required number of soldiers would not be enough. The mass Army that was forming required a new structure to manage the forces and resources the nation was mobilizing for the war effort. Realizing this, GA Marshall reorganized the Army into the Army Ground Forces, Army Air Forces, and Army Service Forces. His foresight organized the Army for the evolving nature of warfare.

7-124. Preparing for combat required more than manning the force. GA Marshall understood that World War I had presented confusing lessons about the future of warfare. His in-theater experience during that war and later reflection distilled a vision of the future. He believed that maneuver of motorized formations spearheaded by tanks and supported logistically by trucks (instead of horse-drawn wagons) would replace the almost siege-like battles of World War I. So while the French trusted the Maginot Line, GA Marshall emphasized the new technologies that would heighten the speed and complexity of the coming conflict.

7-125. Further, GA Marshall championed the common sense training to prepare soldiers to go overseas ready to fight and win. By having new units spend sufficient time on marksmanship, fitness, drill, and fieldcraft, GA Marshall ensured that soldiers and leaders had the requisite competence and confidence to face an experienced enemy.

7-126. Before and during the war, GA Marshall showed a gift for communicating with the American public. He worked closely with the press, frequently confiding in senior newsmen so they would know about the Army's activities and the progress of the war. They responded to his trust by not printing damaging or premature stories. His relaxed manner and complete command of pertinent facts reassured the press, and through it the nation, that America's youth were entrusted to the right person.

7-127. He was equally successful with Congress. GA Marshall understood that getting what he wanted meant asking, not demanding. His humble and respectful approach with lawmakers won his troops what they needed; arrogant demands would have never worked. Because he never sought anything for himself (his five-star rank was awarded over his objections), his credibility soared.

7-128. However, GA Marshall knew how to shift his approach depending on the audience, the environment, and the situation. He refused to be intimidated by leaders such as Prime Minister Winston Churchill, Secretary of War Henry Stimson, or even the president. Though he was always respectful, his integrity demanded that he stand up for his deeply held convictions—and he did, without exception.

7-129. The US role in Europe was to open a major second front to relieve pressure on the Soviet Union and ensure the Allied victory over Germany. GA Marshall had spent years preparing the Army for Operation Overlord, the D-Day invasion that would become the main effort by the Western Allies and the one expected to lead to final victory over Nazi Germany. Many assumed GA Marshall would command it. President Roosevelt might have felt obligated to reward the general's faithful and towering service, but GA Marshall never raised the subject. Ultimately, the president told GA Marshall that it was more important that he lead global resourcing than command a theater of war. GA Eisenhower got the command, while GA Marshall continued to serve on staff.

7-130. GA Marshall didn't request the command that would have placed him alongside immortal combat commanders like Washington, Grant, and Lee. His decision reflects the value of selfless service that kept him laboring for decades without the recognition that came to some of his associates. GA Marshall never attempted to be anyplace but where his country needed him. And there, finally as Army Chief of Staff, GA Marshall served with unsurpassed vision and brilliance, engineering the greatest victory in our nation's history and setting an extraordinary example for those who came after him.

SUMMARY

7-131. Just as GA Marshall prepared for the coming war, strategic leaders today ready the Army for the next conflict. They may not have years before the next D-Day; it could be just hours away. Strategic leaders operate between extremes, balancing a constant awareness of the current national and global situation with a steady focus on the Army's long-term mission and goals.

7-132. Since the nature of future military operations is so unclear, the vision of the Army's strategic leaders is especially crucial. Identifying what's important among the concerns of mission, soldiers, weapons, logistics, and technology produces decisions that determine the structure and capability of tomorrow's Army.

7-133. Within the institution, strategic leaders build support for the end state they desire. That means building a staff that can take broad guidance and turn it into initiatives that move the Army forward. To obtain the required support, strategic leaders also seek to achieve consensus beyond the Army, working with Congress and the other services on budget, force structure, and strategy issues and working with other countries and militaries on shared interests. The way strategic leaders communicate direction to soldiers, DA civilians, and citizens determines the understanding and support for the new ideas.

7-134. Like GA Marshall, today's strategic leaders are deciding how to transform today's force into tomorrow's. These leaders have little guidance. Still, they know that they work to develop the next generation of Army leaders, build the organizations of the future, and resource the systems that will help gain the next success. The way strategic leaders communicate direction to soldiers, DA civilians, and citizens determines the understanding and support for the new ideas. To communicate with these diverse audiences, strategic leaders work through multiple media, adjust the message when necessary, and constantly reinforce Army themes.

7-135. To lead change personally and move the Army establishment toward their concept of the future, strategic leaders transform political and conceptual programs into practical and concrete initiatives. That process increasingly involves leveraging technology and shaping the culture. By knowing themselves, the strategic players, the operational requirements, the geopolitical situation, and the American public, strategic leaders position the force and the nation for success. Because there may be no time for a World War II or Desert Storm sort of buildup, success for Army strategic leaders means being ready to win a variety of conflicts now and remaining ready in the uncertain years ahead.

7-136. Strategic leaders prepare the Army for the future through their leadership. That means **influencing** people—members of the Army, members of other government agencies, and the people of the nation the Army serves—by providing purpose, direction, and motivation. It means **operating** to accomplish today's missions, foreign and domestic. And it means **improving** the institution—making sure its people are trained and that its equipment and organizations are ready for tomorrow's missions, anytime, anywhere.

Appendix A

Roles and Relationships

A-1. When the Army speaks of soldiers, it refers to commissioned officers, warrant officers, noncommissioned officers (NCOs), and enlisted personnel—both men and women. The terms commissioned officer and warrant officer are used when it is necessary to specifically address or refer to a particular group of officers. All Army leaders—soldiers and DA civilians—share the same goal: to accomplish their organization's mission. The roles and responsibilities of Army leaders—commissioned, warrant, noncommissioned, and DA civilian—overlap. Figure A-1 summarizes them.

A-2. Commissioned officers are direct representatives of the President of the United States. Commissions are legal instruments the president uses to appoint and exercise direct control over qualified people to act as his legal agents and help him carry out his duties. The Army retains this direct-agent relationship with the president through its commissioned officers. The commission serves as the basis for a commissioned officer's legal authority. Commissioned officers command, establish policy, and manage Army resources. They are normally generalists who assume progressively broader responsibilities over the course of a career.

A-3. Warrant officers are highly specialized, single-track specialty officers who receive their authority from the Secretary of the Army upon their initial appointment. However, Title 10 USC authorizes the commissioning of warrant officers (WO1) upon promotion to chief warrant officer (CW2). These commissioned warrant officers are direct representatives of the president of the United States. They derive their authority from the same source as commissioned officers but remain specialists, in contrast to commissioned officers, who are generalists. Warrant officers can and do command detachments, units, activities, and vessels as well as lead, coach,

train, and counsel subordinates. As leaders and technical experts, they provide valuable skills, guidance, and expertise to commanders and organizations in their particular field.

A-4. NCOs, the backbone of the Army, train, lead, and take care of enlisted soldiers. They receive their authority from their oaths of office, law, rank structure, traditions, and regulations. This authority allows them to direct soldiers, take actions required to accomplish the mission, and enforce good order and discipline. NCOs represent officer, and sometimes DA civilian, leaders. They ensure their subordinates, along with their personal equipment, are prepared to function as effective unit and team members. While commissioned officers command, establish policy, and manage resources, NCOs conduct the Army's daily business.

A-5. As members of the executive branch of the federal government, DA civilians are part of the Army. They derive their authority from a variety of sources, such as commanders, supervisors, Army regulations, and Title 5 USC. DA civilians' authority is job-related: they normally exercise authority related to their positions. DA civilians fill positions in staff and base sustaining operations that would otherwise have to be filled by officers and NCOs. Senior DA civilians establish policy and manage Army resources, but they do not have the authority to command.

A-6. The complementary relationship and mutual respect between the military and civilian members of the Army is a long-standing tradition. Since the Army's beginning in 1775, military and DA civilian duties have stayed separate, yet necessarily related. Taken in combination, traditions, functions, and laws serve to delineate the particular duties of military and civilian members of the Army.

THE COMMISSIONED OFFICER

- Commands, establishes policy, and manages Army resources.
- Integrates collective, leader, and soldier training to accomplish missions.
- Deals primarily with units and unit operations.
- Concentrates on unit effectiveness and readiness.

THE WARRANT OFFICER

- Provides quality advice, counsel, and solutions to support the command.
- Executes policy and manages the Army's systems.
- Commands special-purpose units and task-organized operational elements.
- Focuses on collective, leader, and individual training.
- Operates, maintains, administers, and manages the Army's equipment, support activities, and technical systems.
- Concentrates on unit effectiveness and readiness.

THE NONCOMMISSIONED OFFICER

- Trains soldiers and conducts the daily business of the Army within established policy.
- Focuses on individual soldier training.
- Deals primarily with individual soldier training and team leading.
- Ensures that subordinate teams, NCOs, and soldiers are prepared to function as effective unit and team members.

THE DEPARTMENT OF THE ARMY CIVILIAN

- Establishes and executes policy, leads people, and manages programs, projects, and Army systems.
- Focuses on integrating collective, leader, and individual training.
- Operates, maintains, administers, and manages Army equipment and support, research, and technical activities.
- Concentrates on DA civilian individual and organizational effectiveness and readiness.

Figure A-1. Roles and Responsibilities of Commissioned, Warrant, Noncommissioned, and DA Civilian Leaders

AUTHORITY

A-7. Authority is the legitimate power of leaders to direct subordinates or to take action within the scope of their positions. Military authority begins with the Constitution, which divides it between Congress and the president. (The Constitution appears in Appendix F.) Congress has the authority to make laws that govern the Army. The president, as commander in chief, commands the armed forces, including the Army. Two types of military authority exist: command and general military.

Command Authority

A-8. Command is the authority that a commander in the armed forces lawfully exercises over subordinates by virtue of rank or assignment. Command includes the authority and responsibility for effectively using available resources to organize, direct, coordinate, employ, and control military forces so that they accomplish assigned missions. It also includes responsibility for health, welfare, morale, and discipline of assigned personnel.

A-9. Command authority originates with the president and may be supplemented by law or regulation. It is the authority that a commander lawfully exercises over subordinates by virtue of rank or assignment. Only commissioned and warrant officers may command Army units and installations. DA civilians may exercise general supervision over an Army installation or activity; however, they act under the authority of a military supervisor. DA civilians do not command. (AR 600-20 addresses command authority in more detail.)

A-10. Army leaders are granted command authority when they fill command-designated positions. These normally involve the direction and control of other soldiers and DA civilians. Leaders in command-designated positions have the inherent authority to issue orders, carry out the unit mission, and care for both military members and DA civilians within the leader's scope of responsibility.

General Military Authority

A-11. General military authority originates in oaths of office, law, rank structure, traditions, and regulations. This broad-based authority also allows leaders to take appropriate corrective actions whenever a member of any armed service, anywhere, commits an act involving a breach of good order or discipline. AR 600-20, paragraph 4-5, states this specifically, giving commissioned, warrant, and noncommissioned officers authority to "quell all quarrels, frays, and disorders among persons subject to military law"—in other words, to maintain good order and discipline.

A-12. All enlisted leaders have general military authority. For example, dining facility managers, platoon sergeants, squad leaders, and tank commanders all use general military authority when they issue orders to direct and control their subordinates. Army leaders may exercise general military authority over soldiers from different units.

A-13. For NCOs, another source of general military authority stems from the combination of the chain of command and the NCO support channel. The chain of command passes orders and policies through the NCO support channel to provide authority for NCOs to do their job.

Delegation of Authority

A-14. Just as Congress and the president cannot participate in every aspect of armed forces operations, most leaders cannot handle every action directly. To meet the organization's goals, officers delegate authority to NCOs and, when appropriate, to DA civilians. These leaders, in turn, may further delegate that authority.

A-15. Unless restricted by law, regulation, or a superior, leaders may delegate any or all of their authority to their subordinate leaders. However, such delegation must fall within the leader's scope of authority. Leaders cannot delegate authority they do not have and subordinate leaders may not assume authority that their superiors do not have, cannot delegate, or have retained. The task or duty to be performed limits the authority of the leader to which it is assigned.

A-16. When a leader is assigned a task or duty, the authority necessary to accomplish it accompanies the assignment. When a leader delegates a task or duty to a subordinate, he delegates the requisite authority as well. However, leaders always retain responsibility for the outcome of any tasks they assign. They must answer for any actions or omissions related to them.

RESPONSIBILITY AND ACCOUNTABILITY

A-17. No definitive lines separate officer, NCO, and DA civilian responsibilities. Officers, NCOs, and DA civilians lead other officers, NCOs, and DA civilians and help them carry out their responsibilities. Commanders set overall policies and standards, but all leaders must provide the guidance, resources, assistance, and supervision necessary for subordinates to perform their duties. Similarly, subordinates must assist and advise their leaders. Mission accomplishment demands that officers, NCOs, and DA civilians work together to advise, assist, and learn from each other. Responsibilities fall into two categories: command and individual.

Command Responsibility

A-18. Command responsibility refers to collective or organizational accountability and includes how well units perform their missions. For example, a company commander is responsible for all the tasks and missions assigned to his company; his leaders hold him accountable for completing them. Military and DA civilian leaders have responsibility for what their sections, units, or organizations do or fail to do.

Individual Responsibility

A-19. All soldiers and DA civilians must account for their personal conduct. Commissioned officers, warrant officers, and DA civilians assume personal responsibility when they take their oath. DA civilians take the same oath as commissioned officers. Soldiers take their initial oath of enlistment. Members of the Army account for their actions to their fellow soldiers or coworkers, the appointed leader, their unit or organization, the Army, and the American people.

COMMUNICATIONS AND THE CHAIN OF COMMAND

A-20. Communication among individuals, teams, units, and organizations is essential to efficient and effective mission accomplishment. As Chapter 4 discusses, two-way communication is more effective than one-way communication. Mission accomplishment depends on information passing accurately to and from subordinates and leaders, up and down the chain of command and NCO support channel, and laterally among adjacent organizations or activities. In garrison operations, organizations working on the same mission or project should be considered "adjacent."

A-21. The Army has only one chain of command. Through this chain of command, leaders issue orders and instructions and convey policies. A healthy chain of command is a two-way communications channel. Its members do more than transmit orders; they carry information from within the unit or organization back up to its leader. They furnish information about how things are developing, notify the leader of problems, and provide requests for clarification and help. Leaders at all levels use the chain of command—their subordinate leaders—to keep their people informed and render assistance. They continually facilitate the process of gaining the necessary clarification and solving problems.

A-22. Beyond conducting their normal duties, NCOs train soldiers and advise commanders on individual soldier readiness and the training needed to ensure unit readiness. Officers and DA civilian leaders should consult their command sergeant major, first sergeant, or NCO assistant, before implementing policy. Commanders, commissioned and warrant officers, DA civilian leaders, and NCOs must continually communicate to avoid duplicating instructions or issuing conflicting orders. Continuous and open lines of communication enable commanders and DA

civilian leaders to freely plan, make decisions, and program future training and operations.

THE NONCOMMISSIONED OFFICER SUPPORT CHANNEL

A-23. The NCO support channel parallels and reinforces the chain of command. NCO leaders work with and support the commissioned and warrant officers of their chain of command. For the chain of command to work efficiently, the NCO support channel must operate effectively. At battalion level and higher, the NCO support channel begins with the command sergeant major, extends through first sergeants and platoon sergeants, and ends with section chiefs, squad leaders, or team leaders. (TC 22-6 discusses the NCO support channel.)

A-24. The connection between the chain of command and NCO support channel is the senior NCO. Commanders issue orders through the chain of command, but senior NCOs must know and understand the orders to issue effective implementing instructions through the NCO support channel. Although the first sergeant and command sergeant major are not part of the

formal chain of command, leaders should consult them on all individual soldier matters.

A-25. Successful leaders have a good relationship with their senior NCOs. Successful commanders have a good leader-NCO relationship with their first sergeants and command sergeants major. The need for such a relationship applies to platoon leaders and platoon sergeants as well as to staff officers and NCOs. Senior NCOs have extensive experience in successfully completing missions and dealing with enlisted soldier issues. Also, senior NCOs can monitor organizational activities at all levels, take corrective action to keep the organization within the boundaries of the commander's intent, or report situations that require the attention of the officer leadership. A positive relationship between officers and NCOs creates conditions for success.

DA CIVILIAN SUPPORT

A-26. The Army employs DA civilians because they possess or develop technical skills that are necessary to accomplish some missions. The specialized skills of DA civilians are essential to victory but, for a variety of reasons, they are difficult to maintain in the uniformed components. The Army expects DA civilian leaders to be more than specialists: they are expected to apply technical, conceptual, and interpersonal skills together to accomplish missions—in a combat theater, if necessary.

A-27. While the command sergeant major is the advocate in a unit for soldier issues, DA civilians have no single advocate. Rather, their own leaders, civilian personnel advisory center, or civilian personnel operations center represent them and their issues to the chain of command. Often the senior DA civilian in an organization or the senior DA civilian in a particular career field has the additional duty of advising and counseling junior DA civilians on job-related issues and career development.

Appendix B

Performance Indicators

B-1. Appendix B is organized around the leadership dimensions that Chapters 1 through 7 discuss and that Figure B-1 shows. This appendix lists indicators for you to use to assess the leadership of yourself and others based on these leadership dimensions. Use it as an assessment and counseling tool, not as a source of phrases for evaluation reports. When you prepare an evaluation, make comments that apply specifically to the individual you are evaluating. Do not limit yourself to the general indicators listed here. Be specific; be precise; be objective; be fair.

Leaders of character and competence act to achieve excellence by providing purpose, direction and motivation.

Values	Attributes	Skills	Actions
"Be"	"Be"	"Know"	"Do"

| | Mental[1] | Interpersonal | Influencing | Operating | Improving |
| Loyalty Duty Respect Selfless Service Honor Integrity Personal Courage | Physical[2] Emotional[3] | Conceptual Technical Tactical | Communicating Decision Making Motivating | Planning/ Preparing Executing Assessing | Developing Building Learning |

[1] The mental attributes of an Army leader are will, self-discipline, initiative, judgment, self-confidence, intelligence, and cultural awareness.

[2] The physical attributes of an Army leader are health fitness, physical fitness, and military and professional bearing.

[3] The emotional attributes of an Army leader are self-control, balance, and stability.

[4] The interpersonal, conceptual, technical, and tactical skills are different for direct, organizational, and strategic leaders.

[5] The influencing, operating, and improving actions are different for direct, organizational, and strategic leaders.

Figure B-1. Leadership Dimensions

VALUES

LOYALTY

B-2. Leaders who demonstrate loyalty—

- Bear true faith and allegiance in the correct order to the Constitution, the Army, and the organization.
- Observe higher headquarters' priorities.
- Work within the system without manipulating it for personal gain.

DUTY

B-3. Leaders who demonstrate devotion to duty—

- Fulfill obligations—professional, legal, and moral.
- Carry out mission requirements.
- Meet professional standards.
- Set the example.
- Comply with policies and directives.
- Continually pursue excellence.

RESPECT

B-4. Leaders who demonstrate respect—

- Treat people as they should be treated.
- Create a climate of fairness and equal opportunity.
- Are discreet and tactful when correcting or questioning others.
- Show concern for and make an effort to check on the safety and well-being of others.
- Are courteous.
- Don't take advantage of positions of authority.

SELFLESS SERVICE

B-5. Leaders who demonstrate selfless service—

- Put the welfare of the nation, the Army, and subordinates before their own.
- Sustain team morale.
- Share subordinates' hardships.
- Give credit for success to others and accept responsibility for failure themselves.

HONOR

B-6. Leaders who demonstrate honor—

- Live up to Army values.
- Don't lie, cheat, steal, or tolerate those actions by others.

INTEGRITY

B-7. Leaders who demonstrate integrity—

- Do what is right legally and morally.
- Possess high personal moral standards.
- Are honest in word and deed.
- Show consistently good moral judgment and behavior.
- Put being right ahead of being popular.

PERSONAL COURAGE

B-8. Leaders who demonstrate personal courage—

- Show physical and moral bravery.
- Take responsibility for decisions and actions.
- Accept responsibility for mistakes and shortcomings.

ATTRIBUTES

MENTAL ATTRIBUTES

B-9. Leaders who demonstrate desirable mental attributes—

- Possess and display will, self-discipline, initiative, judgment, self-confidence, intelligence, common sense, and cultural awareness.
- Think and act quickly and logically, even when there are no clear instructions or the plan falls apart.
- Analyze situations.
- Combine complex ideas to generate feasible courses of action.
- Balance resolve and flexibility.
- Show a desire to succeed; do not quit in the face of adversity.
- Do their fair share.
- Balance competing demands.
- Embrace and use the talents of all members to build team cohesion.

PHYSICAL ATTRIBUTES

B-10. Leaders who demonstrate desirable physical attributes—

- Maintain an appropriate level of physical fitness and military bearing.
- Present a neat and professional appearance.

- Meet established norms of personal hygiene, grooming, and cleanliness.
- Maintain Army height and weight standards (not applicable to DA civilians).
- Render appropriate military and civilian courtesies.
- Demonstrate nonverbal expressions and gestures appropriate to the situation.
- Are personally energetic.
- Cope with hardship.
- Complete physically demanding endeavors.
- Continue to function under adverse conditions.
- Lead by example in performance, fitness, and appearance.

EMOTIONAL ATTRIBUTES

B-11. Leaders who demonstrate appropriate emotional attributes—

- Show self-confidence.
- Remain calm during conditions of stress, chaos, and rapid change.
- Exercise self-control, balance, and stability.
- Maintain a positive attitude.
- Demonstrate mature, responsible behavior that inspires trust and earns respect.

SKILLS

INTERPERSONAL SKILLS

B-12. Leaders who demonstrate interpersonal skills—

- Coach, teach, counsel, motivate, and empower subordinates.
- Readily interact with others.
- Earn trust and respect.
- Actively contribute to problem solving and decision making.
- Are sought out by peers for expertise and counsel

CONCEPTUAL SKILLS

B-13. Leaders who demonstrate conceptual skills—

- Reason critically and ethically.
- Think creatively.
- Anticipate requirements and contingencies.
- Improvise within the commander's intent.
- Use appropriate reference materials.
- Pay attention to details.

TECHNICAL SKILLS

B-14. Leaders who demonstrate technical skills—

- Possess or develop the expertise necessary to accomplish all assigned tasks and functions.
- Know standards for task accomplishment.
- Know the small unit tactics, techniques, and procedures that support the organization's mission.
- Know the drills that support the organization's mission.
- Prepare clear, concise operation orders.
- Understand how to apply the factors of mission, enemy, terrain and weather, troops, time available, and civil considerations (METT-TC) to mission analysis.
- Master basic soldier skills.
- Know how to use and maintain equipment.

- Know how and what to inspect or check.
- Use technology, especially information technology, to enhance communication.

TACTICAL SKILLS

B-15. Leaders who demonstrate tactical skills—

- Know how to apply warfighting doctrine within the commander's intent.
- Apply their professional knowledge, judgment, and warfighting skill at the appropriate leadership level.
- Combine and apply skill with people, ideas, and things to accomplish short-term missions.
- Apply skill with people, ideas, and things to train for, plan, prepare, execute and assess offensive, defensive, stability, and support actions.

ACTIONS

INFLUENCING

B-16. Leaders who influence—

- Use appropriate methods to reach goals while operating and improving.
- Motivate subordinates to accomplish assigned tasks and missions.
- Set the example by demonstrating enthusiasm for—and, if necessary, methods of—accomplishing assigned tasks.
- Make themselves available to assist peers and subordinates.
- Share information with subordinates.
- Encourage subordinates and peers to express candid opinions.
- Actively listen to feedback and act appropriately based on it.
- Mediate peer conflicts and disagreements.
- Tactfully confront and correct others when necessary.
- Earn respect and obtain willing cooperation of peers, subordinates, and superiors.
- Challenge others to match their example.

- Take care of subordinates and their families, providing for their health, welfare, morale, and training.
- Are persuasive in peer discussions and prudently rally peer pressure against peers when required.
- Provide a team vision for the future.
- Shape the organizational climate by setting, sustaining, and ensuring a values-based environment.

Communicating

B-17. Leaders who communicate effectively—

- Display good oral, written, and listening skills.
- Persuade others.
- Express thoughts and ideas clearly to individuals and groups.

B-18. **Oral Communication.** Leaders who effectively communicate orally—

- Speak clearly and concisely.
- Speak enthusiastically and maintain listeners' interest and involvement.

- Make appropriate eye contact when speaking.
- Use gestures that are appropriate but not distracting.
- Convey ideas, feelings, sincerity, and conviction.
- Express well-thought-out and well-organized ideas.
- Use grammatically and doctrinally correct terms and phrases.
- Use appropriate visual aids.
- Act to determine, recognize and resolve misunderstandings.
- Listen and watch attentively; make appropriate notes; convey the essence of what was said or done to others.
- React appropriately to verbal and nonverbal feedback.
- Keep conversations on track.

B-19. **Written Communication.** Leaders who effectively communicate in writing—

- Are understood in a single rapid reading by the intended audience.
- Use correct grammar, spelling, and punctuation.
- Have legible handwriting.
- Put the "bottom line up front."
- Use the active voice.
- Use an appropriate format, a clear organization, and a reasonably simple style.
- Use only essential acronyms and spell out those used.
- Stay on topic.
- Correctly use facts and data.

(DA Pam 600-67 discusses techniques for writing effectively.)

Decision Making

B-20. Leaders who make effective, timely decisions—

- Employ sound judgment and logical reasoning.
- Gather and analyze relevant information about changing situations to recognize and define emerging problems.

- Make logical assumptions in the absence of facts.
- Uncover critical issues to use as a guide in both making decisions and taking advantage of opportunities.
- Keep informed about developments and policy changes inside and outside the organization.
- Recognize and generate innovative solutions.
- Develop alternative courses of action and choose the best course of action based on analysis of their relative costs and benefits.
- Anticipate needs for action.
- Relate and compare information from different sources to identify possible cause-and-effect relationships.
- Consider the impact and implications of decisions on others and on situations.
- Involve others in decisions and keep them informed of consequences that affect them.
- Take charge when in charge.
- Define intent.
- Consider contingencies and their consequences.
- Remain decisive after discovering a mistake.
- Act in the absence of guidance.
- Improvise within commander's intent; handle a fluid environment.

Motivating

B-21. Leaders who effectively motivate—
- Inspire, encourage, and guide others toward mission accomplishment.
- Don't show discouragement when facing setbacks.
- Attempt to satisfy subordinates' needs.
- Give subordinates the reason for tasks.
- Provide accurate, timely, and (where appropriate) positive feedback.
- Actively listen for feedback from subordinates.
- Use feedback to modify duties, tasks, requirements, and goals when appropriate.

- Recognize individual and team accomplishments and reward them appropriately.
- Recognize poor performance and address it appropriately.
- Justly apply disciplinary measures.
- Keep subordinates informed.
- Clearly articulate expectations.
- Consider duty positions, capabilities, and developmental needs when assigning tasks.
- Provide early warning to subordinate leaders of tasks they will be responsible for.
- Define requirements by issuing clear and concise orders or guidance.
- Allocate as much time as possible for task completion.
- Accept responsibility for organizational performance. Credit subordinates for good performance. Take responsibility for and correct poor performance.

OPERATING

B-22. Leaders who effectively operate—

- Accomplish short-term missions.
- Demonstrate tactical and technical competency appropriate to their rank and position.
- Complete individual and unit tasks to standard, on time, and within the commander's intent.

Planning and Preparing

B-23. Leaders who effectively plan—

- Develop feasible and acceptable plans for themselves and others that accomplish the mission while expending minimum resources and posturing the organization for future missions.
- Use forward planning to ensure each course of action achieves the desired outcome.
- Use reverse planning to ensure that all tasks can be executed in the time available and that tasks depending on other tasks are executed in the correct sequence.
- Determine specified and implied tasks and restate the higher headquarters' mission in terms appropriate to the organization.

- Incorporate adequate controls such as time phasing; ensure others understand when actions should begin or end.
- Adhere to the "1/3–2/3 Rule"; give subordinates time to plan.
- Allocate time to prepare and conduct rehearsals.
- Ensure all courses of action accomplish the mission within the commander's intent.
- Allocate available resources to competing demands by setting task priorities based on the relative importance of each task.
- Address likely contingencies.
- Remain flexible.
- Consider SOPs, the factors of METT-TC, and the military aspects of terrain (OCOKA).
- Coordinate plans with higher, lower, adjacent, and affected organizations.
- Personally arrive on time and meet deadlines; require subordinates and their organizations to accomplish tasks on time.
- Delegate all tasks except those they are required to do personally.
- Schedule activities so the organization meets all commitments in critical performance areas.
- Recognize and resolve scheduling conflicts.
- Notify peers and subordinates as far in advance as possible when their support is required.
- Use some form of a personal planning calendar to organize requirements.

Executing

B-24. Leaders who effectively execute—

- Use technical and tactical skills to meet mission standards, take care of people, and accomplish the mission with available resources.
- Perform individual and collective tasks to standard.
- Execute plans, adjusting when necessary, to accomplish the mission.
- Encourage initiative.
- Keep higher and lower headquarters, superiors, and subordinates informed.

- Keep track of people and equipment.
- Make necessary on-the-spot corrections.
- Adapt to and handle fluid environments.
- Fight through obstacles, difficulties, and hardships to accomplish the mission.
- Keep track of task assignments and suspenses; adjust assignments, if necessary; follow up.

Assessing

B-25. Leaders who effectively assess—

- Use assessment techniques and evaluation tools (especially AARs) to identify lessons learned and facilitate consistent improvement.
- Establish and employ procedures for monitoring, coordinating, and regulating subordinates' actions and activities.
- Conduct initial assessments when beginning a new task or assuming a new position.
- Conduct IPRs.
- Analyze activities to determine how desired end states are achieved or affected.
- Seek sustainment in areas when the organization meets the standard.
- Observe and assess actions in progress without oversupervising.
- Judge results based on standards.
- Sort out important actual and potential problems.
- Conduct and facilitate AARs; identify lessons.
- Determine causes, effects, and contributing factors for problems.
- Analyze activities to determine how desired end states can be achieved ethically.

IMPROVING

B-26. Leaders who effectively improve the organization—

- Sustain skills and actions that benefit themselves and each of their people for the future.
- Sustain and renew the organization for the future by managing change and exploiting individual and institutional learning capabilities.

- Create and sustain an environment where all leaders, subordinates, and organizations can reach their full potential.

Developing

B-27. Leaders who effectively develop—

- Strive to improve themselves, subordinates, and the organization.
- Mentor by investing adequate time and effort in counseling, coaching, and teaching their individual subordinates and subordinate leaders.
- Set the example by displaying high standards of duty performance, personal appearance, military and professional bearing, and ethics.
- Create a climate that expects good performance, recognizes superior performance, and doesn't accept poor performance.
- Design tasks to provide practice in areas of subordinate leaders' weaknesses.
- Clearly articulate tasks and expectations and set realistic standards.
- Guide subordinate leaders in thinking through problems for themselves.
- Anticipate mistakes and freely offer assistance without being overbearing.
- Observe, assess, counsel, coach, and evaluate subordinate leaders.
- Motivate subordinates to develop themselves.
- Arrange training opportunities that help subordinates achieve insight, self-awareness, self-esteem, and effectiveness.
- Balance the organization's tasks, goals, and objectives with subordinates' personal and professional needs.
- Develop subordinate leaders who demonstrate respect for natural resources and the environment.
- Act to expand and enhance subordinates' competence and self-confidence.
- Encourage initiative.
- Create and contribute to a positive organizational climate.
- Build on successes.
- Improve weaknesses.

Building

B-28. Leaders who effectively build—

- Spend time and resources improving the organization.

- Foster a healthy ethical climate.

- Act to improve the organization's collective performance.

- Comply with and support organizational goals.

- Encourage people to work effectively with each other.

- Promote teamwork and team achievement.

- Are examples of team players.

- Offer suggestions, but properly execute decisions of the chain of command and NCO support channel—even unpopular ones—as if they were their own.

- Accept and act on assigned tasks.

- Volunteer in useful ways.

- Remain positive when the situation becomes confused or changes.

- Use the chain of command and NCO support channel to solve problems.

- Support equal opportunity.

- Prevent sexual harassment.

- Participate in organizational activities and functions.

- Participate in team tasks and missions without being requested to do so.

- Establish an organizational climate that demonstrates respect for the environment and stewards natural resources.

Learning

B-29. Leaders who effectively learn—

- Seek self-improvement in weak areas.

- Encourage organizational growth.

- Envision, adapt, and lead change.

- Act to expand and enhance personal and organizational knowledge and capabilities.

- Apply lessons learned.

- Ask incisive questions.

- Envision ways to improve.

- Design ways to practice.

- Endeavor to broaden their understanding.

- Transform experience into knowledge and use it to improve future performance.

- Make knowledge accessible to the entire organization.

- Exhibit reasonable self-awareness.

- Take time off to grow and recreate.

- Embrace and manage change; adopt a future orientation.

- Use experience to improve themselves and the organization.

Appendix C

Developmental Counseling

C-1. Subordinate leadership development is one of the most important responsibilities of every Army leader. Developing the leaders who will come after you should be one of your highest priorities. Your legacy and the Army's future rests on the shoulders of those you prepare for greater responsibility.

C-2. Leadership development reviews are a means to focus the growing of tomorrow's leaders. Think of them as AARs with a focus of making leaders more effective every day. These important reviews are not necessarily limited to internal counseling sessions; leadership feedback mechanisms also apply in operational settings such as the CTCs.

C-3. Just as training includes AARs and training strategies to fix shortcomings, leadership development includes performance reviews. These reviews result in agreements between leader and subordinate on a development strategy or plan of action that builds on the subordinate's strengths and establishes goals to improve on weaknesses. Leaders conduct performance reviews and create plans of action during developmental counseling.

C-4. Leadership development reviews are a component of the broader concept of developmental counseling. Developmental counseling is subordinate-centered communication that produces a plan outlining actions that subordinates must take to achieve individual and organizational goals. During developmental counseling, subordinates are not merely passive listeners; they're actively involved in the process. The Developmental Counseling Form (DA Form 4856-E, which is discussed at the end of this appendix) provides a useful framework to prepare for almost any type of counseling. Use it to help you mentally organize issues and isolate important, relevant items to cover during counseling sessions.

C-5. Developmental counseling is a shared effort. As a leader, you assist your subordinates in identifying strengths and weaknesses and creating plans of action. Then you support them throughout the plan implementation and assessment. However, to achieve success, your subordinates must be forthright in their commitment to improve and candid in their own assessment and goal setting.

THE LEADER'S RESPONSIBILITIES

C-6. Organizational readiness and mission accomplishment depend on every member's ability to perform to established standards. Supervisors must mentor their subordinates through teaching, coaching, and counseling. Leaders coach subordinates the same way sports coaches improve their teams: by identifying weaknesses, setting goals, developing and implementing plans of action, and providing oversight and motivation throughout the process. To be effective coaches, leaders must thoroughly understand the strengths, weaknesses, and professional goals of their subordinates. (Chapter 5 discusses coaching.)

C-7. Army leaders evaluate DA civilians using procedures prescribed under the Total Army Performance Evaluation System (TAPES). Although TAPES doesn't address developmental counseling, you can use DA Form 4856-E to counsel DA civilians concerning professional growth and career goals. DA Form 4856-E is not appropriate for documenting counseling concerning DA civilian misconduct or poor performance. The servicing civilian personnel office can provide guidance for such situations.

C-8. Soldiers and DA civilians often perceive counseling as an adverse action. Effective leaders who counsel properly can change that perception. Army leaders conduct counseling to help subordinates become better members of the team, maintain or improve performance, and prepare for the future. Just as no easy answers exist for exactly what to do in all leadership situations, no easy answers exist for exactly what to do in all counseling situations. However, to conduct effective counseling, you should develop a counseling style with the characteristics listed in Figure C-1.

- **Purpose**: Clearly define the purpose of the counseling.

- **Flexibility**: Fit the counseling style to the character of each subordinate and to the relationship desired.

- **Respect**: View subordinates as unique, complex individuals, each with a distinct set of values, beliefs, and attitudes.

- **Communication**: Establish open, two-way communication with subordinates using spoken language, nonverbal actions, gestures, and body language. Effective counselors listen more than they speak.

- **Support**: Encourage subordinates through actions while guiding them through their problems.

Figure C-1. Characteristics of Effective Counseling

THE LEADER AS A COUNSELOR

C-9. Army leaders must demonstrate certain qualities to be effective counselors. These qualities include respect for subordinates, self-awareness and cultural awareness, empathy, and credibility.

RESPECT FOR SUBORDINATES

C-10. As an Army leader, you show respect for subordinates when you allow them to take responsibility for their own ideas and actions. Respecting subordinates helps create mutual respect in the leader-subordinate relationship. Mutual respect improves the chances of changing (or maintaining) behavior and achieving goals.

SELF AWARENESS AND CULTURAL AWARENESS

C-11. As an Army leader, you must be fully aware of your own values, needs, and biases prior to counseling subordinates. Self-aware leaders are less likely to project their biases onto subordinates. Also, aware leaders are more likely to act consistently with their own values and actions.

C-12. Cultural awareness, as discussed in Chapter 2, is a mental attribute. As an Army leader, you need to be aware of the similarities and differences between individuals of different cultural backgrounds and how these factors may influence values, perspectives, and actions. Don't let unfamiliarity with cultural backgrounds hinder you in addressing cultural issues, especially if they generate concerns within the organization or hinder team-building. Cultural awareness enhances your ability to display empathy.

EMPATHY

C-13. Empathy is the action of being understanding of and sensitive to the feelings, thoughts, and experiences of another person to the point that you can almost feel or experience them yourself. Leaders with empathy can put themselves in their subordinate's shoes; they can see a situation from the other person's perspective. By understanding the subordinate's position, you can help a subordinate develop a plan of action that fits the subordinate's personality and needs, one that works for the subordinate. If you don't fully comprehend a situation from your subordinate's point of view, you have less credibility and influence and your subordinate is less likely to commit to the agreed upon plan of action.

CREDIBILITY

C-14. Leaders achieve credibility by being honest and consistent in their statements and actions. To be credible, use a straightforward style with your subordinates. Behave in a manner that your subordinates respect and trust. You can earn credibility by repeatedly demonstrating your willingness to assist a subordinate and being consistent in what you say and do. If you lack credibility with your subordinates you'll find it difficult to influence them.

LEADER COUNSELING SKILLS

C-15. One challenging aspect of counseling is selecting the proper approach to a specific situation. To counsel effectively, the technique you use must fit the situation, your capabilities, and your subordinate's expectations. In some cases, you may only need to give information or listen. A subordinate's improvement may call for just a brief word of praise. Other situations may require structured counseling followed by definite actions.

C-16. All leaders should seek to develop and improve their own counseling abilities. You can improve your counseling techniques by studying human behavior, learning the kinds of problems that affect your subordinates, and developing your interpersonal skills. The techniques needed to provide effective counseling will vary from person to person and session to session. However, general skills that you'll need in almost every situation include active listening, responding, and questioning.

ACTIVE LISTENING

C-17. During counseling, you must actively listen to your subordinate. When you're actively listening, you communicate verbally and nonverbally that you've received the subordinate's message. To fully understand a subordinate's message, you must listen to the words and observe the subordinate's manners. Elements of active listening you should consider include—

- **Eye contact**. Maintaining eye contact without staring helps show sincere interest. Occasional breaks of contact are normal and acceptable. Subordinates may perceive excessive breaks of eye contact, paper shuffling, and clock-watching as a lack of interest or concern. These are guidelines only. Based on cultural background, participants in a particular counseling session may have different ideas about what proper eye contact is.

- **Body posture**. Being relaxed and comfortable will help put the subordinate at ease. However, a too-relaxed position or slouching may be interpreted as a lack of interest.

- **Head nods**. Occasionally nodding your head shows you're paying attention and encourages the subordinate to continue.

- **Facial expressions**. Keep your facial expressions natural and relaxed. A blank look or fixed expression may disturb the subordinate. Smiling too much or frowning may discourage the subordinate from continuing.

- **Verbal expressions**. Refrain from talking too much and avoid interrupting. Let the subordinate do the talking while keeping the discussion on the counseling subject. Speaking only when necessary reinforces the importance of what the subordinate is saying and encourages the subordinate to continue. Silence can also do this, but be careful. Occasional silence may indicate to the subordinate that it's okay to continue talking, but a long silence can sometimes be distracting and make the subordinate feel uncomfortable.

C-18. Active listening also means listening thoughtfully and deliberately to the way a subordinate says things. Stay alert for common themes. A subordinate's opening and closing statements as well as recurring references may indicate the subordinate's priorities. Inconsistencies and gaps may indicate a subordinate's avoidance of the real issue. This confusion and uncertainty may suggest additional questions.

C-19. While listening, pay attention to the subordinate's gestures. These actions complete the total message. By watching the subordinate's actions, you can "see" the feelings behind the words. Not all actions are proof of a subordinate's feelings, but they should be taken into consideration. Note differences between what the subordinate says and does. Nonverbal indicators of a subordinate's attitude include—

- **Boredom.** Drumming on the table, doodling, clicking a ball-point pen, or resting the head in the palm of the hand.
- **Self-confidence.** Standing tall, leaning back with hands behind the head, and maintaining steady eye contact.
- **Defensiveness.** Pushing deeply into a chair, glaring at the leader, and making sarcastic comments as well as crossing or folding arms in front of the chest.
- **Frustration.** Rubbing eyes, pulling on an ear, taking short breaths, wringing the hands, or frequently changing total body position.
- **Interest, friendliness, and openness.** Moving toward the leader while sitting.
- **Openness or anxiety.** Sitting on the edge of the chair with arms uncrossed and hands open.

C-20. Consider these indicators carefully. Although each indicator may show something about the subordinate, don't assume a particular behavior absolutely means something. Ask the subordinate about the indicator so you can better understand the behavior and allow the subordinate to take responsibility for it.

RESPONDING

C-21. Responding skills follow-up on active listening skills. A leader responds to communicate that the leader understands the subordinate. From time to time, check your understanding: clarify and confirm what has been said. Respond to subordinates both verbally and nonverbally. Verbal responses consist of summarizing, interpreting, and clarifying the subordinate's message. Nonverbal responses include eye contact and occasional gestures such as a head nod.

QUESTIONING

C-22. Although questioning is a necessary skill, you must use it with caution. Too many questions can aggravate the power differential between a leader and a subordinate and place the subordinate in a passive mode. The subordinate may also react to excessive questioning as an intrusion of privacy and become defensive. During a leadership development review, ask questions to obtain information or to get the subordinate to think about a particular situation. Generally, the questions should be open-ended so as to evoke more than a yes or no answer. Well-posed questions may help to verify understanding, encourage further explanation, or help the subordinate move through the stages of the counseling session.

COUNSELING ERRORS

C-23. Effective leaders avoid common counseling mistakes. Dominating the counseling by talking too much, giving unnecessary or inappropriate "advice," not truly listening, and projecting personal likes, dislikes, biases, and prejudices all interfere with effective counseling. You should also avoid other common mistakes such as rash judgments, stereotypes, loss of emotional control, inflexible methods of counseling and improper follow-up. To improve your counseling skills, follow the guidelines in Figure C-2.

- Determine the subordinate's role in the situation and what the subordinate has done to resolve the problem or improve performance.
- Draw conclusions based on more than the subordinate's statement.
- Try to understand what the subordinate says and feels; listen to what the subordinate says and how the subordinate says it.
- Show empathy when discussing the problem.
- When asking questions, be sure that you need the information.
- Keep the conversation open-ended; avoid interrupting.
- Give the subordinate your full attention.
- Be receptive to the subordinate's feelings without feeling responsible to save the subordinate from hurting.
- Encourage the subordinate to take the initiative and to say what the subordinate wants to say.
- Avoid interrogating.
- Keep your personal experiences out of the counseling session unless you believe your experiences will really help.
- Listen more; talk less.
- Remain objective.
- Avoid confirming a subordinate's prejudices.
- Help the subordinate help himself.
- Know what information to keep confidential and what to present to the chain of command.

Figure C-2. Guidelines to Improve Counseling

THE LEADER'S LIMITATIONS

C-24. Army leaders can't help everyone in every situation. Even professional counselors can't provide all the help that a person might need. You must recognize your limitations and, when the situation calls for it, refer a subordinate to a person or agency more qualified to help.

C-25. These agencies Figure C-3 lists can help you and your people resolve problems. Although it's generally in an individual's best interest to seek help first from their first-line leaders, leaders must always respect an individual's right to contact most of these agencies on their own.

Activity	Description
Adjutant General	Provides personnel and administrative services support such as orders, ID cards, retirement assistance, deferments, and in- and out-processing.
American Red Cross	Provides communications support between soldiers and families and assistance during or after emergency or compassionate situations.
Army Community Service	Assists military families through their information and referral services, budget and indebtedness counseling, household item loan closet, information on other military posts, and welcome packets for new arrivals.
Army Substance Abuse Program	Provides alcohol and drug abuse prevention and control programs for DA civilians.
Better Opportunities for Single Soldiers (BOSS)	Serves as a liaison between upper levels of command on the installation and single soldiers.
Army Education Center	Provides services for continuing education and individual learning services support.
Army Emergency Relief	Provides financial assistance and personal budget counseling; coordinates student loans through Army Emergency Relief education loan programs.
Career Counselor	Explains reenlistment options and provides current information on prerequisites for reenlistment and selective reenlistment bonuses.
Chaplain	Provides spiritual and humanitarian counseling to soldiers and DA civilians.
Claims Section, SJA	Handles claims for and against the government, most often those for the loss and damage of household goods.
Legal Assistance Office	Provides legal information or assistance on matters of contracts, citizenship, adoption, marital problems, taxes, wills, and powers of attorney.
Community Counseling Center	Provides alcohol and drug abuse prevention and control programs for soldiers.
Community Health Nurse	Provides preventive health care services.
Community Mental Health Service	Provides assistance and counseling for mental health problems.
Employee Assistance Program	Provides health nurse, mental health service, and social work services for DA civilians.
Equal Opportunity Staff Office and Equal Employment Opportunity Office	Provides assistance for matters involving discrimination in race, color, national origin, gender, and religion. Provides, information on procedures for initiating complaints and resolving complaints informally.
Family Advocacy Officer	Coordinates programs supporting children and families including abuse and neglect investigation, counseling, and educational programs.
Finance and Accounting Office	Handles inquiries for pay, allowances, and allotments.
Housing Referral Office	Provides assistance with housing on and off post.
Inspector General	Renders assistance to soldiers and DA civilians. Corrects injustices affecting individuals and eliminates conditions determined to be detrimental to the efficiency, economy, morale, and reputation of the Army. Investigates matters involving fraud, waste, and abuse.
Social Work Office	Provides services dealing with social problems to include crisis intervention, family therapy, marital counseling, and parent or child management assistance.
Transition Office	Provides assistance and information on separation from the Army.

Figure C-3. Support Activities

TYPES OF DEVELOPMENTAL COUNSELING

C-26. You can often categorize developmental counseling based on the topic of the session. The two major categories of counseling are event-oriented and performance/professional growth.

EVENT-ORIENTED COUNSELING

C-27. Event-oriented counseling involves a specific event or situation. It may precede events, such as going to a promotion board or attending a school; or it may follow events, such as a noteworthy duty performance, a problem with performance or mission accomplishment, or a personal problem. Examples of event-oriented counseling include, but are not limited to—

- Specific instances of superior or substandard performance.
- Reception and integration counseling.
- Crisis counseling.
- Referral counseling.
- Promotion counseling.
- Separation counseling.

Counseling for Specific Instances

C-28. Sometimes counseling is tied to specific instances of superior or substandard duty performance. You tell your subordinate whether or not the performance met the standard and what the subordinate did right or wrong. The key to successful counseling for specific performance is to conduct it as close to the event as possible.

C-29. Many leaders focus counseling for specific instances on poor performance and miss, or at least fail to acknowledge, excellent performance. You should counsel subordinates for specific examples of superior as well as substandard duty performance. To measure your own performance and counseling emphasis, you can note how often you document counseling for superior versus substandard performance.

C-30. You should counsel subordinates who don't meet the standard. If the subordinate's performance is unsatisfactory because of a lack of knowledge or ability, you and the subordinate should develop a plan to improve the subordinate's skills. Corrective training may be required at times to ensure the subordinate

knows and achieves the standard. Once the subordinate can achieve the standard, you should end the corrective training.

C-31. When counseling a subordinate for a specific performance, take the following actions:

- Tell the subordinate the purpose of the counseling, what was expected, and how the subordinate failed to meet the standard.
- Address the specific unacceptable behavior or action, not the person's character.
- Tell the subordinate the effect of the behavior, action, or performance on the rest of the organization.
- Actively listen to the subordinate's response.
- Remain unemotional.
- Teach the subordinate how to meet the standard.
- Be prepared to do some personal counseling, since a failure to meet the standard may be related to or the result of an unresolved personal problem.
- Explain to the subordinate what will be done to improve performance (plan of action). Identify your responsibilities in implementing the plan of action; continue to assess and follow up on the subordinate's progress. Adjust plan of action as necessary.

Reception and Integration Counseling

C-32. As the leader, you must counsel new team members when they arrive at your organization. This reception and integration counseling serves two purposes. First, it identifies and helps fix any problems or concerns that new members may have, especially any issues resulting from the new duty assignment. Second, it lets them know the organizational standards and how they fit into the team. It clarifies job titles and sends the message that the chain of command cares. Reception and integration counseling should begin immediately upon arrival so new team members can quickly become integrated into the organization. (Figure C-4 gives some possible discussion points.)

- Organizational standards.
- Chain of command.
- NCO support channel (who and how used).
- On-and-off duty conduct.
- Personnel/personal affairs/initial clothing issue.
- Organizational history, organization, and mission.
- Soldier programs within the organization, such as soldier of the month/quarter/year and Audie Murphy.
- Off limits and danger areas.
- Functions and locations of support activities (see Figure C-3).
- On- and off-post recreational, educational, cultural, and historical opportunities.
- Foreign nation or host nation orientation.
- Other areas the individual should be aware of, as determined by the leader.

Figure C-4. Reception and Integration Counseling Points

Crisis Counseling

C-33. You may conduct crisis counseling to get a subordinate through the initial shock after receiving negative news, such as notification of the death of a loved one. You may assist the subordinate by listening and, as appropriate, providing assistance. Assistance may include referring the subordinate to a support activity or coordinating external agency support. Crisis counseling focuses on the subordinate's immediate, short-term needs.

Referral Counseling

C-34. Referral counseling helps subordinates work through a personal situation and may or may not follow crisis counseling. Referral counseling may also act as preventative counseling before the situation becomes a problem. Usually, the leader assists the subordinate in identifying the problem and refers the subordinate to the appropriate resource, such as Army Community Services, a chaplain, or an alcohol and drug counselor. (Figure C-3 lists support activities.)

Promotion Counseling

C-35. Leaders must conduct promotion counseling for all specialists and sergeants who are eligible for advancement without waivers but not recommended for promotion to the next higher grade. Army regulations require that soldiers within this category receive initial (event-oriented) counseling when they attain full eligibility and then periodic (performance/personal growth) counseling thereafter.

Adverse Separation Counseling

C-36. Adverse separation counseling may involve informing the soldier of the administrative actions available to the commander in the event substandard performance continues and of the consequences associated with those administrative actions (see AR 635-200).

C-37. Developmental counseling may not apply when an individual has engaged in more serious acts of misconduct. In those situations, you should refer the matter to the commander and the servicing staff judge advocate. When the leader's rehabilitative efforts fail, counseling with a view towards separation fills an administrative prerequisite to many administrative discharges and serves as a final warning to the soldier to improve performance or face discharge. In many situations, it may be beneficial to involve the chain of command as soon as you determine that adverse separation counseling might be required. A unit first sergeant or commander should be the person who informs the soldier of the notification requirements outlined in AR 635-200.

PERFORMANCE AND PROFESSIONAL GROWTH COUNSELING

Performance Counseling

C-38. During performance counseling, you conduct a review of a subordinate's duty performance during a certain period. You and the subordinate jointly establish performance objectives and standards for the next period. Rather than dwelling on the past, you should focus the session on the subordinate's strengths, areas needing improvement, and potential.

C-39. Performance counseling is required under the officer, NCO, and DA civilian evaluation reporting systems. The OER process requires periodic performance counseling as part of the OER Support Form requirements. Mandatory, face-to-face performance counseling between the rater and the rated NCO is required under the NCOERS. TAPES includes a combination of both of these requirements.

C-40. Counseling at the beginning of and during the evaluation period facilitates a subordinate's involvement in the evaluation process. Performance counseling communicates standards and is an opportunity for leaders to establish and clarify the expected values, attributes, skills, and actions. Part IVb (Leader Attributes/Skills/Actions) of the OER Support Form (DA Form 67-9-1) serves as an excellent tool for leaders doing performance counseling. For lieutenants and warrant officers one, the major performance objectives on the OER Support Form are used as the basis for determining the developmental tasks on the Junior Officer Developmental Support Form (DA Form 67-9-1a). Quarterly face-to-face performance and developmental counseling is required for these junior officers as outlined in AR 623-105.

C-41. As an Army leader, you must ensure you've tied your expectations to performance objectives and appropriate standards. You must establish standards that your subordinates can work towards and must teach them how to achieve the standards if they are to develop.

Professional Growth Counseling

C-42. Professional growth counseling includes planning for the accomplishment of individual and professional goals. You conduct this counseling to assist subordinates in achieving organizational and individual goals. During the counseling, you and your subordinate conduct a review to identify and discuss the subordinate's strengths and weaknesses and create a plan of action to build upon strengths and overcome weaknesses. This counseling isn't normally event-driven.

C-43. As part of professional growth counseling, you may choose to discuss and develop a "pathway to success" with the subordinate. This future-oriented counseling establishes short- and long-term goals and objectives. The discussion may include opportunities for civilian or military schooling, future duty assignments, special programs, and reenlistment options. Every person's needs are different, and leaders must apply specific courses of action tailored to each individual.

C-44. Career field counseling is required for lieutenants and captains before they're considered for promotion to major. Raters and senior raters, in conjunction with the rated officer, need to determine where the officer's skills best fit the needs of the Army. During career field counseling, consideration must be given to the rated officer's preference and his abilities (both performance and academic). The rater and senior rater should discuss career field designation with the officer prior to making a recommendation on the rated officer's OER.

C-45. While these categories can help you organize and focus counseling sessions, they should not be viewed as separate, distinct, or exhaustive. For example, a counseling session that focuses on resolving a problem may also address improving duty performance. A session focused on performance may also include a discussion on opportunities for professional growth. Regardless of the topic of the counseling session, leaders should follow the same basic format to prepare for and conduct it.

APPROACHES TO COUNSELING

C-46. An effective leader approaches each subordinate as an individual. Different people and different situations require different counseling approaches. Three approaches to counseling include nondirective, directive, and combined. These approaches differ in the techniques used, but they all fit the definition of counseling and contribute to its overall purpose. The major difference between the approaches is the degree to which the subordinate participates and interacts during a counseling session. Figure C-5 summarizes the advantages and disadvantages of each approach.

NONDIRECTIVE

C-47. The nondirective approach is preferred for most counseling sessions. Leaders use their experienced insight and judgment to assist subordinates in developing solutions. You should partially structure this type of counseling by telling the subordinate about the counseling process and explaining what you expect.

C-48. During the counseling session, listen rather than make decisions or give advice. Clarify what's said. Cause the subordinate to bring out important points, so as to better understand the situation. When appropriate, summarize the discussion. Avoid providing solutions or rendering opinions; instead, maintain a focus on individual and organizational goals and objectives. Ensure the subordinate's plan of action supports those goals and objectives.

DIRECTIVE

C-49. The directive approach works best to correct simple problems, make on-the-spot corrections, and correct aspects of duty performance. The leader using the directive style does most of the talking and tells the subordinate what to do and when to do it. In contrast to the nondirective approach, the leader directs a course of action for the subordinate.

C-50. Choose this approach when time is short, when you alone know what to do, or if a subordinate has limited problem-solving skills. It's also appropriate when a subordinate needs guidance, is immature, or is insecure.

COMBINED

C-51. In the combined approach, the leader uses techniques from both the directive and nondirective approaches, adjusting them to articulate what's best for the subordinate. The combined approach emphasizes the subordinate's planning and decision-making responsibilities.

C-52. With your assistance, the subordinate develops the subordinate's own plan of action. You should listen, suggest possible courses, and help analyze each possible solution to determine its good and bad points. You should then help the subordinate fully understand all aspects of the situation and encourage the subordinate to decide which solution is best.

	Advantages	Disadvantages
Nondirective	• Encourages maturity. • Encourages open communication. • Develops personal responsibility.	• More time-consuming. • Requires greatest counselor skill.
Directive	• Quickest method. • Good for people who need clear, concise direction. • Allows counselors to actively use their experience.	• Doesn't encourage subordinates to be part of the solution. • Tends to treat symptoms, not problems. • Tends to discourage subordinates from talking freely. • Solution is the counselor's, not the subordinate's.
Combined	• Moderately quick. • Encourages maturity. • Encourages open communication. • Allows counselors to actively use their experience.	• May take too much time for some situations.

Figure C-5. Counseling Approach Summary Chart

COUNSELING TECHNIQUES

C-53. As an Army leader, you may select from a variety of techniques when counseling subordinates. These counseling techniques, when appropriately used, cause subordinates to do things or improve upon their performance. You can use these methods during scheduled counseling sessions or while simply coaching a subordinate. Counseling techniques you can use during the nondirective or combined approaches include—

- **Suggesting alternatives**. Discuss alternative actions that the subordinate may take, but both you and the subordinate decide which course of action is most appropriate.

- **Recommending**. Recommend one course of action, but leave the decision to accept the recommended action to the subordinate.

- **Persuading**. Persuade the subordinate that a given course of action is best, but leave the decision to the subordinate. Successful persuasion depends on the leader's credibility, the subordinate's willingness to listen, and their mutual trust.

- **Advising**. Advise the subordinate that a given course of action is best. This is the strongest form of influence not involving a command.

C-54. Some techniques you can use during the directive approach to counseling include—

- **Corrective training**. Teach and assist the subordinate in attaining and maintaining the standards. The subordinate completes corrective training when the subordinate attains the standard.

- **Commanding**. Order the subordinate to take a given course of action in clear, exact words. The subordinate understands that he has been given a command and will face the consequences for failing to carry it out.

THE COUNSELING PROCESS

C-55. Effective leaders use the counseling process. It consists of four stages:

- Identify the need for counseling.
- Prepare for counseling.
- Conduct counseling.
- Follow up.

IDENTIFY THE NEED FOR COUNSELING

C-56. Quite often organizational policies, such as counseling associated with an evaluation or counseling required by the command, focus a counseling session. However, you may conduct developmental counseling whenever the need arises for focused, two-way communication aimed at subordinate development. Developing subordinates consists of observing the subordinate's performance, comparing it to the standard, and then providing feedback to the subordinate in the form of counseling.

PREPARE FOR COUNSELING

C-57. Successful counseling requires preparation. To prepare for counseling, do the following:

- Select a suitable place.
- Schedule the time.
- Notify the subordinate well in advance.
- Organize information.
- Outline the counseling session components.
- Plan your counseling strategy.
- Establish the right atmosphere.

Select a Suitable Place

C-58. Schedule counseling in an environment that minimizes interruptions and is free from distracting sights and sounds.

Schedule the Time

C-59. When possible, counsel a subordinate during the duty day. Counseling after duty hours may be rushed or perceived as unfavorable. The length of time required for counseling depends on the complexity of the issue. Generally a counseling session should last less than an hour. If you need more time, schedule a second session. Additionally, select a time free from competition with other activities and consider what has been planned after the counseling session. Important events can distract a subordinate from concentrating on the counseling.

Notify the Subordinate Well in Advance

C-60. For a counseling session to be a subordinate-centered, two-person effort, the subordinate must have time to prepare for it. The subordinate should know why, where, and when the counseling will take place. Counseling following a specific event should happen as close to the event as possible. However, for performance or professional development counseling, subordinates may need a week or more to prepare or review specific products, such as support forms or counseling records.

Organize Information

C-61. Solid preparation is essential to effective counseling. Review all pertinent information. This includes the purpose of the counseling, facts and observations about the subordinate, identification of possible problems, main points of discussion, and the development of a plan of action. Focus on specific and objective behaviors that the subordinate must maintain or improve as well as a plan of action with clear, obtainable goals.

Outline the Components of the Counseling Session

C-62. Using the information obtained, determine what to discuss during the counseling session. Note what prompted the counseling, what you aim to achieve, and what your role as a counselor is. Identify possible comments or questions to help you keep the counseling session subordinate-centered and help the subordinate progress through its stages. Although you never know what a subordinate will say or do during counseling, a written outline helps organize the session and enhances the chance of positive results. (Figure C-6 is one example of a counseling outline prepared by a platoon leader about to conduct an initial NCOER counseling session with a platoon sergeant.)

Type of counseling: Initial NCOER counseling for SFC Taylor, a recently promoted new arrival to the unit.

Place and time: The platoon office, 1500 hours, 9 October.

Time to notify the subordinate: Notify SFC Taylor one week in advance of the scheduled counseling session.

Subordinate preparation: Have SFC Taylor put together a list of goals and objectives he would like to complete over the next 90 to 180 days. Review the values, attributes, skills, and actions from FM 22-100.

Counselor preparation:

● Review the NCO Counseling Checklist/Record (DA Form 2166-8-1).

● Update or review SFC Taylor's duty description and fill out the rating chain and duty description on the working copy of the NCOER (DA Form 2166-8, Parts II and III).

● Review each of the values and responsibilities in Part IV of the NCOER and the values, attributes, skills and actions in FM 22-100. Think of how each applies to SFC Taylor and the platoon sergeant position.

● Review the actions you consider necessary for a success or excellence in each value and responsibility.

● Make notes in blank spaces in Part IV of the NCOER to assist when counseling.

Role as counselor: Help SFC Taylor to understand the expectations and standards associated with the platoon sergeant position. Assist SFC Taylor in developing the values, attributes, skills, and actions that will enable him to achieve his performance objectives, consistent with those of the platoon and company. Resolve any aspects of the job that aren't clearly understood.

Session outline: Complete an outline following the counseling session components in Figure C-7 and based on the draft duty description on the NCOER, ideally at least two to three days prior to the actual counseling session.

Figure C-6. Example of a Counseling Outline

Plan Counseling Strategy

C-63. As many approaches to counseling exist as there are leaders. The directive, nondirective, and combined approaches to counseling were addressed earlier. Use a strategy that suits your subordinates and the situation.

Establish the Right Atmosphere

C-64. The right atmosphere promotes two-way communication between a leader and subordinate. To establish a relaxed atmosphere, you may offer the subordinate a seat or a cup of coffee. You may want to sit in a chair facing the subordinate since a desk can act as a barrier.

C-65. Some situations make an informal atmosphere inappropriate. For example, during counseling to correct substandard performance, you may direct the subordinate to remain standing while you remain seated behind a desk. This formal atmosphere, normally used to give specific guidance, reinforces the leader's rank, position in the chain of command, and authority.

CONDUCT THE COUNSELING SESSION

C-66. Be flexible when conducting a counseling session. Often counseling for a specific incident occurs spontaneously as leaders encounter subordinates in their daily activities. Such counseling can occur in the field, motor pool, barracks—wherever subordinates perform their duties. Good leaders take advantage of naturally occurring events to provide subordinates with feedback.

C-67. Even when you haven't prepared for formal counseling, you should address the four basic components of a counseling session. Their purpose is to guide effective counseling rather than mandate a series of rigid steps. Counseling sessions consist of—

- Opening the session.
- Discussing the issues.
- Developing the plan of action.
- Recording and closing the session.

Ideally, a counseling session results in a subordinate's commitment to a plan of action. Assessment of the plan of action (discussed below) becomes the starting point for follow-up counseling. (Figure C-7 is an example of a counseling session.)

Open the Session

C-68. In the session opening, state the purpose of the session and establish a subordinate-centered setting. Establish the preferred setting early in the session by inviting the subordinate to speak. The best way to open a counseling session is to clearly state its purpose. For example, an appropriate purpose statement might be: "The purpose of this counseling is to discuss your duty performance over the past month and to create a plan to enhance performance and attain performance goals." If applicable, start the counseling session by reviewing the status of the previous plan of action.

C-69. You and the subordinate should attempt to develop a mutual understanding of the issues. You can best develop this by letting the subordinate do most of the talking. Use active listening; respond, and question without dominating the conversation. Aim to help the subordinate better understand the subject of the counseling, for example, duty performance, a problem situation and its impact, or potential areas for growth.

C-70. Both you and the subordinate should provide examples or cite specific observations to reduce the perception that either is unnecessarily biased or judgmental. However, when the issue is substandard performance, you should make clear how the performance didn't meet the standard. The conversation, which should be two-way, then addresses what the subordinate needs to do to meet the standard. It's important that you define the issue as substandard performance and don't allow the

subordinate to define the issue as an unreasonable standard—unless you consider the standard negotiable or are willing to alter the conditions under which the subordinate must meet the standard.

Develop a Plan of Action

C-71. A plan of action identifies a method for achieving a desired result. It specifies what the subordinate must do to reach the goals set during the counseling session. The plan of action must be specific: it should show the subordinate how to modify or maintain his behavior. It should avoid vague intentions such as "Next month I want you to improve your land navigation skills." The plan must use concrete and direct terms. For example, you might say: "Next week you'll attend the map reading class with 1st Platoon. After the class, SGT Dixon will coach you through the land navigation course. He will help you develop your skill with the compass. I will observe you going through the course with SGT Dixon, and then I will talk to you again and determine where and if you still need additional training." A specific and achievable plan of action sets the stage for successful development.

Record and Close the Session

C-72. Although requirements to record counseling sessions vary, a leader always benefits by documenting the main points of a counseling session. Documentation serves as a reference to the agreed upon plan of action and the subordinate's accomplishments, improvements, personal preferences, or problems. A complete record of counseling aids in making recommendations for professional development, schools, promotions, and evaluation reports.

C-73. Additionally, Army regulations require written records of counseling for certain personnel actions, such as a barring a soldier from reenlisting, processing a soldier for administrative separation, or placing a soldier in the overweight program. When a soldier faces involuntary separation, the leader must take special care to maintain accurate counseling records. Documentation of substandard actions conveys a strong corrective message to subordinates.

C-74. To close the session, summarize its key points and ask if the subordinate understands the plan of action. Invite the subordinate to review the plan of action and what's expected of you, the leader. With the subordinate, establish any follow-up measures necessary to support the successful implementation of the plan of action. These may include providing the subordinate with resources and time, periodically assessing the plan, and following through on referrals. Schedule any future meetings, at least tentatively, before dismissing the subordinate.

FOLLOW UP

Leader's Responsibilities

C-75. The counseling process doesn't end with the counseling session. It continues through implementation of the plan of action and evaluation of results. After counseling, you must support subordinates as they implement their plans of action. Support may include teaching, coaching, or providing time and resources. You must observe and assess this process and possibly modify the plan to meet its goals. Appropriate measures after counseling include follow-up counseling, making referrals, informing the chain of command, and taking corrective measures.

Assess the Plan of Action

C-76. The purpose of counseling is to develop subordinates who are better able to achieve personal, professional, and organizational goals. During the assessment, review the plan of action with the subordinate to determine if the desired results were achieved. You and the subordinate should determine the date for this assessment during the initial counseling session. The assessment of the plan of action provides useful information for future follow-up counseling sessions.

Open the Session
- Establish a relaxed environment. Explain to SFC Taylor that the more one discusses and understands Army values and leader attributes, skills, and actions, the easier it is to develop and incorporate them into an individual leadership style.
- State the purpose of the counseling session. Explain that the initial counseling is based on leader actions (what SFC Taylor needs to do to be a successful platoon sergeant) and not on professional developmental needs (what SFC Taylor needs to do to develop further as an NCO).
- Come to an agreement on the duty description, the meaning of each value and responsibility, and the standards for success and excellence for each value and responsibility. Explain that subsequent counseling will focus on SFC Taylor's developmental needs as well as how well SFC Taylor is meeting the jointly agreed upon performance objectives. Instruct SFC Taylor to perform a self-assessment during the next quarter to identify his developmental needs.
- Ensure SFC Taylor knows the rating chain. Resolve any questions that SFC Taylor has about the job. Discuss the team relationship that exists between a platoon leader and a platoon sergeant and the importance of two-way communication between them.

Discuss the Issue
- Jointly review the duty description on the NCOER, including the maintenance, training, and taking care of soldiers responsibilities. Mention that the duty description can be revised as necessary. Highlight areas of special emphasis and appointed duties.
- Discuss the meaning of each value and responsibility on the NCOER. Discuss the values, attributes, skills, and actions outlined in FM 22-100. Ask open-ended questions to see if SFC Taylor can relate these items to his role as a platoon sergeant.

Figure C-7. Example of a Counseling Session

- Explain that even though the developmental tasks focus on developing leader actions, character development forms the basis for leadership development. Character and actions can't be viewed as separate; they're closely linked. In formulating the plan of action to accomplish major performance objectives, the proper values, attributes, and skills form the basis for the plan. As such, character development must be incorporated into the plan of action.

Assist in Developing a Plan of Action (During the Counseling Session)

- Ask SFC Taylor to identify actions that will facilitate the accomplishment of the major performance objectives. Categorize each action into one of the values or responsibilities listed on the NCOER.

- Discuss how each value and responsibility applies to the platoon sergeant position. Discuss specific examples of success and excellence in each value and responsibility block. Ask SFC Taylor for suggestions to make the goals more objective, specific, and measurable.

- Ensure that SFC Taylor has at least one example of a success or excellence bullet listed under each value and responsibility.

- Discuss SFC Taylor's promotion goals and ask him what he considers to be his strengths and weakness. Obtain copies of the last two master sergeant selection board results and match his goals and objectives to these.

Close the Session

- Check SFC Taylor's understanding of the duty description and performance objectives.

- Stress the importance of teamwork and two-way communication.

- Ensure SFC Taylor understands that you expect him to assist in your development as a platoon leader. This means that both of you have the role of teacher and coach.

- Remind SFC Taylor to perform a self-assessment during the next quarter.

- Set a tentative date during the next quarter for the routinely scheduled follow-up counseling.

Notes on Strategy

- Facilitate answering any questions SFC Taylor may have.

- Expect SFC Taylor to be uncomfortable with the terms and the developmental process; respond in a way that encourages participation throughout the session.

Figure C-7. Example of a Counseling Session (continued)

SUMMARY

C-77. This appendix has discussed developmental counseling. Developmental counseling is subordinate-centered communication that outlines actions necessary for subordinates to achieve individual and organizational goals and objectives. It can be either event-oriented or focused on personal and professional development. Figure C-8 summarizes the major aspects of developmental counseling and the counseling process.

Leaders must demonstrate these qualities to counsel effectively:	The Counseling Process

• Respect for subordinates.
• Self and cultural awareness.
• Credibility.
• Empathy.
 	
Leaders must possess these counseling skills:
• Active listening.
• Responding.
• Questioning.
 	
Effective leaders avoid common counseling mistakes. Leaders should avoid the influence of—
• Personal bias.
• Rash judgments.
• Stereotyping.
• Losing emotional control.
• Inflexible counseling methods.
• Improper follow up.
	1. **Identify the need for counseling.**
	2. **Prepare for counseling.**
	  • Select a suitable place.
	  • Schedule the time.
	  • Notify the subordinate well in advance.
	  • Organize information.
	  • Outline the components of the counseling session.
	  • Plan counseling strategy.
	  • Establish the right atmosphere.
	3. **Conduct the counseling session.**
	  • Open the session.
	  • Discuss the issue.
	  • Develop a plan of action (to include the leader's responsibilities).
	  • Record and close the session.
	4. **Follow up.**
	  • Support plan of action implementation
	  • Assess the plan of action.

Figure C-8. A Summary of Developmental Counseling

THE DEVELOPMENTAL COUNSELING FORM

C-78. The Developmental Counseling Form (DA Form 4856-E) is designed to help Army leaders conduct and record counseling sessions. Figure C-9 shows a completed DA Form 4856-E documenting the counseling of a young soldier with financial problems. While this is an example of a derogatory counseling , you can see that it is still developmental. Leaders must decide when counseling, additional training, rehabilitation, reassignment, or other developmental options have been exhausted. If the purpose of a counseling session is not developmental, refer to paragraphs C-36 and C-37. Figure C-10 shows a routine performance/professional growth counseling for a unit first sergeant. Figure C-11 shows a blank form with instructions on how to complete each block.

DEVELOPMENTAL COUNSELING FORM
For use of this form see FM 22-100

DATA REQUIRED BY THE PRIVACY ACT OF 1974

AUTHORITY: 5 USC 301, Departmental Regulations; 10 USC 3013, Secretary of the Army and E.O. 9397 (SSN)
PRINCIPAL PURPOSE: To assist leaders in conducting and recording counseling data pertaining to subordinates.
ROUTINE USES: For subordinate leader development IAW FM 22-100. Leaders should use this form as necessary.
DISCLOSURE: Disclosure is voluntary.

PART I - ADMINISTRATIVE DATA

Name (Last, First, MI) *Lloyd, Andrew*	Rank / Grade *PFC*	Social Security No. *123-45-6789*	Date of Counseling *28 March 1997*
Organization *2nd Platoon, B Battery, 1 - 1 ADA Bn*		Name and Title of Counselor *SGT Mark Levy, Squad Leader*	

PART II - BACKGROUND INFORMATION

Purpose of Counseling: (Leader states the reason for the counseling, e.g. performance/professional or event-oriented counseling and includes the leader's facts and observations prior to the counseling):

The purpose of this counseling is to inform PFC Lloyd of his responsibility to manage his financial affairs and the potential consequences of poorly managing finances and to help PFC Lloyd develop a plan of action to resolve his financial problems.

Facts: The battery commander received notice of delinquent payment on PFC Lloyd's Deferred Payment Plan (DPP).

A payment of $86.00 is 45 days delinquent

PART III - SUMMARY OF COUNSELING
Complete this section during or immediately subsequent to counseling.

Key Points of Discussion:

PFC Lloyd, late payments on a DPP account reflect a lack of responsibility and poor managing of finances. You should know that the letter of lateness has been brought to the attention of the battery commander, the first sergeant, and the platoon sergeant. They're all questioning your ability to manage your personal affairs. I also remind you that promotions and awards are based more than on just performing MOS-related duties; soldiers must act professionally and responsibly in all areas. Per conversation with PFC Lloyd, the following information was obtained:

He didn't make the DPP payment due to a lack of funds in his checking account. His most recent long distance phone bill was over $220 due to calling his house concerning his grandmother's failing health. PFC Lloyd stated that he wanted to pay for the phone calls himself in order not to burden his parents with the expense of collect calls. He also stated that his calling had tapered down considerably and he expects this month's phone bill to be approximately $50. We made an appointment at ACS and ACS came up with the following information:

PFC Lloyd's monthly obligations: Car payment: $330
  Car insurance: $138
  Rent including utilities: $400
  Other credit cards: $0
Total monthly obligations: $868.00
Monthly take-home pay: $1232.63

We discussed that with approximately $364 available for monthly living expenses, a phone bill in excess of $200 will severely affect PFC Lloyd's financial stability and can't continue. We discussed the need for PFC Lloyd to establish a savings account to help cover emergency expenses. PFC Lloyd agreed that his expensive phone bill and his inability to make the DPP payment is not responsible behavior. He confirmed that he wants to get his finances back on track and begin building a savings account.

OTHER INSTRUCTIONS
This form will be destroyed upon: reassignment (other than rehabilitative transfers), separation at ETS, or retirement. For separation requirements and notification of loss of benefits/consequences, see local directives and AR 635-200.

DA FORM 4856-E, JUN 99   EDITION OF JUN 85 IS OBSOLETE

Figure C-9. Example of a Developmental Counseling Form—Event Counseling

Plan of Action: (Outlines actions that the subordinate will do after the counseling session to reach the agreed upon goals(s). The actions must be specific enough to modify or maintain the subordinate's behavior and include a specific time line for implementation and assessment (Part IV below)).

Based on our discussion, PFC Lloyd will be able to resume normal payment on his DPP account next month (assuming that his phone bill is approximately $50). PFC Lloyd agreed to contact the DPP office and provide a partial payment of $20 immediately. He agreed to exercise self-restraint and not make long distance calls as frequently. He decided that his goal is to make one ten-minute phone call every two weeks. He will write letters to express concern over his grandmother's condition and ask his parents to do the same to keep him informed. His long-term goal is to establish a savings account with a goal of contributing $50 a month.

PFC Lloyd also agreed to attend the check cashing class at ACS on 2, 9, and 16 April.

Assessment date: 27 June

Session Closing: (The leader summarizes the key points of the session and checks if the subordinate understands the plan of action. The subordinate agrees/disagrees and provides remarks if appropriate).

Individual counseled: I agree/ ~~disagree~~ with the information above

Individual counseled remarks:

Signature of Individual Counseled: *Andrew Lloyd* _____ Date: *28 March 1997* _____

Leader Responsibilities: (Leader's responsibilities in implementing the plan of action).

PFC Lloyd will visit the DPP office to make an immediate partial payment of $20 and will give me a copy of the receipt as soon as the payment is made. PFC Lloyd will also provide me with a copy of the next month's phone bill and DPP payment receipt.
PFC Lloyd's finances will be a key topic of discussion at his next monthly counseling session.

Signature of Counselor: *Mark Levy* _____ Date: *28 March 1997* _____

PART IV - ASSESSMENT OF THE PLAN OF ACTION

Assessment (Did the plan of action achieve the desired results? This section is completed by both the leader and the individual counseled and provides useful information for follow-up counseling):

Counselor: _____ Individual Counseled: _____ Date of Assessment: _____

Note: Both the counselor and the individual counseled should retain a record of the counseling.

DA FORM 4856-E (Reverse)

Figure C-9 (continued). Example of a Developmental Counseling Form—Event Counseling

---

DEVELOPMENTAL COUNSELING FORM
For use of this form see FM 22-100

DATA REQUIRED BY THE PRIVACY ACT OF 1974

AUTHORITY: 5 USC 301, Departmental Regulations; 10 USC 3013, Secretary of the Army and E.O. 9397 (SSN)
PRINCIPAL PURPOSE: To assist leaders in conducting and recording counseling data pertaining to subordinates.
ROUTINE USES: For subordinate leader development IAW FM 22-100. Leaders should use this form as necessary.
DISCLOSURE: Disclosure is voluntary.

PART I - ADMINISTRATIVE DATA

Name (Last, First, MI) *McDonald, Stephen*	Rank / Grade *1SG*	Social Security No. *333-33-3333*	Date of Counseling *13 March 1998*
Organization *D Company, 3–95ᵗʰ IN*		Name and Title of Counselor *CPT Peterson, Company Commander*	

PART II - BACKGROUND INFORMATION

Purpose of Counseling: (Leader states the reason for the counseling, e.g. performance/professional or event-oriented counseling and includes the leader's facts and observations prior to the counseling):

- *To discuss duty performance for the period 19 Dec 97 to 11 March 1998.*

- *To discuss short-range professional growth goals/plan for next year.*

- *Talk about long-range professional growth (2-5 years) goals.*

PART III - SUMMARY OF COUNSELING
Complete this section during or immediately subsequent to counseling.

Key Points of Discussion:

- *Performance (sustain):*

- *Emphasized safety and knowledge of demolition, tactical proficiency on the Platoon Live Fire Exercises.*

- *Took charge of company defense during the last major field training exercise; outstanding integration and use of engineer, heavy weapons, and air defense artillery assets. Superb execution of defense preparations and execution.*

- *No dropped white cycle taskings.*

- *Good job coordinating with battalion adjutant on legal and personnel issues.*

- *Continue to take care of soldiers, keep the commander abreast of problems.*

- *Focused on subordinate NCO development; right man for the right job.*

Improve:

- *Get NCODPs on the calendar.*

- *Hold NCOs to standard on sergeants time training.*

OTHER INSTRUCTIONS
This form will be destroyed upon: reassignment (other than rehabilitative transfers), separation at ETS, or upon retirement. For separation requirements and notification of loss of benefits/consequences see local directives and AR 635-200.

DA FORM 4856-E, JUN 99   EDITION OF JUN 85 IS OBSOLETE

**Figure C-10. Example of a Developmental Counseling Form—
Performance/Professional Growth Counseling**

Plan of Action: (Outlines actions that the subordinate will do after the counseling session to reach the agreed upon goals(s). The actions must be specific enough to modify or maintain the subordinate's behavior and include a specific time line for implementation and assessment (Part IV below)).

- _Developmental Plan (next year):_
- _Develop a yearlong plan for NCODPs; coordinate to place on the calendar and training schedules._
- _Resume civilian education; correspondence courses._
- _Develop a company soldier of the month competition._
- _Assist the company XO in modularizing the supply room for quick, efficient load-outs._
- _Put in place a program to develop Ranger School candidates._

Long-range goals (2 to 5 years):
- _Earn bachelor's degree._
- _Attend and graduate the Sergeant Majors Academy._

Session Closing: (The leader summarizes the key points of the session and checks if the subordinate understands the plan of action. The subordinate agrees/disagrees and provides remarks if appropriate).

Individual counseled: I agree/ ~~disagree~~ with the information above

Individual counseled remarks:

Signature of Individual Counseled: 1SG McDonald _____ Date: _13 March 1998_

Leader Responsibilities: (Leader's responsibilities in implementing the plan of action).

- _Coordinate with the 1SG on scheduling of NCODPs and soldier of the month boards._
- _Have the XO meet with the 1SG on developing a plan for modularizing and improving the supply room._
- _Provide time for Ranger candidate program._

Signature of Counselor: _Mark Levy_ _____ Date: _28 March 1997_

PART IV - ASSESSMENT OF THE PLAN OF ACTION

Assessment (Did the plan of action achieve the desired results? This section is completed by both the leader and the individual counseled and provides useful information for follow-up counseling):

1SG McDonald has enrolled in an associates degree program at the University of Kentucky. The supply room received all green evaluations during the last command inspection. Five of seven Ranger applicants successfully completed Ranger School, exceeding the overall course completion rate of 39%. Monthly soldier of the month boards proved to be impractical because of the OPTEMPO; however, the company does now hold quarterly boards during the white cycle. Brigade command sergeant major commented favorably on the last company NCODP he attended and gave the instructor a brigade coin.

Counselor: _CPT Peterson_ _____ Individual Counseled: _1SG McDonald_ _____ Date of Assessment: _1 Aug 98_ _____

Note: Both the counselor and the individual counseled should retain a record of the counseling.

DA FORM 4856-E (Reverse)

**Figure C-10 (continued). Example of a Developmental Counseling Form—
Performance/Professional Growth Counseling**

DEVELOPMENTAL COUNSELING FORM
For use of this form see FM 22-100

DATA REQUIRED BY THE PRIVACY ACT OF 1974

AUTHORITY: 5 USC 301, Departmental Regulations; 10 USC 3013, Secretary of the Army and E.O. 9397 (SSN)
PRINCIPAL PURPOSE: To assist leaders in conducting and recording counseling data pertaining to subordinates.
ROUTINE USES: For subordinate leader development IAW FM 22-100. Leaders should use this form as necessary.
DISCLOSURE: Disclosure is voluntary.

PART I - ADMINISTRATIVE DATA

Name (Last, First, MI)	Rank / Grade	Social Security No.	Date of Counseling
Organization		Name and Title of Counselor	

PART II - BACKGROUND INFORMATION

Purpose of Counseling: (Leader states the reason for the counseling, e.g. performance/professional or event-oriented counseling and includes the leader's facts and observations prior to the counseling):

See paragraph C-68, Open the Session

The leader should annotate pertinent, specific, and objective facts and observations made. If applicable, the leader and subordinate start the counseling session by reviewing the status of the previous plan of action.

PART III - SUMMARY OF COUNSELING
Complete this section during or immediately subsequent to counseling.

Key Points of Discussion:

See paragraphs C-69 and C-70, Discuss the Issues.

The leader and subordinate should attempt to develop a mutual understanding of the issues. Both the leader and the subordinate should provide examples or cite specific observations to reduce the perception that either is unnecessarily biased or judgmental.

OTHER INSTRUCTIONS
This form will be destroyed upon: reassignment (other than rehabilitative transfers), separation at ETS, or upon retirement. For separation requirements and notification of loss of benefits/consequences see local directives and AR 635-200.

DA FORM 4856-E, JUN 99   EDITION OF JUN 85 IS OBSOLETE

Figure C-11. Guidelines on Completing a Developmental Counseling Form

Plan of Action: (Outlines actions that the subordinate will do after the counseling session to reach the agreed upon goals(s). The actions must be specific enough to modify or maintain the subordinate's behavior and include a specific time line for implementation and assessment (Part IV below)).

See paragraph C-71, Develop a Plan of Action

The plan of action specifies what the subordinate must do to reach the goals set during the counseling session. The plan of action must be specific and should contain the outline, guideline(s), and time line that the subordinate follows. A specific and achievable plan of action sets the stage for successful subordinate development.

Remember, event-oriented counseling with corrective training as part of the plan of action can't be tied to a specified time frame. Corrective training is complete once the subordinate attains the standard.

Session Closing: (The leader summarizes the key points of the session and checks if the subordinate understands the plan of action. The subordinate agrees/disagrees and provides remarks if appropriate).

Individual counseled: I agree/ <u>disagree</u> with the information above

Individual counseled remarks:

See paragraph C-72 through C-74, Close the Session

Signature of Individual Counseled:_____ Date: _____

Leader Responsibilities: (Leader's responsibilities in implementing the plan of action).

See paragraph C76, Leader's Responsibilities

To accomplish the plan of action, the leader must list the resources necessary and commit to providing them to the soldier.

Signature of Counselor: _____ Date:_____

PART IV - ASSESSMENT OF THE PLAN OF ACTION

Assessment (Did the plan of action achieve the desired results? This section is completed by both the leader and the individual counseled and provides useful information for follow-up counseling):

The assessment of the plan of action provides useful information for future follow-up counseling. This block should be completed prior to the start of a follow-up counseling session. During an event-oriented counseling session, the counseling session is not complete until this block is completed.

During performance/professional growth counseling, this block serves as the starting point for future counseling sessions. Leaders must remember to conduct this assessment based on resolution of the situation or the established time line discussed in the plan of action block above.

Counselor:_____ Individual Counseled: _____ Date of Assessment: _____

Note: Both the counselor and the individual counseled should retain a record of the counseling.

DA FORM 4856-E (Reverse)

Figure C-11 (continued). Guidelines on Completing a Developmental Counseling Form

Appendix D

A Leader Plan of Action and the ECAS

D-1. By completing a set of tasks (shown in Figure D-1), leaders can improve, sustain, or reinforce a standard of performance within their organizations. Leaders may complete some or all of the sub-tasks shown in Figure D-1, depending on the situation.

D-2. A leader plan of action (developed in step 3) identifies specific leader actions necessary to achieve improvement. It is similar to the individual plan of action that Appendix C discusses.

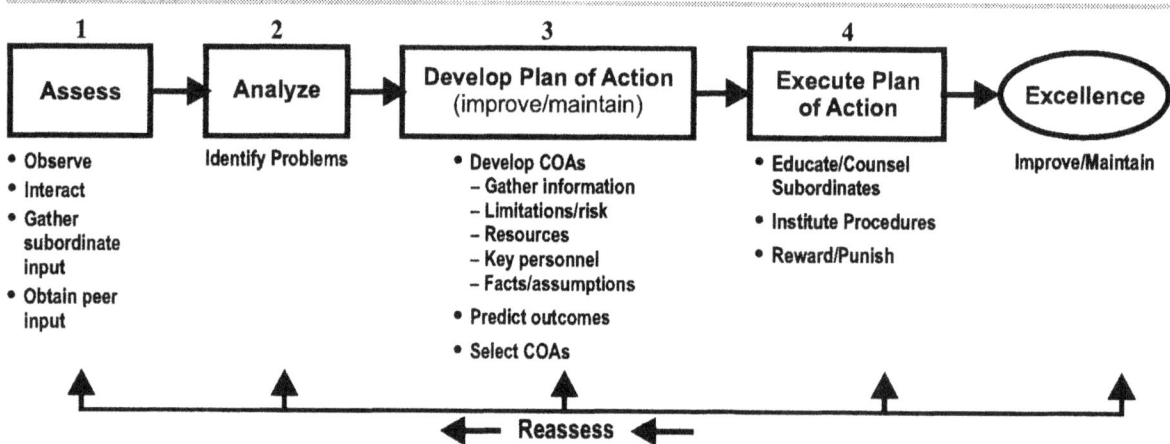

Figure D-1. The Leader Plan of Action Development Process

D-3. Begin your plan of action by assessing your unit (Step 1). Observe, interact, and gather feedback from others; or conduct formal assessments of the workplace. Then analyze the information you gathered to identify what needs improvement (Step 2). Once you have identified what needs improvement, begin to develop courses of action to make the improvements.

D-4. In Step 3, you develop your plan of action. First, develop and consider several possible courses of action to correct the weaknesses you identified. Gather important information, assess the limitations and risks associated with the various courses, identify available key personnel and resources, and verify facts and assumptions. Attempt to predict the outcome for each possible course of action. Based on your predictions, select several leader actions to deal with the problems.

D-5. Execute your plan of action (Step 4) by educating, training, or counseling your subordinates; instituting new policies or procedures; and revising or enforcing proper systems of rewards and punishment. Your organization moves towards excellence by improving substandard or weak areas and maintaining conditions that meet or exceed the standard. Finally, periodically reassesses your unit to identify new matters of concern or to evaluate the effectiveness of the leader actions.

D-6. You can use this process for many areas of interest within your organization. A case study demonstrating how to use an ECAS to prepare a leader plan of action follows. It includes a description of how one leader gathered information to complete the survey. (You can obtain the form used to conduct an ECAS through Training Support Centers by ordering GTA 22-6-1.)

PREPARATION OF AN ECAS

D-7. 2LT Christina Ortega has been a military police platoon leader for almost eight months. When she first came to the platoon, it was a well-trained, cohesive group. Within two months of her taking charge, she and her platoon deployed on a six-month rotation to support operations in Bosnia. The unit performed well, and she quickly earned a reputation as a leader with high standards for herself and her unit. Now redeployed, she must have her platoon ready in two months for a rotation at the Combat Maneuver Training Center (CMTC). She realizes that within that time she must get the unit's equipment ready for deployment, train her soldiers on different missions they will encounter at the CMTC, and provide them some much needed and deserved time off.

D-8. As 2LT Ortega reflects on her first eight months of leadership, she remembers how she took charge of the platoon. She spoke individually with the leaders in the platoon about her expectations and gathered information about her subordinates. She stayed up all night completing the leadership philosophy memorandum that she gave to every member of her platoon. After getting her feet on the ground and getting to know her soldiers, she assessed the platoon's ethical climate using the ECAS. Her unit's overall ECAS score was very good. She committed herself to maintaining that positive ethical climate by continuing the established policies and by monitoring the climate periodically.

D-9. Having completed a major deployment and received a recent influx of some new soldiers, 2LT Ortega decides to complete another ECAS. She heads to the unit motor pool to observe her soldiers preparing for the next day's training exercise. The platoon is deploying to the local training area for the "best squad" competition prior to the ARTEP evaluation at the CMTC. "The best squad competition has really become a big deal in the company," she thinks. "Squad rivalry is fierce, and the squad leaders seem to be looking for an edge so they can come out on top and win the weekend pass that goes to the winning squad."

D-10. She talks to as many of her soldiers as she can, paying particular attention to the newest members of the unit. One new soldier, a vehicle driver for SSG Smith, the 2nd Squad Leader, appears very nervous and anxious. During her conversation with the soldier, 2LT Ortega discovers some disturbing information.

D-11. The new soldier, PFC O'Brien, worries about his vehicle's maintenance and readiness for the next day. His squad leader has told him to "get the parts no matter what." PFC O'Brien says that he admires SSG Smith because he realizes that SSG Smith just wants to perform well and keep up the high standards of his previous driver. He recounts that SSG Smith has vowed to win the next day's land navigation competition. "SSG Smith even went so far as to say that he knows we'll win because he already knows the location of the points for the course. He saw them on the XO's desk last night and wrote them on his map."

D-12. 2LT Ortega thanks the soldier for talking honestly with her and immediately sets him straight on the proper and improper way to get repair parts. By the time she leaves, PFC O'Brien knows that 2LT Ortega has high standards and will not tolerate improper means of meeting them. Meanwhile, 2LT Ortega heads back toward the company headquarters to find the XO.

D-13. She finds the XO busily scribbling numbers and dates on pieces of paper. He is obviously involved and frantic. He looks up at her and manages a quick "Hi, Christina," before returning to his task. The battalion XO apparently did not like the way the unit status report (USR) portrayed the status of the maintenance in the battalion and refused to send that report forward. Not completely familiar with the USR, 2LT Ortega goes to the battalion motor officer to get some more information. After talking to a few more people in her platoon, 2LT Ortega completes the ECAS shown in Figure D-2.

GTA 22-6-1

Ethical
Climate
Assessment
Survey

An ethical climate is one in which our stated Army values are routinely articulated, supported, practiced and respected. The Ethical Climate of an organization is determined by a variety of factors, including the *individual character* of unit members, the *policies and practices* within the organization, the *actions of unit leaders*, and *environmental and mission factors*. Leaders should periodically assess their unit's ethical climate and take appropriate actions to maintain the high ethical standards expected of all Army organizations. This survey will assist you in making these assessments and in identifying the actions necessary to accomplish this vital leader function. FM 22-100, Army Leadership, provides specific leader actions necessary to sustain or improve your ethical climate, as necessary.

DISTRIBUTION RESTRICTION STATEMENT:
Approved for public release: distribution is unlimited
References: FM 22-100
DISTRIBUTION: U.S. Army Training Support Centers (TSCs)
HEADQUARTERS, DEPARTMENT OF THE ARMY
OCTOBER 1997

DISTRIBUTION RESTRICTION STATEMENT:
Approved for public release: distribution is unlimited
References: FM 22-100
DISTRIBUTION: U.S. Army Training Support Centers (TSCs)
HEADQUARTERS, DEPARTMENT OF THE ARMY
OCTOBER 1997

E. We maintain an organizational creed, motto, and/or philosophy that is consistent with Army values. — 4

F. We submit unit reports that reflect accurate information. — 3

G. We ensure unit members are aware of, and are comfortable using, the various channels available to report unethical behavior. — 4

H. We treat fairly those individuals in our unit who report unethical behavior. — 5

I. We hold accountable (i.e., report and/or punish) members of our organization who behave unethically. — 4

Section II Total — 31

*Use the following scale for questions in Section III.

Never	Hardly Ever	Sometimes	Almost Always	Always
1	2	3	4	5

III. Unit Leader Actions - *"What do I do?"* This section focuses on what you do as the leader of your organization to encourage an ethical climate.

A. I discuss Army values in orientation programs when I welcome new members to my organization. — 5

B. I routinely assess the ethical climate of my unit (i.e., sensing sessions, climate surveys, etc.). — 5

C. I communicate my expectations regarding ethical behavior in my unit, and require subordinates to perform tasks in an ethical manner. — 5

D. I encourage discussions of ethical issues in After Action Reviews, training meetings, seminars, and workshops. — 3

E. I encourage unit members to raise ethical questions and concerns to the chain of command or other individuals, if needed (i.e., chaplain, IG, etc.). — 5

F. I consider ethical behavior in performance evaluations, award and promotion recommendations, and adverse personnel actions. — 4

G. I include maintaining a strong ethical climate as one of my unit's goals and objectives. — 5

Section III Total — 32

INSTRUCTIONS

Answer the questions in this survey according to how you currently perceive your unit and your own leader actions, NOT according to how you would prefer them to be or how you think they should be. This information is for your use, (not your chain of command's) to determine if you need to take action to improve the Ethical Climate in your organization. Use the following scale for all questions in Sections I and II.

Strongly Disagree	Disagree	Neither Agree nor Disagree	Agree	Strongly Agree
1	2	3	4	5

I. Individual Character - *"Who are we?"* This section focuses on your organization's members' commitment to Army values. Please answer the following questions based on your observations of the ethical commitment in your unit. (This means your *immediate* unit. If you are a squad leader, it means you and your squad. If you are a civilian supervisor, it means you and your section.)

A. In general, the members of my unit demonstrate a commitment to Army values (honor, selfless service, integrity, loyalty, courage, duty and respect). — 4

B. The members of my unit typically accomplish a mission by "doing the right thing" rather than compromising Army values. — 2

C. I understand, and I am committed to, the Army's values as outlined in FM 22-100, Army Leadership. — 5

Section I Total — 11

II. Unit/Workplace Policies & Practices - *"What do we do?"* This section focuses on what you, and the leaders who report to you, do to maintain an ethical climate in your workplace. (This does not mean your superiors. Their actions will be addressed in Section IV).

A. We provide clear instructions which help prevent unethical behavior. — 2

B. We promote an environment in which subordinates can learn from their mistakes. — 5

C. We maintain appropriate, not dysfunctional, levels of stress and competition in our unit. — 1

D. We discuss ethical behavior and issues during regular counseling sessions. — 3

IV. Environmental/Mission Factors - *"What surrounds us?"* This section focuses on the external environment surrounding your organization. Answer the following questions to assess the impact of these factors on the ethical behavior in your organization.

Use the following scale for all questions in Section IV. ***Note: the scale is reversed for this section (Strongly Agree is scored as a "1", not a "5") ***

Strongly Agree	Agree	Neither Agree nor Disagree	Disagree	Strongly Disagree
1	2	3	4	5

A. My unit is currently under an excessive amount of stress (i.e., inspections, limited resources, frequent deployments, training events, deadlines, etc.). — 1

B. My higher unit leaders foster a 'zero defects' outlook on performance, such that they do not tolerate mistakes. — 1

C. My higher unit leaders over-emphasize competition between units. — 1

D. My higher unit leaders appear to be unconcerned with unethical behavior as long as the mission is accomplished. — 2

E. I do not feel comfortable bringing up ethical issues with my supervisors. — 5

F. My peers in my unit do not seem to take ethical behavior very seriously. — 1

Section IV Total — 11

Place the Total Score from each section in the spaces below:
(A score of 1 or 2 on any question requires some immediate leader action.)

Section I - Individual Character Total Score — 11
Section II - Leader Action Total Score — 31
Section III - Unit Policies and Procedures Total Score — 32
Section IV - Environmental/Mission Factors Total Score — 11

| ECAS Total Score (I + II + III + IV) | 85 |

25 - 75	76-100	101 - 125
Take *Immediate* Action to Improve Ethical Climate	Take Actions to Improve Ethical Climate	Maintain a Healthy Ethical Climate

Figure D-2. Example of an Ethical Climate Assessment Survey

PREPARATION OF A LEADER PLAN OF ACTION

D-14. 2LT Ortega looks at her ECAS score and determines that she needs to take action to improve the ethical climate in her platoon. To help determine where she should begin, 2LT Ortega looks at the scores for each question. She knows that any question receiving a "1" or "2" must be addressed immediately in her plan of action. As 2LT Ortega reviews the rest of the scores for her unit, she identifies additional problems to correct. Furthermore, she decides to look at a few actions in which her unit excels and to describe ways to sustain the performance. As she continues to develop the leader plan of action, she looks at each subject she has identified. She next develops the plan shown in Figure D-3 to correct the deficiencies. At the bottom of the form, she lists at least two actions she plans to take to maintain the positive aspects of her platoon's ethical climate.

D-15. 2LT Ortega has already completed the first three steps (assess, analyze, and develop a plan of action) specified in Figure D-1. When she takes action to implement the plan she will have completed the process. She must then follow up to ensure her actions have the effects she intended.

Actions to *correct* negative aspects of the ethical climate in the organization

Problem: Dysfunctional competition/stress in the unit (the competition is causing some members of the unit to seek ways to gain an unfair advantage over others) [ECAS question # II.C., IV.A. & IV.C.]
Action:

- Postpone the platoon competition; focus on the readiness of equipment and soldier preparation rather than competition.
- Build some time in the long-range calendar to allow soldiers time to get away from work and relax.
- Focus on the group's accomplishment of the mission (unit excellence). Reward the platoon, not squads, for excellent performance. Reward teamwork.

Problem: Battalion XO "ordering" the changing of reports [IV B., D. & F.]
Action:

- Go see the company XO first and discuss what he should do.
- If the XO won't deal with it, see the commander myself to raise the issue.

Problem: Squad leader's unethical behavior [I.B. & II.A.]
Action:

- Reprimand the squad leader for getting the land navigation points unfairly.
- Counsel the squad leader on appropriate ways to give instructions and accomplish the mission without compromising values.

Problem: Unclear instructions given by the squad leader ("get the parts no matter what") [II.A.]
Action:

- Have the platoon sergeant give a class (NCODP) on proper guidelines for giving instructions and appropriate ethical considerations when asking subordinates to complete a task.
- Have the platoon sergeant counsel the squad leader(s) on the importance of using proper supply procedures.

Problem: Company XO "changing report" to meet battalion XO's needs [IV.B. & F.]
Action:

- Have an informal discussion with the company XO about correct reporting or see the company commander to raise the issue about the battalion XO.

Actions to *maintain* positive aspects of the ethical climate in the organization

Maintain: Continue to hold feedback (sensing) sessions and conduct ECAS assessments to maintain a feel for how the platoon is accomplishing its mission. [II.D. & G.; III.A. & B.]

Maintain: Continue to reward people who perform to high standards without compromising values. Punish those caught compromising them. [III.E. & F.]

Figure D-3. Example of a Leader Plan of Action

Appendix E

Character Development

E-1. Everyone who becomes part of America's Army, soldier or DA civilian, has character. On the day a person joins the Army, leaders begin building on that character. Army values emphasize the relationship between character and competence. Although competence is a fundamental attribute of Army leaders, character is even more critical. This appendix discusses the actions Army leaders take to develop their subordinates' character.

E-2. Army leaders are responsible for refining the character of soldiers and DA civilians. How does the Army as an institution ensure proper character development? What should leaders do to inculcate Army values in their subordinates?

E-3. Leaders teach Army values to every new member of the Army. Together with the leader attributes described in Chapter 2, Army values establish the foundation of leaders of character. Once members learn these values, their leaders ensure adherence. Adhering to the principles Army values embody is essential, for the Army cannot tolerate unethical behavior. Unethical behavior destroys morale and cohesion; it undermines the trust and confidence essential to teamwork and mission accomplishment.

E-4. Ethical conduct must reflect beliefs and convictions, not just fear of punishment. Over time, soldiers and DA civilians adhere to Army values because they want to live ethically and profess the values because they know it's right to do so. Once people believe and demonstrate Army values, they are persons of character. Ultimately, Army leaders are charged with the with the essential role of developing character in others. Figure E-1 shows the leader actions that support character development.

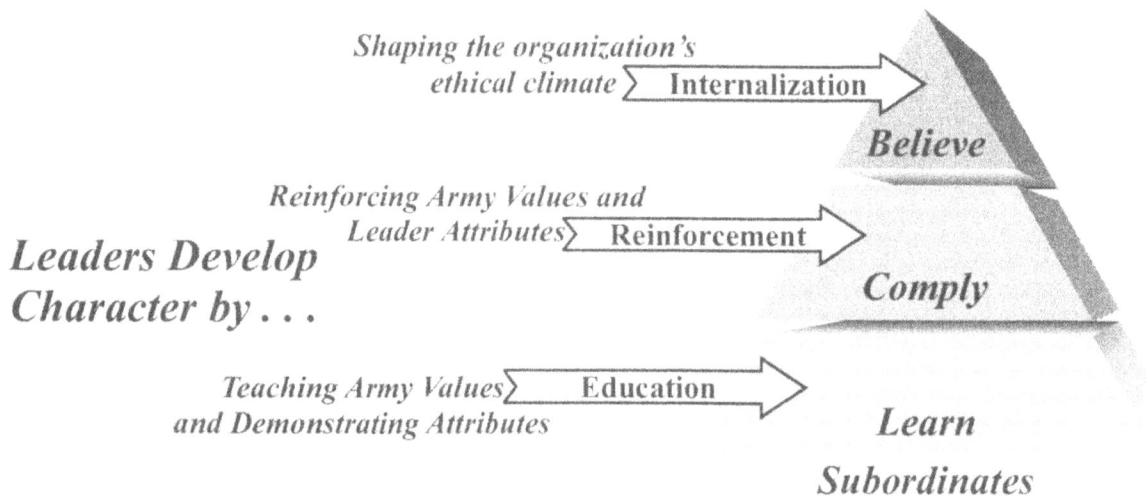

Figure E-1. Character Development

LEADERS TEACH VALUES; SUBORDINATES LEARN THE CULTURE

E-5. Army leaders must teach their subordinates moral principles, ethical theory, Army values, and leadership attributes. Through their leaders' programs, soldiers and DA civilians develop character through education, experience, and reflection. By educating their subordinates and setting the example, Army leaders enable their subordinates to make ethical decisions that in turn contribute to excellence. Subordinates gain deeper understanding from experiencing, observing, and reflecting on the aspects of Army leadership under the guidance of their leaders.

E-6. Inculcating Army values doesn't end with basic training. All Army leaders should seek to deepen subordinates' understanding of the ethical aspects of character through classes, informal discussions, one-on-one coaching, and formal developmental counseling. Army leaders can also improve their own understanding through study, reflection, and discussions with peers and superiors.

LEADERS REINFORCE VALUES; SUBORDINATES COMPLY

E-7. Leaders reinforce and discipline behavior to guide subordinates' development. To help subordinates live according to Army values, leaders enforce rules, policies, and regulations. Still, soldiers and DA civilians of character do more than merely comply with established institutional rules. Acting correctly but without complete understanding or sound motivation is not good enough in America's values-based Army. People of character behave correctly through correct understanding and personal desire. Understanding comes from training and self-development. Personal desire comes from a person's realization that Army values are worth adopting and living by and from that person's decision to do just that.

E-8. Character stems from a thorough understanding of Army values; however, this understanding must go beyond knowing the one-line definitions. Individuals must also know why Army values are important and how to apply them to everyday Army life. Leaders can promote Army values by setting the example themselves and pointing out other examples of Army values in both normal and exceptional activities. Army leaders can use unit histories and traditions, prominent individuals, and recent events to bring Army values to life and explain why adhering to them is important.

LEADERS SHAPE THE ETHICAL CLIMATE; SUBORDINATES INTERNALIZE ARMY VALUES

E-9. Doing the right thing is good. Doing the right thing for the right reason and with the right intention is better. People of character must possess the desire to act ethically in all situations. One of the Army leader's primary responsibilities is to maintain an ethical climate that supports development of such a character. When an organization's ethical climate nurtures ethical behavior, over time, people think, feel, and act ethically—they internalize the aspects of sound character.

E-10. Leaders should influence others' character development and foster correct actions through role modeling, teaching, and coaching. Army leaders seek to build a climate in which subordinates and organizations can reach their full potential. Together, these actions promote organizational excellence.

E-11. Army leaders can use the ECAS to assess ethical aspects of their own character and actions, the workplace, and the external environment. Once they have done their assessment, leaders prepare and carry out a plan of action. The plan of action focuses on solving ethical problems within the leaders' span of influence; leaders pass ethical problems they cannot change to higher headquarters. E-12. Becoming a person of character and a leader of character is a career-long process involving both self-development and developmental counseling. While individuals are responsible for their own character development, leaders are responsible for encouraging, supporting, and assessing the efforts of their subordinates. Leaders of character can develop only through continual study, reflection, experience, and feedback.

Appendix F

The Constitution of the United States

As a member of the Army, you have taken an oath to "support and defend the Constitution of the United States against all enemies, foreign and domestic [and to] bear true faith and allegiance to the same." But what is this document that you have sworn to protect? In essence, the Constitution is a blueprint establishing the powers and responsibilities of the three branches of the United States government as well as the rights of American citizens. Especially important for the Army are those provisions in the Constitution that place fundamental military authority in Congress and the president. As part of that authority, Congress has the power to "provide for the common Defense," which includes the power to "raise and support Armies," and the president is the commander in chief of the armed forces. So the Constitution establishes the critical principle that America's military leaders are subordinate to the nation's civilian authorities. Given the importance of that concept to our system of government, this appendix contains a copy of the US Constitution. As you read it, you will see that, although the Constitution was written over 200 years ago, it remains relevant today for Army leaders and all Americans.

THE PREAMBLE

We the People of the United States, in Order to form a more perfect Union, establish Justice, insure domestic Tranquility, provide for the common defence, promote the general Welfare, and secure the Blessings of Liberty to ourselves and our Posterity, do ordain and establish this Constitution for the United States of America.

THE CONSTITUTION

ARTICLE I

Section 1

All legislative Powers herein granted shall be vested in a Congress of the United States, which shall consist of a Senate and House of Representatives.

Section 2

The House of Representatives shall be composed of Members chosen every second Year by the People of the several States, and the Electors in each State shall have the Qualifications requisite for Electors of the most numerous Branch of the State Legislature.

No Person shall be a Representative who shall not have attained to the Age of twenty five Years, and been seven Years a Citizen of the United States, and who shall not, when elected, be an Inhabitant of that State in which he shall be chosen.

Representatives and direct Taxes shall be apportioned among the several States which may be included within this Union, according to their respective Numbers, which shall be determined by adding to the whole Number of free Persons, including those bound to Service for a Term of Years, and excluding Indians not taxed, three fifths of all other Persons. The actual Enumeration shall be made within three Years after the first Meeting of the Congress of the United States, and within every subsequent Term of ten Years, in such Manner as they shall by Law direct. The Number of Representatives shall not exceed one for every thirty Thousand, but each State shall have at Least one Representative; and until such enumeration shall be made, the State of New Hampshire shall be entitled to chuse three, Massachusetts eight, Rhode-Island and Providence Plantations one, Connecticut five, New-York six, New Jersey four, Pennsylvania eight, Delaware one, Maryland six, Virginia ten, North Carolina five, South Carolina five, and Georgia three.

When vacancies happen in the Representation from any State, the Executive Authority thereof shall issue Writs of Election to fill such Vacancies.

The House of Representatives shall chuse their Speaker and other Officers; and shall have the sole Power of Impeachment.

Section 3

The Senate of the United States shall be composed of two Senators from each State, chosen by the Legislature thereof, for six Years; and each Senator shall have one Vote.

Immediately after they shall be assembled in Consequence of the first Election, they shall be divided as equally as may be into three Classes. The Seats of the Senators of the first Class shall be vacated at the Expiration of the second Year, of the second Class at the Expiration of the fourth Year, and of the third Class at the Expiration of the sixth Year, so that one third may be chosen every second Year; and if Vacancies happen by Resignation, or otherwise, during the Recess of the Legislature of any State, the Executive thereof may make temporary Appointments until the next Meeting of the Legislature, which shall then fill such Vacancies.

No Person shall be a Senator who shall not have attained to the Age of thirty Years, and been nine Years a Citizen of the United States, and who shall not, when elected, be an Inhabitant of that State for which he shall be chosen.

The Vice President of the United States shall be President of the Senate, but shall have no Vote, unless they be equally divided.

The Senate shall chuse their other Officers, and also a President pro tempore, in the Absence of the Vice President, or when he shall exercise the Office of President of the United States.

The Senate shall have the sole Power to try all Impeachments. When sitting for that Purpose, they shall be on Oath or Affirmation. When the President of the United States is tried, the Chief Justice shall preside: And no Person shall be convicted without the Concurrence of two thirds of the Members present.

Judgment in Cases of Impeachment shall not extend further than to removal from Office, and disqualification to hold and enjoy any Office of honor, Trust or Profit under the United States: but the Party convicted shall nevertheless be liable and subject to Indictment, Trial, Judgment and Punishment, according to Law.

Section 4

The Times, Places and Manner of holding Elections for Senators and Representatives, shall be prescribed in each State by the Legislature thereof; but the Congress may at any time by Law make or alter such Regulations, except as to the Places of chusing Senators.

The Congress shall assemble at least once in every Year, and such Meeting shall be on the first Monday in December, unless they shall by Law appoint a different Day.

Section 5

Each House shall be the Judge of the Elections, Returns and Qualifications of its own Members, and a Majority of each shall constitute a Quorum to do Business; but a smaller Number may adjourn from day to day, and may be authorized to compel the Attendance of absent Members, in such Manner, and under such Penalties as each House may provide.

Each House may determine the Rules of its Proceedings, punish its Members for disorderly Behaviour, and, with the Concurrence of two thirds, expel a Member.

Each House shall keep a Journal of its Proceedings, and from time to time publish the same, excepting such Parts as may in their Judgment require Secrecy; and the Yeas and Nays of the Members of either House on any question shall, at the Desire of one fifth of those Present, be entered on the Journal.

Neither House, during the Session of Congress, shall, without the Consent of the other, adjourn for more than three days, nor to any other Place than that in which the two Houses shall be sitting.

Section 6

The Senators and Representatives shall receive a Compensation for their Services, to be ascertained by Law, and paid out of the Treasury of the United States. They shall in all Cases, except Treason, Felony and Breach of the Peace, be privileged from Arrest during their Attendance at the Session of their respective Houses, and in going to and returning from the same; and for any Speech or Debate in either House, they shall not be questioned in any other Place.

No Senator or Representative shall, during the Time for which he was elected, be appointed to any civil Office under the Authority of the United States, which shall have been created, or the Emoluments whereof shall have been encreased during such time; and no Person holding any Office under the United States, shall be a Member of either House during his Continuance in Office.

Section 7

All Bills for raising Revenue shall originate in the House of Representatives; but the Senate may propose or concur with Amendments as on other Bills.

Every Bill which shall have passed the House of Representatives and the Senate, shall, before it become a Law, be presented to the President of the United States; If he approve he shall sign it, but if not he shall return it, with his Objections to that House in which it shall have originated, who shall enter the Objections at large on their Journal, and proceed to reconsider it. If after such Reconsideration two thirds of that House shall agree to pass the Bill, it shall be sent, together with the Objections, to the other House, by which it shall likewise be reconsidered, and if approved by two thirds of that House, it shall become a Law. But in all such Cases the Votes of both Houses shall be determined by Yeas and Nays, and the Names of the Persons voting for and against the Bill shall

be entered on the Journal of each House respectively. If any Bill shall not be returned by the President within ten Days (Sundays excepted) after it shall have been presented to him, the Same shall be a Law, in like Manner as if he had signed it, unless the Congress by their Adjournment prevent its Return, in which Case it shall not be a Law.

Every Order, Resolution, or Vote to which the Concurrence of the Senate and House of Representatives may be necessary (except on a question of Adjournment) shall be presented to the President of the United States; and before the Same shall take Effect, shall be approved by him, or being disapproved by him, shall be re-passed by two thirds of the Senate and House of Representatives, according to the Rules and Limitations prescribed in the Case of a Bill.

Section 8

The Congress shall have Power To lay and collect Taxes, Duties, Imposts and Excises, to pay the Debts and provide for the common Defence and general Welfare of the United States; but all Duties, Imposts and Excises shall be uniform throughout the United States;

To borrow Money on the credit of the United States;

To regulate Commerce with foreign Nations, and among the several States, and with the Indian Tribes;

To establish an uniform Rule of Naturalization, and uniform Laws on the subject of Bankruptcies throughout the United States;

To coin Money, regulate the Value thereof, and of foreign Coin, and fix the Standard of Weights and Measures;

To provide for the Punishment of counterfeiting the Securities and current Coin of the United States;

To establish Post Offices and post Roads;

To promote the Progress of Science and useful Arts, by securing for limited Times to Authors and Inventors the exclusive Right to their respective Writings and Discoveries;

To constitute Tribunals inferior to the supreme Court;

To define and punish Piracies and Felonies committed on the high Seas, and Offences against the Law of Nations;

To declare War, grant Letters of Marque and Reprisal, and make Rules concerning Captures on Land and Water;

To raise and support Armies, but no Appropriation of Money to that Use shall be for a longer Term than two Years;

To provide and maintain a Navy;

To make Rules for the Government and Regulation of the land and naval Forces;

To provide for calling forth the Militia to execute the Laws of the Union, suppress Insurrections and repel Invasions;

To provide for organizing, arming, and disciplining, the Militia, and for governing such Part of them as may be employed in the Service of the United States, reserving to the States respectively, the Appointment of the Officers, and the Authority of training the Militia according to the discipline prescribed by Congress;

To exercise exclusive Legislation in all Cases whatsoever, over such District (not exceeding ten Miles square) as may, by Cession of particular States, and the Acceptance of Congress, become the Seat of the Government of the United States, and to exercise like Authority over all Places purchased by the Consent of the Legislature of the State in which the Same shall be, for the Erection of Forts, Magazines, Arsenals, dock-Yards, and other needful Buildings;—And

To make all Laws which shall be necessary and proper for carrying into Execution the foregoing Powers, and all other Powers vested by this Constitution in the Government of the United States, or in any Department or Officer thereof.

Section 9

The Migration or Importation of such Persons as any of the States now existing shall think proper to admit, shall not be prohibited by the Congress prior to the Year one thousand eight hundred and eight, but a Tax or duty may be imposed on such Importation, not exceeding ten dollars for each Person.

The Privilege of the Writ of Habeas Corpus shall not be suspended, unless when in Cases of Rebellion or Invasion the public Safety may require it.

No Bill of Attainder or ex post facto Law shall be passed.

No Capitation, or other direct, Tax shall be laid, unless in Proportion to the Census or Enumeration herein before directed to be taken.

No Tax or Duty shall be laid on Articles exported from any State.

No Preference shall be given by any Regulation of Commerce or Revenue to the Ports of one State over those of another: nor shall Vessels bound to, or from, one State, be obliged to enter, clear, or pay Duties in another.

No Money shall be drawn from the Treasury, but in Consequence of Appropriations made by Law; and a regular Statement and Account of the Receipts and Expenditures of all public Money shall be published from time to time.

No Title of Nobility shall be granted by the United States: And no Person holding any Office of Profit or Trust under them, shall, without the Consent of the Congress, accept of any present, Emolument, Office, or Title, of any kind whatever, from any King, Prince, or foreign State.

Section 10

No State shall enter into any Treaty, Alliance, or Confederation; grant Letters of Marque and Reprisal; coin Money; emit Bills of Credit; make any Thing but gold and silver Coin a Tender in Payment of Debts; pass any Bill of Attainder, ex post facto Law, or Law impairing the Obligation of Contracts, or grant any Title of Nobility.

No State shall, without the Consent of the Congress, lay any Imposts or Duties on Imports or Exports, except what may be absolutely necessary for executing its inspection Laws: and the net Produce of all Duties and Imposts, laid by any State on Imports or Exports, shall be for the Use of the Treasury of the United States; and all such Laws shall be subject to the Revision and Controul of the Congress.

No State shall, without the Consent of Congress, lay any Duty of Tonnage, keep Troops, or Ships of War in time of Peace, enter into any Agreement or Compact with another State, or with a foreign Power, or engage in War, unless actually invaded, or in such imminent Danger as will not admit of delay.

ARTICLE II

Section 1

The executive Power shall be vested in a President of the United States of America. He shall hold his Office during the Term of four Years, and, together with the Vice President, chosen for the same Term, be elected, as follows:

Each State shall appoint, in such Manner as the Legislature thereof may direct, a Number of Electors, equal to the whole Number of Senators and Representatives to which the State may be entitled in the Congress: but no Senator or Representative, or Person holding an Office of Trust or Profit under the United States, shall be appointed an Elector.

The Electors shall meet in their respective States, and vote by Ballot for two Persons, of whom one at least shall not be an Inhabitant of the same State with themselves. And they shall make a List of all the Persons voted for, and of the Number of Votes for each; which List they shall sign and certify, and transmit sealed to the Seat of the Government of the United States, directed to the President of the Senate. The President of the Senate shall, in the Presence of the Senate and House of Representatives, open all the Certificates, and the Votes shall then be counted. The Person having the greatest Number of Votes shall be the President, if such Number be a Majority of the whole Number of Electors appointed; and if there be more than one who have such Majority, and have an equal Number of Votes, then the House

of Representatives shall immediately chuse by Ballot one of them for President; and if no Person have a Majority, then from the five highest on the List the said House shall in like Manner chuse the President. But in chusing the President, the Votes shall be taken by States, the Representation from each State having one Vote; A quorum for this Purpose shall consist of a Member or Members from two thirds of the States, and a Majority of all the States shall be necessary to a Choice. In every Case, after the Choice of the President, the Person having the greatest Number of Votes of the Electors shall be the Vice President. But if there should remain two or more who have equal Votes, the Senate shall chuse from them by Ballot the Vice President.

The Congress may determine the Time of chusing the Electors, and the Day on which they shall give their Votes; which Day shall be the same throughout the United States.

No Person except a natural born Citizen, or a Citizen of the United States, at the time of the Adoption of this Constitution, shall be eligible to the Office of President; neither shall any Person be eligible to that Office who shall not have attained to the Age of thirty five Years, and been fourteen Years a Resident within the United States.

In Case of the Removal of the President from Office, or of his Death, Resignation, or Inability to discharge the Powers and Duties of the said Office, the Same shall devolve on the Vice President, and the Congress may by Law provide for the Case of Removal, Death, Resignation or Inability, both of the President and Vice President, declaring what Officer shall then act as President, and such Officer shall act accordingly, until the Disability be removed, or a President shall be elected.

The President shall, at stated Times, receive for his Services, a Compensation, which shall neither be encreased nor diminished during the Period for which he shall have been elected, and he shall not receive within that Period any other Emolument from the United States, or any of them.

Before he enter on the Execution of his Office, he shall take the following Oath or Affirmation:—"I do solemnly swear (or affirm) that I will faithfully execute the Office of President of the United States, and will to the best of my Ability, preserve, protect and defend the Constitution of the United States."

Section 2

The President shall be Commander in Chief of the Army and Navy of the United States, and of the Militia of the several States, when called into the actual Service of the United States; he may require the Opinion, in writing, of the principal Officer in each of the executive Departments, upon any Subject relating to the Duties of their respective Offices, and he shall have Power to grant Reprieves and Pardons for Offences against the United States, except in Cases of Impeachment.

He shall have Power, by and with the Advice and Consent of the Senate, to make Treaties, provided two thirds of the Senators present concur; and he shall nominate, and by and with the Advice and Consent of the Senate, shall appoint Ambassadors, other public Ministers and Consuls, Judges of the supreme Court, and all other Officers of the United States, whose Appointments are not herein otherwise provided for, and which shall be established by Law: but the Congress may by Law vest the Appointment of such inferior Officers, as they think proper, in the President alone, in the Courts of Law, or in the Heads of Departments.

The President shall have Power to fill up all Vacancies that may happen during the Recess of the Senate, by granting Commissions which shall expire at the End of their next Session.

Section 3

He shall from time to time give to the Congress Information of the State of the Union, and recommend to their Consideration such Measures as he shall judge necessary and expedient; he may, on extraordinary Occasions, convene both Houses, or either of them, and in Case of Disagreement between them, with Respect to the Time of Adjournment, he may adjourn them to such Time as he shall think proper; he shall receive Ambassadors and other public Ministers; he shall take Care that the Laws be faithfully executed, and shall Commission all the Officers of the United States.

Section 4

The President, Vice President and all civil Officers of the United States, shall be removed from Office on Impeachment for, and Conviction of, Treason, Bribery, or other high Crimes and Misdemeanors.

ARTICLE III

Section 1

The judicial Power of the United States, shall be vested in one supreme Court, and in such inferior Courts as the Congress may from time to time ordain and establish. The Judges, both of the supreme and inferior Courts, shall hold their Offices during good Behaviour, and shall, at stated Times, receive for their Services, a Compensation, which shall not be diminished during their Continuance in Office.

Section 2

The judicial Power shall extend to all Cases, in Law and Equity, arising under this Constitution, the Laws of the United States, and Treaties made, or which shall be made, under their Authority;—to all Cases affecting Ambassadors, other public Ministers and Consuls;—to all Cases of admiralty and maritime Jurisdiction;—to Controversies to which the United States shall be a Party;—to Controversies between two or more States;—between a State and Citizens of another State;—between Citizens of different States, —between Citizens of the same State claiming Lands under Grants of different States, and between a State, or the Citizens thereof, and foreign States, Citizens or Subjects.

In all Cases affecting Ambassadors, other public Ministers and Consuls, and those in which a State shall be Party, the supreme Court shall have original Jurisdiction. In all the other Cases before mentioned, the supreme Court shall have appellate Jurisdiction, both as to Law and Fact, with such Exceptions, and under such Regulations as the Congress shall make.

The Trial of all Crimes, except in Cases of Impeachment, shall be by Jury; and such Trial shall be held in the State where the said Crimes shall have been committed; but when not committed within any State, the Trial shall be at such Place or Places as the Congress may by Law have directed.

Section 3

Treason against the United States, shall consist only in levying War against them, or in adhering to their Enemies, giving them Aid and Comfort. No Person shall be convicted of Treason unless on the Testimony of two Witnesses to the same overt Act, or on Confession in open Court.

The Congress shall have Power to declare the Punishment of Treason, but no Attainder of Treason shall work Corruption of Blood, or Forfeiture except during the Life of the Person attainted.

ARTICLE IV
Section 1

Full Faith and Credit shall be given in each State to the public Acts, Records, and judicial Proceedings of every other State. And the Congress may by general Laws prescribe the Manner in which such Acts, Records and Proceedings shall be proved, and the Effect thereof.

Section 2

The Citizens of each State shall be entitled to all Privileges and Immunities of Citizens in the several States.

A Person charged in any State with Treason, Felony, or other Crime, who shall flee from Justice, and be found in another State, shall on Demand of the executive Authority of the State from which he fled, be delivered up, to be removed to the State having Jurisdiction of the Crime.

No Person held to Service or Labour in one State, under the Laws thereof, escaping into another, shall, in Consequence of any Law or Regulation therein, be discharged from such Service or Labour, but shall be delivered up on Claim of the Party to whom such Service or Labour may be due.

Section 3

New States may be admitted by the Congress into this Union; but no new State shall be formed or erected within the Jurisdiction of any other State; nor any State be formed by the Junction of two or more States, or Parts of States, without the Consent of the Legislatures of the States concerned as well as of the Congress.

The Congress shall have Power to dispose of and make all needful Rules and Regulations respecting the Territory or other Property belonging to the United States; and nothing in this Constitution shall be so construed as to Prejudice any Claims of the United States, or of any particular State.

Section 4

The United States shall guarantee to every State in this Union a Republican Form of Government, and shall protect each of them against Invasion; and on Application of the Legislature, or of the Executive (when the Legislature cannot be convened) against domestic Violence.

ARTICLE V

The Congress, whenever two thirds of both Houses shall deem it necessary, shall propose Amendments to this Constitution, or, on the Application of the Legislatures of two thirds of the several States, shall call a Convention for proposing Amendments, which, in either Case, shall be valid to all Intents and Purposes, as Part of this Constitution, when ratified by the Legislatures of three fourths of the several States, or by Conventions in three fourths thereof, as the one or the other Mode of Ratification may be proposed by the Congress; Provided that no Amendment which may be made prior to the Year One thousand eight hundred and eight shall in any Manner affect the first and fourth Clauses in the Ninth Section of the first Article; and that no State, without its Consent, shall be deprived of its equal Suffrage in the Senate.

ARTICLE VI

All Debts contracted and Engagements entered into, before the Adoption of this Constitution, shall be as valid against the United States under this Constitution, as under the Confederation.

This Constitution, and the Laws of the United States which shall be made in Pursuance thereof; and all Treaties made, or which shall be made, under the Authority of the United States, shall be the supreme Law of the Land; and the Judges in every State shall be bound thereby, any Thing in the Constitution or Laws of any State to the Contrary notwithstanding.

The Senators and Representatives before mentioned, and the Members of the several State Legislatures, and all executive and judicial

Officers, both of the United States and of the several States, shall be bound by Oath or Affirmation, to support this Constitution; but no religious Test shall ever be required as a Qualification to any Office or public Trust under the United States.

ARTICLE VII

The Ratification of the Conventions of nine States, shall be sufficient for the Establishment of this Constitution between the States so ratifying the same.

Done in Convention by the Unanimous Consent of the States present the Seventeenth Day of September in the Year of our Lord one thousand seven hundred and Eighty seven and of the Independence of the United States of America the Twelfth. In witness whereof We have hereunto subscribed our Names,

G. WASHINGTON—President. And deputy from Virginia

Delaware
Geo: Read
Gunning Bedford jun
John Dickinson
Richard Bassett
Jaco: Broom
South Carolina
J. Rutledge
Charles Cotesworth
Pinckney
Charles Pinckney
Pierce Butler
Georgia
William Few
Abr Baldwin
New Hampshire
John Langdon
Nicholas
Massachusetts
Nathaniel Gorham
Rufus King
Connecticut
Wm: Saml. Johnson
Roger Sherman

Maryland
James McHenry
Dan of St Thos. Jenifer
Danl Carroll.
Virginia
John Blair---
James Madison Jr.
New York
Alexander Hamilton
New Jersy
Wil: Livingston
David Brearley
Wm. Paterson
Jona: Dayton
Pennsylvania
B Franklin
Thomas Mifflin
Robt Morris
Geo. Clymer
Thos. FitzSimons
Jared Ingersoll
James Wilson
Gouv Morris

Attest William Jackson
Secretary

North Carolina
Wm. Blount
Richd. Dobbs Spaight
Hu Williamson

THE PREAMBLE TO THE BILL OF RIGHTS

Congress of the United States begun and held at the City of New York, on Wednesday the fourth of March, one thousand seven hundred and eighty-nine.

The Conventions of a number of the States, having at the time of their adopting the Constitution expressed a desire in order to prevent misconstruction or abuse of its powers, that further declaratory and restrictive clauses should be added: And as extending the ground of public confidence in the Government will best ensure the beneficent ends of its institution.

Resolved by the Senate and House of Representatives of the United States of America in Congress assembled, two thirds of both Houses concurring that the following Articles be proposed to the Legislatures of the several states as Amendments to the Constitution of the United States, all or any of which articles, when ratified by three fourths of the said Legislatures to be valid to all intents and purposes as part of the said Constitution. viz.

Articles in addition to, and Amendment of the Constitution of the United States of America, proposed by Congress and Ratified by the Legislatures of the several States, pursuant to the fifth Article of the original Constitution.

THE BILL OF RIGHTS

AMENDMENT I

Congress shall make no law respecting an establishment of religion, or prohibiting the free exercise thereof; or abridging the freedom of speech, or of the press; or the right of the people peaceably to assemble, and to petition the Government for a redress of grievances.

AMENDMENT II

A well regulated Militia being necessary to the security of a free State, the right of the people to keep and bear Arms, shall not be infringed.

AMENDMENT III

No Soldier shall, in time of peace be quartered in any house, without the consent of the Owner, nor in time of war, but in a manner to be prescribed by law.

AMENDMENT IV

The right of the people to be secure in their persons, houses, papers, and effects, against unreasonable searches and seizures, shall not be violated, and no Warrants shall issue, but upon probable cause, supported by Oath or affirmation, and particularly describing the place to be searched, and the persons or things to be seized.

AMENDMENT V

No person shall be held to answer for a capital, or otherwise infamous crime, unless on a presentment or indictment of a Grand Jury, except in cases arising in the land or naval forces, or in the Militia, when in actual service in time of War or public danger; nor shall any person be subject for the same offence to be twice put in jeopardy of life or limb; nor shall be compelled in any criminal case to be a witness against himself, nor be deprived of life, liberty, or property, without due process of law; nor shall private property be taken for public use, without just compensation.

AMENDMENT VI

In all criminal prosecutions, the accused shall enjoy the right to a speedy and public trial, by an impartial jury of the State and district wherein the crime shall have been committed, which district shall have been previously ascertained by law, and to be informed of the nature and cause of the accusation; to be confronted with the witnesses against him; to have compulsory process for obtaining witnesses in his favor, and to have the Assistance of Counsel for his defence.

AMENDMENT VII

In suits at common law, where the value in controversy shall exceed twenty dollars, the right of trial by jury shall be preserved, and no fact tried by a jury, shall be otherwise reexamined in any Court of the United States, than according to the rules of the common law.

AMENDMENT VIII

Excessive bail shall not be required, nor excessive fines imposed, nor cruel and unusual punishments inflicted.

AMENDMENT IX

The enumeration in the Constitution, of certain rights, shall not be construed to deny or disparage others retained by the people.

AMENDMENT X

The powers not delegated to the United States by the Constitution, nor prohibited by it to the States, are reserved to the States respectively, or to the people.

AMENDMENTS 11 THROUGH 27

AMENDMENT XI

Passed by Congress March 4, 1794. Ratified February 7, 1795.

The Judicial power of the United States shall not be construed to extend to any suit in law or equity, commenced or prosecuted against one of the United States by Citizens of another State, or by Citizens or Subjects of any Foreign State.

Note: *Article III, section 2, of the Constitution was modified by amendment 11.*

AMENDMENT XII

Passed by Congress December 9, 1803. Ratified June 15, 1804.

The Electors shall meet in their respective states and vote by ballot for President and Vice-President, one of whom, at least, shall not be an inhabitant of the same state with themselves; they shall name in their ballots the person voted for as President, and in distinct ballots the person voted for as Vice-President, and they shall make distinct lists of all persons voted for as President, and of all persons voted for as Vice-President, and of the number of votes for each, which lists they shall sign and certify, and transmit sealed to the seat of the government of the United States, directed to the President of the Senate; the President of the Senate shall, in the presence of the Senate and House of Representatives, open all the certificates and the votes shall then be counted;

The person having the greatest number of votes for President, shall be the President, if such number be a majority of the whole number of Electors appointed; and if no person have such majority, then from the persons having the highest numbers not exceeding three on the list of those voted for as President, the House of Representatives shall choose immediately, by ballot, the President. But in choosing the President, the votes shall be taken by states, the representation from each state having one vote; a quorum for this purpose shall consist of a member or members from two-thirds of the states, and a majority of all the states shall be necessary to a choice. [And if the House of Representatives shall not choose a President whenever the right of choice shall devolve upon them, before the fourth day of March next following, then the Vice-President shall act as President, as in case of the death or other constitutional disability of the President.]* The person having the greatest number of votes as Vice-President, shall be the Vice-President, if such number be a majority of the whole number of Electors appointed, and if no person have a majority, then from the two highest numbers on the list, the Senate shall choose the Vice-President; a quorum for the purpose shall consist of two-thirds of the whole number of Senators, and a majority of the whole number shall be necessary to a choice. But no person constitutionally ineligible to the office of President shall be eligible to that of Vice-President of the United States.

* Superseded by section 3 of the 20th amendment.

Note: *A portion of Article II, section 1 of the Constitution was superseded by the 12th amendment.*

AMENDMENT XIII

Passed by Congress January 31, 1865. Ratified December 6, 1865.

Section 1

Neither slavery nor involuntary servitude, except as a punishment for crime whereof the party shall have been duly convicted, shall exist within the United States, or any place subject to their jurisdiction.

Section 2

Congress shall have power to enforce this article by appropriate legislation.

Note: *A portion of Article IV, section 2, of the Constitution was superseded by the 13th amendment.*

AMENDMENT XIV

Passed by Congress June 13, 1866. Ratified July 9, 1868.

Section 1

All persons born or naturalized in the United States, and subject to the jurisdiction thereof, are citizens of the United States and of the State wherein they reside. No State shall make or enforce any law which shall abridge the privileges or immunities of citizens of the United States; nor shall any State deprive any person of life, liberty, or property, without due process of law; nor deny to any person within its jurisdiction the equal protection of the laws.

Section 2

Representatives shall be apportioned among the several States according to their respective numbers, counting the whole number of persons in each State, excluding Indians not taxed. But when the right to vote at any election for the choice of electors for President and Vice-President of the United States, Representatives in Congress, the Executive and Judicial officers of a State, or the members of the Legislature thereof, is denied to any of the male inhabitants of such State, being twenty-one years of age,* and citizens of the United States, or in any way abridged, except for participation in rebellion, or other crime, the basis of representation therein shall be reduced in the proportion which the number of such male citizens shall bear to the whole number of male citizens twenty-one years of age in such State.

Section 3

No person shall be a Senator or Representative in Congress, or elector of President and Vice-President, or hold any office, civil or military, under the United States, or under any State, who, having previously taken an oath, as a member of Congress, or as an officer of the United States, or as a member of any State legislature, or as an executive or judicial officer of any State, to support the Constitution of the United States, shall have engaged in insurrection or rebellion against the same, or given aid or comfort to the enemies thereof. But Congress may by a vote of two-thirds of each House, remove such disability.

Section 4

The validity of the public debt of the United States, authorized by law, including debts incurred for payment of pensions and bounties for services in suppressing insurrection or rebellion, shall not be questioned. But neither the United States nor any State shall assume or pay any debt or obligation incurred in aid of insurrection or rebellion against the United States, or any claim for the loss or emancipation of any slave; but all such debts, obligations and claims shall be held illegal and void.

Section 5

The Congress shall have the power to enforce, by appropriate legislation, the provisions of this article.

* Changed by section 1 of the 26th amendment.

Note: *Article I, section 2, of the Constitution was modified by section 2 of the 14th amendment.*

AMENDMENT XV

Passed by Congress February 26, 1869. Ratified February 3, 1870.

Section 1

The right of citizens of the United States to vote shall not be denied or abridged by the United States or by any State on account of race, color, or previous condition of servitude.

Section 2

The Congress shall have the power to enforce this article by appropriate legislation.

AMENDMENT XVI

Passed by Congress July 2, 1909. Ratified February 3, 1913.

The Congress shall have power to lay and collect taxes on incomes, from whatever source derived, without apportionment among the several States, and without regard to any census or enumeration.

Note*: Article I, section 9, of the Constitution was modified by amendment 16.*

AMENDMENT XVII

Passed by Congress May 13, 1912. Ratified April 8, 1913.

The Senate of the United States shall be composed of two Senators from each State, elected by the people thereof, for six years; and each Senator shall have one vote. The electors in each State shall have the qualifications requisite for electors of the most numerous branch of the State legislatures.

When vacancies happen in the representation of any State in the Senate, the executive authority of such State shall issue writs of election to fill such vacancies: Provided, That the legislature of any State may empower the executive thereof to make temporary appointments until the people fill the vacancies by election as the legislature may direct.

This amendment shall not be so construed as to affect the election or term of any Senator chosen before it becomes valid as part of the Constitution.

Note*: Article I, section 3, of the Constitution was modified by the 17th amendment.*

AMENDMENT XVIII

Passed by Congress December 18, 1917. Ratified January 16, 1919. Repealed by amendment 21.

Section 1

After one year from the ratification of this article the manufacture, sale, or transportation of intoxicating liquors within, the importation thereof into, or the exportation thereof from the United States and all territory subject to the jurisdiction thereof for beverage purposes is hereby prohibited.

Section 2

The Congress and the several States shall have concurrent power to enforce this article by appropriate legislation.

Section 3

This article shall be inoperative unless it shall have been ratified as an amendment to the Constitution by the legislatures of the several States, as provided in the Constitution, within seven years from the date of the submission hereof to the States by the Congress.

AMENDMENT XIX

Passed by Congress June 4, 1919. Ratified August 18, 1920.

The right of citizens of the United States to vote shall not be denied or abridged by the United States or by any State on account of sex.

Congress shall have power to enforce this article by appropriate legislation.

AMENDMENT XX

Passed by Congress March 2, 1932. Ratified January 23, 1933.

Section 1

The terms of the President and the Vice President shall end at noon on the 20th day of January, and the terms of Senators and Representatives at noon on the 3d day of January, of the years in which such terms would have ended if this article had not been ratified; and the terms of their successors shall then begin.

Section 2

The Congress shall assemble at least once in every year, and such meeting shall begin at noon on the 3d day of January, unless they shall by law appoint a different day.

Section 3

If, at the time fixed for the beginning of the term of the President, the President elect shall have died, the Vice President elect shall become President. If a President shall not have been chosen before the time fixed for the beginning of his term, or if the President elect shall have failed to qualify, then the Vice President elect shall act as President until a President shall have qualified; and the Congress may by law provide for the case wherein neither a President elect nor a Vice President shall have qualified, declaring who shall then act as President, or the manner in which one who is to act shall be selected, and such person shall act accordingly until a President or Vice President shall have qualified.

Section 4

The Congress may by law provide for the case of the death of any of the persons from whom the House of Representatives may choose a President whenever the right of choice shall have devolved upon them, and for the case of the death of any of the persons from whom the Senate may choose a Vice President whenever the right of choice shall have devolved upon them.

Section 5

Sections 1 and 2 shall take effect on the 15th day of October following the ratification of this article.

Section 6

This article shall be inoperative unless it shall have been ratified as an amendment to the Constitution by the legislatures of three-fourths of the several States within seven years from the date of its submission.

Note: Article I, section 4, of the Constitution was modified by section 2 of this amendment. In addition, a portion of the 12th amendment was superseded by section 3.

AMENDMENT XXI

Passed by Congress February 20, 1933. Ratified December 5, 1933.

Section 1

The eighteenth article of amendment to the Constitution of the United States is hereby repealed.

Section 2

The transportation or importation into any State, Territory, or Possession of the United States for delivery or use therein of intoxicating liquors, in violation of the laws thereof, is hereby prohibited.

Section 3

This article shall be inoperative unless it shall have been ratified as an amendment to the Constitution by conventions in the several States, as provided in the Constitution, within seven years from the date of the submission hereof to the States by the Congress.

AMENDMENT XXII

Passed by Congress March 21, 1947. Ratified February 27, 1951.

Section 1

No person shall be elected to the office of the President more than twice, and no person who has held the office of President, or acted as President, for more than two years of a term to which some other person was elected President shall be elected to the office of President more than once. But this Article shall not apply to any person holding the office of President when this Article was proposed by Congress, and shall not prevent any person who may be holding the office of President, or acting as President, during the term within which this Article becomes operative from holding the office of President or acting as President during the remainder of such term.

Section 2

This article shall be inoperative unless it shall have been ratified as an amendment to the Constitution by the legislatures of three-fourths of the several States within seven years from the date of its submission to the States by the Congress.

AMENDMENT XXIII

Passed by Congress June 16, 1960. Ratified March 29, 1961.

Section 1

The District constituting the seat of Government of the United States shall appoint in such manner as Congress may direct:

A number of electors of President and Vice President equal to the whole number of Senators and Representatives in Congress to which the District would be entitled if it were a State, but in no event more than the least populous State; they shall be in addition to those appointed by the States, but they shall be considered, for the purposes of the election of President and Vice President, to be electors appointed by a State; and they shall meet in the District and perform such duties as provided by the twelfth article of amendment.

Section 2

The Congress shall have power to enforce this article by appropriate legislation.

AMENDMENT XXIV

Passed by Congress August 27, 1962. Ratified January 23, 1964.

Section 1

The right of citizens of the United States to vote in any primary or other election for President or Vice President, for electors for President or Vice President, or for Senator or Representative in Congress, shall not be denied or abridged by the United States or any State by reason of failure to pay poll tax or other tax.

Section 2

The Congress shall have power to enforce this article by appropriate legislation.

AMENDMENT XXV

Passed by Congress July 6, 1965. Ratified February 10, 1967.

Section 1

In case of the removal of the President from office or of his death or resignation, the Vice President shall become President.

Section 2

Whenever there is a vacancy in the office of the Vice President, the President shall nominate a Vice President who shall take office upon confirmation by a majority vote of both Houses of Congress.

Section 3

Whenever the President transmits to the President pro tempore of the Senate and the Speaker of the House of Representatives his written declaration that he is unable to discharge the powers and duties of his office, and until he transmits to them a written declaration to the contrary, such powers and duties shall be discharged by the Vice President as Acting President.

Section 4

Whenever the Vice President and a majority of either the principal officers of the executive departments or of such other body as Congress may by law provide, transmit to the President pro tempore of the Senate and the Speaker of the House of Representatives their written declaration that the President is unable to discharge the powers and duties of his office, the Vice President shall immediately assume the powers and duties of the office as Acting President.

Thereafter, when the President transmits to the President pro tempore of the Senate and the Speaker of the House of Representatives his written declaration that no inability exists, he shall resume the powers and duties of his office unless the Vice President and a majority of either the principal officers of the executive department or of such other body as Congress may by law provide, transmit within four days to the President pro tempore of the Senate and the Speaker of the House of Representatives their written declaration that the President is unable to discharge the powers and duties of his office. Thereupon Congress shall decide the issue, assembling within forty-eight hours for that purpose if not in session. If the Congress, within twenty-one days after receipt of the latter written declaration, or, if Congress is not in session, within twenty-one days after Congress is required to assemble, determines by two-thirds vote of both Houses that the President is unable to discharge the powers and duties of his office, the Vice President shall continue to discharge the same as Acting President; otherwise, the President shall resume the powers and duties of his office.

Note: Article II, section 1, of the Constitution was affected by the 25th amendment.

AMENDMENT XXVI

Passed by Congress March 23, 1971. Ratified July 1, 1971.

Section 1

The right of citizens of the United States, who are eighteen years of age or older, to vote shall not be denied or abridged by the United States or by any State on account of age.

Section 2

The Congress shall have power to enforce this article by appropriate legislation.

Note: Amendment 14, section 2, of the Constitution was modified by section 1 of the 26th amendment.

AMENDMENT XXVII

Originally proposed Sept. 25, 1789. Ratified May 7, 1992.

No law, varying the compensation for the services of the Senators and Representatives, shall take effect, until an election of representatives shall have intervened.

Source Notes

This section lists sources by page number. Where material appears in a paragraph, both the page number and paragraph number are listed. Quotations are identified by the quoted person's name. Boldface indicates titles of examples.

PART I—THE LEADER, LEADERSHIP, AND THE HUMAN DIMENSION
Chapter 1—The Army Leadership Framework

1-1   Douglas MacArthur: *A Soldier Speaks: Public Papers and Speeches of General of the Army Douglas MacArthur*, ed. Vorin E. Whan Jr. (New York: Frederick A. Praeger, Publishers, 1965), 354, 356.

1-1   J. Lawton Collins: in *The Infantry School Quarterly* (April 1953): 30.

1-2   Edward C. Meyer: in *The Chiefs of Staff, United States Army: On Leadership and the Profession of Arms* (Pentagon, Washington, D.C.: The Information Management Support Center, 24 March 1997), 10 (hereafter referred to as *Chiefs of Staff*).

1-2   1-3, "The Creed of the Noncommissioned Officer": *TC 22-6, The Army Noncommissioned Officer Guide*, (23 November 1990), inside front cover (hereafter cited as *TC 22-6*).

1-8   **COL Chamberlain at Gettysburg**: John J. Pullen, *The Twentieth Maine* (1957; reprint, Dayton, Ohio: Press of Morningside Bookshop, 1980), 114-125 (hereafter cited as Pullen); "The Alabamians drove the Maine men...": Geoffrey C. Ward, *The Civil War: An Illustrated History* (New York: Knopf, 1990), 220 (hereafter cited as Ward).

1-10   Douglas E. Murray: in *ARMY Magazine* 39, no. 12 (December 1989): 39.

1-13   "More than anything else...": *TRADOC Pam 525-100-2, Leadership and Command on the Battlefield: Battalion and Company* (Fort Monroe, Va., 10 June 1993), 43 (hereafter cited as *TRADOC Pam 525-100-2*).

1-14   George C. Marshall: in *Selected Speeches and Statements of General of the Army George C. Marshall*, ed. H.A. DeWeerd (Washington, D.C.: The *Infantry Journal*, 1945), 176.

1-16   **Small Unit Leaders' Initiative in Normandy**: Stephen Ambrose, *D-Day June 6, 1944: The Climactic Battle of World War II* (New York: Simon & Schuster, 1994), 235-36 (hereafter cited as Ambrose, *D-Day*); "This certainly wasn't the way...": Sam Gibbons memoir (New Orleans: Eisenhower Center, University of New Orleans (hereafter cited as Eisenhower Center)).

1-17   1-69, "When I became Chief of Staff...": Edward C. Meyer, "A Return to Basics," *Military Review* 60, no. 4 (July 1980): 4.

1-18   George Bush: in *Quotes for the Military Writer/Speaker* (Department of the Army: Chief of Public Affairs, 1989), 6 (hereafter cited as *Military Quotes 1989*).

Chapter 2—The Leader and Leadership: What a Leader Must Be, Know, and Do

2-1   *Oath of Enlistment*: AR 601-280 (29 Sep 1995), 72; 10 USC 502.

2-1   *Oath of Office*: DA Form 71, December 1988; Standard Form 61, June 1986; 5 USC 3331. The oath administered to DA civilians omits the words *"having been appointed a [rank] in the United States Army."*

2-2   Julius W. Gates, "The Thunder of a Mighty Fighting Force," *ARMY Magazine* 38, no. 10 (October 1988): 41.

2-3   S. L. A. Marshall, *Men Against Fire: The Problem of Battle Command in Future War* (Peter Smith: Gloucester, Mass., 1978), 200 (hereafter cited as S. L. A. Marshall).

2-4   John A. Wickham Jr.: *Collected Works of the Thirtieth Chief of Staff, United States Army* (Washington, D.C.: Department of the Army, 1988), 191 (hereafter cited as Wickham).

2-4   **Duty in Korea**: *Highlights in the History of the Army Nurse Corps* (Washington, D.C.: US Army Center for Military History, 1996), 23.

2-5   John M. Schofield: in *Manual for Noncommissioned Officers and Privates of Infantry of the Army of the United States* (West Point, N.Y.: US Military Academy Library Special Collections, 1917), 12.

2-6   Omar N. Bradley: in *Military Review* 28, no. 2 (May 1948): 62.

2-6   **GA Marshall Continues to Serve**: David McCullough, *Truman* (New York: Simon & Schuster, 1992), 475, 532-35.

2-7   Thomas J. Jackson: in Robert Debs Heinl, *Dictionary of Military and Naval Quotations* (Annapolis: US Naval Institute Press, 1988), 151 (hereafter cited as Heinl).

2-8   **MSG Gordon and SFC Shugart in Somalia**: Mark Bowden, "Blackhawk Down," *Philadelphia Inquirer* [hereafter cited as Bowden], Chapter 8 (November 23, 1997) [http://phillynews.com/packages/somalia/nov 23/default23.asp] and official sources.

2-8   J. Lawton Collins, *Lightning Joe: An Autobiography* (Baton Rouge: Louisiana State University Press, 1979), 444.

2-9   2-31, "Integrity has three parts...": Stephen L. Carter, *Integrity* (New York: Basic Books, 1996), 7.

2-9   William Connelly, "NCOs: It's Time to Get Tough," *ARMY Magazine* 31, no. 10 (October 1981): 31.

2-10   **WO1 Thompson at My Lai**: James S. Olson and Randy Roberts, *My Lai: A Brief History with Documents* (Boston: Bedford Books, 1998), 159, 909-92. See also W.R. Peters, *The My Lai Inquiry* (New York: W.W. Norton, 1979), 66-76.

2-11   Dandridge M. Malone, *Small Unit Leadership: A Commonsense Approach* (Novato, Calif.: Presidio Press, 1983), 29.

2-12   William G. Bainbridge, "First, and Getting Firster: The NCO and Moral Discipline," *ARMY Magazine* 25, no. 10 (October 1975): 24.

2-12   John T. Nelson, II: "*Auftragstatik*: A Case for Decentralized Combat Leadership," in *The Challenge of Military Leadership,* ed. Lloyd J. Matthews and Dale E. Brown (Washington, D.C.: Pergamon-Brassey's International Defense Publishers, Inc., 1989), 33.

2-13   **The Quick Reaction Platoon**: interview with LTC J. Baughman, School for Command Preparation, US Army Command and General Staff College, Fort Leavenworth, Kans., 1998.

2-14   2-56: "It is not genius...": Napoleon Bonaparte, in Heinl, 239.

2-14   2-58: "see though the forests...": Pullen, 111.

2-16   Geoffrey C. Ward: in Ward, 184.

2-16   George S. Patton Jr., *War as I Knew It* (Boston: Houghton-Mifflin Co., 1947), 402 (hereafter cited as Patton).

2-16   George C. Marshall: in Forrest C. Pogue, *George C. Marshall: Ordeal and Hope 1939-1942* (New York: Viking Press, 1966), 97.

2-17   Julius W. Gates, "From the Top," *Army Trainer* 9, no. 1 (Fall 1989): 5.

2-17   Aristotle: *Nicomachean Ethics,* trans. Martin Ostwald (New York: Macmillan Publishing Co., 1962), 100.

2-17   2-75: The division of emotional attributes into self-control, balance, and stability is based on Daniel Goleman, *Emotional Intelligence* (New York: Bantam Books, 1995).

2-18   Theresa Kristek: in Donna Miles, "The Women of Just Cause," *Soldiers* (March 1990), 21 (hereafter cited as "Just Cause").

2-18   *Noncommissioned Officer's Manual*: James E. Mose, *Noncommissioned Officer's Manual* (Menasha, Wis.: George Banta Publishing Co., 1917), 23.

2-18   **BG Thomas Jackson at First Bull Run**: quotations from William C. Davis, *Battle at Bull Run: A History of the First Major Campaign of the Civil War* (Baton Rouge: Louisiana State University Press, 1977), 196-97.

2-19   Margaret Chase Smith: speech to graduating women naval officers at Naval Station, Newport, Rhode Island, 1952 (Skowhegan, Maine: Margaret Chase Smith Library).

2-19   **Character and Prisoners**: TRADOC Pam 525-100-4, *Leadership and Command on the Battlefield: Noncommissioned Officer Corps* (Fort Monroe, Va., 1994), 26 (hereafter cited as *TRADOC Pam 525-100-4*).

2-20   **GA Eisenhower's Message**: GA Eisenhower's handwritten statement from the Harry Butcher diary, 6/20/44 (Eisenhower Center).

2-22   **The Qualification Report**: This vignette is based on an actual incident.

2-24   Omar N. Bradley, "American Military Leadership," *Army Information Digest* 8, no. 2 (February 1953): 5.

Chapter 3—The Human Dimension

3-1   "The Creed of the Noncommissioned Officer": *TC 22-6,* inside front cover.

3-1   3-4: Wickham, 310-11.

3-3   "NSDQ": Bowden, Chapter 29 (December 14, 1997) [http://phillynews.com/packages/somalia/dec14/default14.asp].

3-3   Audie Murphy: in Harold B. Simpson, *Audie Murphy: American Soldier* (Dallas, Tex.: Alcor Publishing Co., 1982), 271.

3-3   Richard A. Kidd, "NCOs Make It Happen," *ARMY Magazine* 44, no. 10 (October 1994): 31-36.

3-4   **The 96th Division on Leyte**: Richard Gerhardt, interview by Ed Ruggero, 1998.

3-5 **The K Company Visit**: Harold P. Leinbaugh and John D. Campbell, *The Men of Company K: The Autobiography of a World War II Rifle Company* (New York: William Morrow, 1985), 167-68 (hereafter cited as Leinbaugh & Campbell).

3-6 George S. Patton Jr.: Patton, 340.

3-6 **Task Force Ranger in Somalia**: Bowden, Chapter 1 (November 16, 1997) [http://phillynews.com/packages/somalia/nov16/default16.asp].

3-7 3-31: Bowden, Chapter 10 (November 25, 1997) [http://phillynews.com/packages/somalia/nov25/default25.asp].

3-7 Marie Bezubic: in "Just Cause," 23.

3-14 "When you're first sergeant...": Michelle McCormick, *Polishing Up the Brass: Honest Observations on Modern Military Life* (Harrisburg, Pa.: Stackpole, 1988), 102.

3-19 3-92: "Great events sometimes turn...": Pullen, 128.

PART II—DIRECT LEADERSHIP
Chapter 4—Direct Leadership Skills

4-2 James J. Karolchyk, "Leading by Example," *EurArmy* (January 1986): 25-26.

4-4 "If a squad leader...": *TRADOC Pam 525-100-2*, 35.

4-5 Randolph S. Hollingsworth: in *The Noncommissioned Officer Corps on Leadership, the Army and America: Quotes for Winners*, 2d ed. (Washington D.C.: The Information Management Support Center, January 1998), 18 (hereafter cited as *Quotes for Winners*).

4-8 4-28: "comes from a civilization..." and "Artist John Wolfe...": S. L. A. Marshall, 78.

4-11 "The first thing ...": *TRADOC Pam 525-100-4*, 5.

4-11 **Technical Skill into Combat Power**: Stephen Ambrose, *Citizen Soldiers: The US Army from the Normandy Beaches to the Bulge to the Surrender of Germany, June 7, 1944–May 7, 1945* (New York: Simon & Schuster, 1997), 64 (hereafter cited as Ambrose, *Citizen Soldiers*).

4-12 "I felt we had to...": *TRADOC Pam 525-100-2*, 47.

4-13 **Task Force Kingston**: Martin Blumenson, "Task Force Kingston," *ARMY Magazine* (April 1964): 50-60.

Chapter 5—Direct Leadership Actions

5-2 Daniel E. Wright: in *Quotes for Winners*, 18.

5-11 William G. Bainbridge, "Quality, Training and Motivation," *ARMY Magazine*, (October 1976): 28.

5-13 John D. Woodyard, "Are You a Whetstone?" *NCO Journal* (Summer 1993): 18.

5-14 Glen E. Morrell, "Looking to the Future" *Sergeants' Business* (March 1986): 7.

5-15 William A. Connelly, "Keep Up with Change in the '80s" *ARMY Magazine* (October 1982): 29.

5-16 Richard A. Kidd, "NCOs Make It Happen," *ARMY Magazine*, 44, no. 10 (October 1994): 34.

5-19 **Trust Earned**: Henry Berry, *Make the Kaiser Dance: Living Memories of the Doughboy* (New York: Arbor House, 1978), 416–419.

5-22 **Replacements in the ETO**: Ambrose, *Citizen Soldiers,* 276; "We discovered..." George Wilson, *If You Survive* (New York: Ivy Books, 1987), 214; "We were just numbers...": Leinbaugh & Campbell, 91.

5-24 **SGT York**: David D. Lee, *Sergeant York: An American Hero* (Lexington: University Press of Kentucky, 1985), 33-38.

5-25 Omar N. Bradley: in *Quotes for the Military Writer* (Washington, D.C.: Department of the Army, Office of the Chief of Information, August, 1972), 19-1 (hereafter referred to as *Military Quotes 1972*).

5-27 5-133: Theodore Roosevelt: in John C. Maxwell, *Leadership 101–Inspirational Quotes and Insights for Leaders* (Tulsa, Okla.: Honor Books, 1994), 52 (hereafter cited as Maxwell).

PART III—ORGANIZATIONAL AND STRATEGIC LEADERSHIP
Chapter 6—Organizational Leadership

6-4 **Knowing Your People**: Adolf von Schell, *Battle Leadership* (1933; reprint, Quantico, Va.: The Marine Corps Association, 1988), 11-12.

6-9 6-30: General Order No. 1664, HQ, US Army Vietnam, APO, San Francisco 09307, 13 April 1967 (Washington, D.C.: Army Nurse Corps Archives, US Army Center for Military History).

6-10 Carl F. Vuono: in *Military Quotes 1989*, 13.

6-11 **GEN Grant at Vicksburg**: *Joint Military Operations Collection* (15 July 1997), Chapter I.

6-13 Edward C. Meyer: in *Chiefs of Staff*, 7.

6-17 George C. Marshall: in *Military Quotes 1972*, 13-1.

6-19 George S. Patton Jr.: Patton, 354.

6-22   Lucian K. Truscott, *Command Missions: A Personal Story* (New York: Dutton, 1954), 556.
6-23   George S. Patton Jr.: Patton, 357.
6-24   6-107–6-108: Gary B. Griffin, "The Directed Telescope: A Traditional Element of Effective Command" (Fort Leavenworth, Kans.: US Army Command and General Staff College, Combat Studies Institute, 1991), 26-32.
6-25   John O. Marsh Jr.: in *Military Quotes 1989*, 38.
6-26   Woodrow Wilson: in *Quotes for the Military Writer/Speaker* (Department of the Army: Chief of Public Affairs, 1982), 64.
6-27   Manton S. Eddy: in *Military Review* (March 1948): 44.
6-29   The discussion of GEN Ridgway is condensed from Jack J. Gifford, "Invoking Force of Will to Move the Force," in *Studies in Battle Command* by the Combat Studies Institute (Fort Leavenworth: Kans.: US Army Command and General Staff College, 1995), 143-46.

Chapter 7—Strategic Leadership

7-5   **Allied Command During the Battle of the Bulge**: J.D. Morelock *Generals of the Ardennes: American Leadership in the Battle of the Bulge* (Washington, D.C.: National Defense University Press, 1994), 65.
7-5   7-24–7-25, "But MG Chamberlain sensed...": Pullen, 272-73.
7-7   William J. Crowe Jr.: William J. Crowe Jr. with David Chanoff, *The Line of Fire: From Washington to the Gulf, the Politics and Battles of the New Military* (New York: Simon & Shuster, 1993), 54.
7-9   **Strategic Flexibility in Haiti**: *Joint Military Operations Collection* (15 July 1997), Chapter VII.
7-10   James Thomas Flexner, *George Washington in the American Revolution (1775-1783)* (Boston: Little, Brown, & Co., 1968), 535.
7-12   **Show of Force in the Philippines**: Bob Woodward, *The Commanders*, (New York: Pocket Books, 1991), 120-25.
7-13   Arleigh A. Burke: in Karel Montor and others, *Naval Leadership: Voices of Experience* (Annapolis: US Naval Institute Press, 1987), 16.
7-14   7-64: Gordon R. Sullivan and Michael V. Harper. *Hope is Not A Method* (New York: Times Business, 1996), 90-91.
7-16   **The D-Day Decision**: Ambrose, *D-Day*, 189; quotations from interview by Walter Cronkite for CBS TV (Eisenhower Center).
7-17   Antoine-Henri de Jomini, *The Art of War*, trans. G.H. Mendell and W.P. Craighill (1862; reprint, Westport Conn.: Greenwood Press, Publishers, n.d.)162-63.
7-19   Dwight D. Eisenhower: in Maxwell, 52.
7-25   7-113: John P. Kotter, *Leading Change* (Boston: Harvard Business School Press, 1996), 20-26.
7-25   Douglas MacArthur: in *Military Quotes 1972*, 18-3.

Glossary

The glossary lists acronyms and abbreviations used in this manual. AR 310-25, JP 1-02, and FM 101-5-1 define most standard Army and joint terms. AR 310-50 lists authorized abbreviations and brevity codes.

1LT	first lieutenant
2LT	second lieutenant
.50 cal	machine gun, caliber 50
5 USC	Title 5, United States Code
10 USC	Title 10, United States Code
AAR	after-action review
ACS	Army Community Service
ADM	admiral
AO	area of operations
APFT	Army Physical Fitness Test
BOSS	Better Opportunities for Single Soldiers
CCIR	commander's critical information requirements
CINC	commander in chief of a combatant command
CMTC	Combat Maneuver Training Center
COL	colonel
comp.	compiler
CPL	corporal
CPT	captain
CSM	command sergeant major
CTC	combat training center
CW2	chief warrant officer, W-2
DA	Department of the Army
DA civilian	a civilian employee of the Department of the Army
DA Pam	Department of the Army Pamphlet
DCPDS	Defense Civilian Personnel Data System (the automated personnel system used to manage DA and DOD civilians)
D-Day	6 June 1944, the date of the Allied invasion of Normandy, France, during World War II
D-day	the execution date of any military operation
DOD	Department of Defense
DOTLMS	doctrine, training, leader development, organization, materiel, and soldiers
DPP	deferred payment plan

ECAS	ethical climate assessment survey
ed.	editor, edition
EFMB	Expert Field Medical Badge
EO	equal opportunity
ETO	European Theater of Operations (used during World War II)
EUCOM	United States European Command
FM	field manual
GA	General of the Army
GEN	general
GI	general issue (refers to equipment; sometimes used during World War II to refer to soldiers)
GTA	graphic training aid
HQ	headquarters
HQDA	Headquarters, Department of the Army
Humvee	phonetic spelling of the acronym HMMWV, high-mobility, multipurpose wheeled vehicle
IFOR	the NATO Implementation Force sent to Bosnia to implement the 1995 Dayton peace accords
IPR	in-process review
JOPES	Joint Operation Planning and Execution System
JP	Joint Publication
JSPS	Joint Strategic Planning System
JTF	joint task force
LDRSHIP	an aid for remembering Army values (loyalty, duty, respect, selfless service, honor, integrity, personal courage)
LTC	lieutenant colonel
LTG	lieutenant general
MAJ	major
MDMP	military decision-making process
METL	mission essential task list
METT-TC	mission, enemy, terrain and weather, troops, time available, civil considerations (in tactics, the major factors considered during mission analysis)
MOS	military occupational specialty
MOUT	military operations on urbanized terrain
NATO	North Atlantic Treaty Organization
NCO	noncommissioned officer
NCODP	noncommissioned officer development program (used to designate either a unit program or training conducted as part of that program)
NCOER	noncommissioned officer evaluation report

NCOERS	Noncommissioned Officer Evaluation Reporting System
OCOKA	observation, concealment, obstacles, key terrain, and avenues of approach (in tactics, the military aspects of terrain considered during a mission analysis)
OER	officer evaluation report
OERS	Officer Evaluation Reporting System
OPTEMPO	operational tempo
PMCS	preventive maintenance checks and services
PPBS	Planning, Programming and Budgeting System
PT	physical training
PVT	private
PW	prisoner of war
PX	post exchange
QM	quartermaster
R&R	rest and recuperation
ROE	rules of engagement
RPG	rocket propelled grenade (a man-portable, single-shot, antitank rocket)
SFC	sergeant first class
SGT	sergeant
SHAEF	Supreme Headquarters, Allied Expeditionary Force (the highest Allied headquarters in the ETO during World War II)
SIDPERS	Standard Installation/Division Personnel System
SJA	staff judge advocate
SOP	standing operating procedures
SSG	staff sergeant
TAPES	Total Army Performance Evaluation System (the system used to evaluate DA civilians' performance)
TLP	troop leading procedures
TOC	tactical operations center
TOW	tube-launched, optically-tracked, wire-guided (refers to the Army's heavy, antitank missile system)
UCMJ	Uniform Code of Military Justice
UN	United Nations
US	United States
USAREUR	United States Army, Europe
USC	United States Code
USR	unit status report (also called the "2715 report"; a report of unit personnel, equipment, training and overall readiness submitted monthly by active component units and quarterly by reserve component units)
vols.	volumes

WAAC   Women's Army Auxiliary Corps

WO1   warrant officer

XO   Executive Officer

Bibliography

The Army uses the world wide web to disseminate information. All major commands and schools maintain web pages. Since uniform resource locators (URLs) change from time to time, this bibliography does not list any. If you do not know the URL for an organization's web page, you can usually find it by accessing the page for the major command or installation the organization belongs to and looking for a link to the organization's page. The Center for Army Leadership maintains a page under the US Army Command and General Staff College at Fort Leavenworth, KS. The US Army Training and Doctrine Command home page contains a link to Fort Leavenworth's home page. The library or learning center at your installation will provide you with free Internet access for official business. In addition, DA Pam 25-30 (available on CD-ROM) lists all Army publications and blank forms. Use it to find Army publications related to specific topics. The US Army Publishing Agency (USAPA) home page has a link to an online extract of DA Pam 25-30 which can link you to full-text copies of selected publications.

SOURCES USED

JOINT AND MULTISERVICE PUBLICATIONS

DOD Directive 5500.7. *Joint Ethics Regulation*. (n.d.).

DOD GEN–36A. *The Armed Forces Officer*. Armed Forces Information Service. Washington, D.C.: 1988.

JP 0-2. *Unified Action Armed Forces (UNAAF)*. 24 February 1995.

JP 1-02. *Department of Defense Dictionary of Military and Associated Terms*. 23 March 1994.

JP 3-0. Doctrine for Joint Operations. 1 February 1995.

USSOCOM Publication 1. *Special Operations in Peace and War*. (n.d.).

ARMY PUBLICATIONS
Army Regulation (AR)

AR 5-1. *Army Management Philosophy*. 12 June 1992.

AR 200-1. *Environmental Protection and Enhancement*. 21 February 1997.

AR 310-25. *Dictionary of Unites States Army Terms*. 21 May 1986.

AR 600-20. *Army Command Policy*. 30 March 1988.

AR 600-100. *Army Leadership*. 17 September 1993.

AR 623-105. *Officer Evaluation Reporting System*. 1 October 1997.

AR 635-200. *Enlisted Personnel*. 26 June 1996.

AR 690-950. *Career Management*. 18 August 1988.

AR 690-400. *Total Army Performance Evaluation System*. 22 May 1993.

Department of the Army Pamphlet (DA Pam)

DA Pam 10-1. *Organization of the US Army*. 14 June 1994.

DA Pam 350-58. *Leader Development for America's Army*. 13 October 1994.

DA Pam 600-3. *Commissioned Officer Development*. 01 October 1998.

DA Pam 600-11. *Warrant Officer Professional Development*. 30 December 1996.

DA Pam 600-25. *US Army Noncommissioned Officer Professional Development Guide*. 30 April 1987.

DA Pam 600-67. *Effective Writing for Army Leaders*. 2 June 1986.

DA Pam 600-69. *Unit Climate Profile Commander's Handbook*. 01 October 1986.

DA Pam 623-205. *The Noncommissioned Officer Evaluation Reporting System*. 29 January 1988.

DA Pam 672-20. *Decorations, Awards, and Honors, Incentive Awards*. 1 July 1993.

DA Pam 690-46. *Mentoring for Civilian Members of the Force*. 31 July 1995.

DA Pam 690-400. *Total Army Performance Evaluation System (TAPES)*. 1 June 1993.

Field Manual (FM)

FM 21-20. *Physical Fitness Training*. 30 September 1992.

FM 22-9. *Soldier Performance in Continuous Operations*. 12 December 1991.

FM 22-51. *Leaders' Manual for Combat Stress Control*. 29 September 1994.

FM 25-100. *Training the Force*. 15 November 1988.

FM 25-101. *Battle Focused Training*. 30 September 1990.

FM 27-10. *The Law of Land Warfare*. 18 July 1956.

FM 100-1. *The Army*. 14 June 1994.

FM 100-5. *Operations*. 14 June 1993.

FM 100-6. *Information Operations*. 27 August 1996.

FM 100-7. *Decisive Force: the Army in Theater Operations*. 31 May 1995.

FM 100-34. *Command and Control*. TBP

FM 100-40. *Tactics*. TPB

Training Circular (TC)

TC 22-6. *The Army Noncommissioned Officer Guide*. 23 November 1990.

TC 25-20. *A Leader's Guide to After-Actions Reviews*. 30 September 1993.

TC 25-30. *A Leader's Guide to Company Training Meetings*. 27 April 1994.

US Army Training and Doctrine Command Publications

TRADOC Reg 351-10. *Institutional Leader Education and Training*. 1 May 1995.

TRADOC Pam 525-5. *Force XXI Operations*. 01 August 1994.

TRADOC Pam 525-100-1. *Leadership and Command on the Battlefield*. 1992.

TRADOC Pam 525-100-2. *Leadership and Command on the Battlefield, Battalion and Company.* 1993.

TRADOC Pam 525-100-4. *Leadership and Command on the Battlefield, Noncommissioned Officer Corps.* 1994.

MISCELLANEOUS GOVERNMENT PUBLICATIONS

GTA 22-6-1. *Ethical Climate Assessment Survey.* October 1997.

10 USC, secs. 801–946. *The Uniform Code of Military Justice.* 1950.

CIVILIAN TRAINING AND LEADER DEVELOPMENT POLICY
Laws

DOD 1400.25-M. *Civilian Personnel Manual CPM Basic Installment #8, Training.* 1 October 1985.

Executive Order 11348. *Training Programs.* 20 April 1967.

Federal Workforce Restructuring Act of 1994. *The Government Employee Training Act (GETA).* 1994.

5 USC, Chapter 41. *Training,* 1994.

5 CFR, Part 410/412. 12 December 1996.

Memoranda

HQ TRADOC, TAPC-CPP-T. *Mandatory Supervisory Training.* 29 May 1992.

HQ TRADOC, TAPC-CPP-T. *Supervisory Training Policy.* 29 June 1993.

HQ TRADOC, TAPC-CPP-T. *Mandatory New Manager's Training.* 22 September 1994.

SAMR-CPP-MP. *Action Officer Development Course.* 10 July 1996.

DEPARTMENT OF THE ARMY FORMS

DA Form 67-9. Officer Evaluation Report. October 1997.

DA Form 67-9-1. Office Evaluation Report Support Form. October 1997.

DA Form 67-9-1a. Junior Officer Developmental Support Form. October 1997.

DA Form 2028. Recommended Changes to Publications and Blank Forms. February 1974.

DA Form 2166-8. NCO Evaluation Report. October 1999.

DA Form 2166-8-1. NCO Counseling Checklist/Record. October 1999.

DA Form 4856-E. Developmental Counseling Form. June 1999.

DA Form 7222. Senior System Civilian Evaluation Report. May 1993.

DA Form 7222-1. Senior System Civilian Evaluation Report Support Form. May 1993.

DA Form 7223. Base System Civilian Evaluation Report. May 1993.

DA Form 7223-1. Base System Civilian Performance Counseling Checklist/Record. May 1993.

NONMILITARY PUBLICATIONS

Ambrose, Stephen E. *Citizen Soldiers: The US Army from the Normandy Beaches to the Bulge to the Surrender of Germany, June 7, 1944–May 7, 1945.* New York: Simon & Schuster, 1997.

———. *D-Day June 6, 1944: The Climactic Battle of World War II.* New York: Simon & Schuster, 1994.

Aristotle. "Nichomachean Ethics." In *The Basic Works of Aristotle.* New York: Random House, 1941.

Arms, Larry, and Jaime Cavazos. "The Army's SMAs from the Beginning to the Present." *NCO Journal* (Summer 1994): 9-13.

Associates, Department of Behavioral Sciences and Leadership, United States Military Academy. *Leadership in Organizations.* Garden City Park, N.Y.: Avery Publishing Group, Inc., 1988.

Axinn, Sidney. *A Moral Military.* Philadelphia: Temple University Press, 1989.

Bass, Bernard M. *A New Paradigm of Leadership: An Inquiry into Transformational Leadership.* Washington, D.C.: US Army Research Institute for the Behavioral and Social Sciences, 1996.

———. *Leadership and Performance Beyond Expectations.* New York: The Free Press, 1985.

———. *Transformational Leadership: Industrial, Military, and Educational Impact.* Mahwah, N.J.: Lawrence Erlbaum Associates, Inc., 1998.

Bentham, Jeremy. *The Principles of Morals and Legislation.* New York: Macmillan, 1948.

Berry, Henry. *Make the Kaiser Dance: Living Memories of the Doughboy.* New York: Arbor House, 1978.

Blumenson, Martin, and James L. Stokesbury. *Masters of the Art of Command.* Boston: Houghton Mifflin Company, 1975.

Chilcoat, Richard A. *Strategic Art: The New Discipline for 21st Century Leaders.* Carlisle Barracks, Pa.: US Army War College, Strategic Studies Institute, 1995.

Christopher, Paul. *The Ethics of War and Peace: An Introduction to Legal and Moral Issues.* Englewood Cliffs, N. J.: Prentice Hall, 1994.

Clement, S. D., and D. B. Ayers. *A Matrix of Organizational Leadership Dimensions.* Leadership Monograph Series, no. 8. Indianapolis: US Army Administration Center, 1976.

Collins, James C., and Jerry I. Porras. *Built To Last.* New York: Harper Business, 1994.

Combat Studies Institute. *Studies in Battle Command.* Fort Leavenworth, Kans.: US Army Command and General Staff College, 1995.

DePree, Max. *Leadership Is an Art.* New York: Dell Publishing, 1989.

Donagan, Alan. *A Theory of Morality.* Chicago: University of Chicago Press, 1977.

Doughty, Robert A. *The Seeds of Disaster: The Development of French Army Doctrine 1919-1939.* Hamden, Conn.: Archon Books, 1985.

Eichelberger, Robert L. *Our Jungle Road to Tokyo.* New York: The Viking Press, 1950.

Fall, Bernard B. *Street Without Joy.* Harrisburg, Pa.: Stackpole Books, March 1994.

Fitton, Robert A., ed., *Leadership Quotations from the Military Tradition.* Boulder, Colo.: Westview Press, 1990.

Frankena, William. *Ethics.* Englewood Cliffs, N. J.: Prentice Hall, 1973.

Fussell, Paul. *Doing Battle: The Making of a Skeptic.* Boston: Little, Brown & Co, 1996.

Gardner, John W. *On Leadership.* New York: The Free Press, 1990.

Goleman, Daniel. *Emotional Intelligence.* New York: Bantam Books, 1995.

Graen, George, and Chun Hui, "Development of Leaders for Dealing with 'Third Cultures': How to Manage Cross-Cultural Partners." Paper presented at the 1996 Army: "Leadership Challenges of the 21st Century Army" Symposium, Cantigny, Wheaton, Ill., March 1996.

————. "US Army Leadership in Century XXI: Challenges and Implications for Training." Paper presented at the 1996 Army: "Leadership Challenges of the 21st Century Army" Symposium, Cantigny, Wheaton, Ill., March 1996.

Griffin, Gary B. "The Directed Telescope: A Traditional Element of Effective Command." Fort Leavenworth, Kans.: US Army Command and General Staff College, Combat Studies Institute, 1991.

Hammer, Michael, and James Champy. *Reengineering the Corporation*. New York: Harper Business, 1993.

Hartle, Anthony. *Moral Issues in Military Decision Making*. Lawrence, Kans.: University of Kansas Press, 1989.

Haslam, Diana. *The Effects of Continuous Operations Upon Military Performance of the Infantryman (Exercise "Early Call II")*. UK Army Personnel Research Establishment, 1978.

Heifetz, Ronald A. *Leadership Without Easy Answers*. Cambridge: Harvard University Press, 1994.

Heinl, Robert D. *Dictionary of Military and Naval Quotations*. Annapolis: US Naval Institute Press, 1988.

Heller, Charles E., and William A. Stofft. *America's First Battles–1776-1965*. Lawrence, Kans.: University of Kansas Press, 1986.

Hunt, James G. (Jerry), and Robert L. Phillips, eds. "Executive Summary of the 1996 Army Symposium: 'Leadership Challenges of the 21st Century Army.'" Paper presented at the US Army Research Institute for the Behavioral and Social Sciences, Cantigny, Wheaton, Ill., May 1996.

Infantry in Battle. 2nd ed. Garrett & Massie: Richmond, Va., 1939. Reprint, Fort Leavenworth, Kans.: US Army Command and General Staff College, 1981.

Jones, Charles "T," and R. Manning Ancell, eds. *Four Star Leadership for Leaders*. Mechanicsburg, Pa.: Executive Books, 1997.

Kant, Immanuel. *Groundwork of the Metaphysic of Morals*. New York: Harper & Row, 1964.

————. *The Metaphysical Elements of Justice*. New York: Macmillan, 1965.

Kilmann, Ralph H., Mary J. Sexton, and Roy Serpa. "Issues in Understanding and Changing Culture." *California Management Review*, 28 (1986): 87-94.

Kotter, John P. *Leading Change*. Boston: Harvard Business School Press, 1996.

Kouzes, James M., and Barry Z. Posner. *The Leadership Challenge*. San Francisco: Jossey-Bass, 1987.

Kuhn, Thomas S. *The Structure of Scientific Revolutions*. Chicago: University of Chicago Press, 1974.

Lane, Larry. "Mentors and Shapers." *Soldiers* 53 (March 1998): 16-19.

Leckie, William H. *The Buffalo Soldiers: A Narrative of the Negro Cavalry in the West*. Norman, Okla.: University of Oklahoma Press, 1967.

Lee, Blaine. *The Power Principle: Influence with Honor*. New York: Simon & Schuster, 1997.

Malone, Dandridge M. (Mike). *Small Unit Leadership: A Commonsense Approach*. Novato, Calif.: Presidio Press, 1983.

Marshall, S. L. A. *Men Against Fire: The Problem of Battle Command in Future War*. Gloucester, Mass.: Peter Smith, 1947.

Maxwell, John C. *Leadership 101–Inspirational Quotes and Insights for Leaders*. Tulsa, Okla.: Honor Books, 1994.

Mill, J. S. *Utilitarianism, On Liberty, and Considerations on Representative Government*. London: Everyman's Library, 1988.

Montor, Karel, and others. *Naval Leadership: Voices of Experiences*. Annapolis: US Naval Institute Press, 1987.

Moore, Harold G., and Joseph L. Galloway. *We Were Soldiers Once...And Young*. New York: Random House, 1992.

Morden, Bettie J. *The Women's Army Corps, 1945-1978*. Army Historical Series. Washington, D.C.: Center of Military History, 1990.

Morelock, J. D. *Generals of the Ardennes: American Leadership in the Battle of the Bulge*. Washington, D.C.: National Defense University Press, 1994.

O'Toole, James. *Leading Change*. New York: Jossey-Bass, Inc., 1996.

Palmer, Bruce, Jr. *The 25-Year War: America's Military Role in Vietnam*. New York: Simon & Schuster, 1985.

Phillips, Robert L., and James G. Hunt, eds. *Strategic Leadership: A Multiorganizational-Level Perspective*. West Pamort, Conn.: Quorum Books, 1992.

Puryear, Edgar F. Jr. *Nineteen Stars*. Novato, Calif.: Presidio Press, 1971.

Rawls, John. *A Theory of Justice*. Cambridge: Harvard University Press, 1971.

Remarque, Erich Marie. *All Quiet on the Western Front*. Englewood Cliffs, N. J.: Prentice Hall, 1996.

Rogers, Robert. J. "A Study of Leadership in the First Infantry Division During World War II: Terry de la Mesa Allen and Clarence Ralph Huebner," Master's thesis, US Army Command and General Staff College, Fort Leavenworth, Kans., 1965.

Rost, Joseph C. *Leadership for the Twenty-First Century*. New York: Praeger, 1991.

Schön, Donald. *The Reflective Practitioner*. New York: Basic Books, 1983.

Senge, Peter M. *The Fifth Discipline: The Art and Practice of the Learning Organization*. New York: Doubleday Currency, 1990.

Senge, Peter M., Art Kleiner, Charlotte Roberts, Richard R. Ross, and Bryan J. Smith. *The Fifth Discipline Fieldbook: Strategies and Tools for Building a Learning Organization*. New York: Doubleday Currency, 1994.

Sheehan, Neil. *A Bright Shining Lie: John Paul Vann and America in Vietnam*. New York: Vintage Books, 1989.

Smircich, Linda. "Concepts of Culture and Organizational Analysis." *Administrative Science Quarterly*, 28 (1983): 339-358.

Solomon, Robert C. "Ethical Leadership, Emotions, and Trust: Beyond 'Charisma.'" Paper presented at the Kellogg Leadership Studies Project, Ethics and Leadership Working Papers, Ethics and Leadership Focus Group, October 1996. Originally published in *Ethics and Excellence: Cooperation and Integrity in Business* (New York: Oxford University Press, Inc., 1993).

Spears, John A., Emil K. Kleuver, William L. Lynch, Michael T. Matthies, and Thomas L. Owens. *Striking a Balance in Leader Development: A Case for Conceptual Competence*. National Security Program Discussion Paper Series, no. 92-02. Cambridge: John F. Kennedy School of Government at Harvard University, 1992.

Sullivan, Gordon R., and Michael V. Harper. *Hope Is Not A Method*. New York: Times Business, 1996.

Swain, Richard M. *Lucky War: Third Army in Desert Storm*. Fort Leavenworth, Kans.: US Army Command and General Staff College Press, 1994.

Taylor, Robert L., and William E. Rosenbach. *Military Leadership: In Pursuit of Excellence*, 3rd ed. Boulder, Colo.: Westview Press, 1996.

Thompson, Dennis M., and John H. Grubbs. "'Embracing Other Cultures: West Point and Beyond,' from *Embracing Cultural Geography: An Army Imperative.*" *Assembly* 55 (May-June 1997).

Toner, James H. *True Faith and Allegiance: The Burden of Military Ethics*. Lexington, Ky.: University of Kentucky Press, 1995.

Truscott, L. K., Jr. *Command Missions*. New York: E. P. Dutton & Co., 1954.

Vandiver, Frank. *Facing Armageddon: The First World War Experienced*. Edited by Hugh Cecil and Peter H. Liddle. Great Britain: Pen and Sword Paperback, 1996.

Von Schell, Adolf. *Battle Leadership*. Fort Benning, Columbus, Ga.: The *Benning Herald*, 1933.

Wakin, Malham M., ed. *War, Morality, and the Military Profession*. 2nd ed. Boulder, Colo.: Westview, 1986.

Ward, Geoffrey C. *The Civil War: An Illustrated History*. New York: Knopf, 1990.

Watkins, Karen E., and Victoria J. Marsick. *Sculpting The Learning Organization*. New York: Jossey-Bass, 1993.

Whiting, Charles. *HERO: The Life and Death of Audie Murphy*. Chelsea, Mich.: Scarborough House Publishers, 1990.

Yukl, Gary. *Leadership in Organizations*. 3rd ed. Englewood Cliffs, N. J.: Prentice Hall, 1994.

SUGGESTED READINGS FOR DIRECT LEADERS

Bass, Bernard M. *Stogdill's Handbook of Leadership*. New York: Free Press, 1981.

Bennis, Warren, and Burt Nanus. *Leaders: The Strategies for Taking Charge*. New York: Harper & Row, 1985.

Blanchard, Kenneth H., Patricia Zigarmi, and Drea Zigarmi. *Leadership and the One Minute Manager*. New York: Morrow, 1985.

Burns, James MacGregor. Leadership. New York: Harper & Row, 1978.

Chamberlain, Joshua Lawrence. *The Passing of the Armies*. Dayton, Ohio: Press of Morningside Bookshop, 1981.

Clarke, Bruce C. *Guidelines for the Leader and the Commander*. Harrisburg, Pa.: Stackpole Books. 1973.

Clausewitz, Carl von. *On War*. Edited and translated by Michael Howard and Peter Paret. Princeton, N.J.: Princeton University Press, 1976.

Collins, Arthur S., Jr. *Common Sense Training*. San Rafael, Calif.: Presidio Press, 1978.

Crane, Stephen. *The Red Badge of Courage*. Logan, Iowa: Perfection Form, 1979.

Forester, C. S. *Rifleman Dodd*. Garden City, N.Y.: Sun Dial Press, 1944.

Gabriel, Richard A. *To Serve with Honor: A Treatise on Military Ethics and the Way of the Soldier*. Westport, Conn.: Greenwood Press, 1982.

Holmes, Richard. *Acts of War: The Behavior of Men in Battle*. New York: Free Press, 1985.

Jacobs, Bruce. *Heroes of the Army: The Medal of Honor and its Winners*. New York: W.W. Norton & Co., 1956.

Keegan, John. *The Face of Battle (A Study of Agincourt, Waterloo, and the Somme)*. New York: Viking Press, 1976. Reprint, New York: Vintage Books, 1977, and New York: Penguin Books, 1978.

Kellett, Anthony. *Combat Motivation: The Behavior of Soldiers in Battle*. Boston: Kluwer-Nijhoff Publishing, 1982.

MacDonald, Charles B. *The Battle of the Huertgen Forest*. New York: J. P. Lippincott Co., 1963.

Matthews, Lloyd J. *The Challenge of Military Leadership*. New York: Pergamon-Brassey's International Defense Publishers, Inc., 1989.

Myrer, Anton. *Once an Eagle*. New York: Dell Publishing Co., 1970.

Newman, Aubrey S. *Follow Me*. San Francisco: Presidio Press, 1981.

Norton, Oliver Willcox. *The Attack and Defense of Little Round Top*. Dayton, Ohio: Press of Morningside Bookshop, 1978.

Nye, Roger H. *The Challenge of Command: Reading for Military Excellence*. Garden City, N.Y.: Avery Publishing Co., 1986.

Peters, Thomas J., and Nancy Austin. *A Passion for Excellence, The Leadership Difference*. New York: Random House, 1985.

Pullen, John J. *The Twentieth Maine*. Philadelphia: J.B. Lippincott Co., 1957. Reprint, Dayton, Ohio: Press of Morningside Bookshop, 1980.

Sajer, Guy. *The Forgotten Soldier*. New York: Harper & Row, 1971.

Shaara, Michael. *The Killer Angels*. New York: Ballantine Books, 1975.

Smith, Perry M. *Taking Charge: A Practical Guide for Leaders*. Washington, D.C.: National Defense University Press, 1986.

Stockdale, James B. *A Vietnam Experience: Ten Years of Reflection*. Stanford, Calif.: Hoover Press, 1984.

Waterman, Robert H., and Thomas J. Peters. *In Search of Excellence*. New York: Harper & Row, 1982.

ADDITIONAL READINGS
FOR ORGANIZATIONAL AND STRATEGIC LEADERS

Ardant du Picq, Charles Jean Jacques Joseph. *Battle Studies: Ancient and Modern*. Translated by John W. Greely and Robert C. Cotton. Harrisburg, Pa.: Military Service Publishing Co., 1947.

Belasco, Hames A. *Flight of the Buffalo: Soaring to Excellence, Learning to Let Employees Lead*. New York: Warner Books, 1993.

Blair, Clay. *The Forgotten War*. New York: Doubleday, 1987.

Clancy, Tom. *Into the Storm*. New York: G. P. Putnam's Sons, 1997.

Dixon, Norman. *On the Psychology of Military Incompetence*. New York: Basic Books, 1976.

Dupuy, Trevor N. *A Genius for War: The German Army and General Staff, 1807-1945*. Englewood Cliffs, N. J.: Prentice-Hall, 1977.

Fehrenback, R. R. *This Kind of War: A Study in Unpreparedness*. New York: Macmillan Co., 1963.

Flexner, James T. *George Washington in the American Revolution*. Boston: Little, Brown & Co., 1968.

Freeman, Douglas Southall. *Lee's Lieutenants: A Study in Command*. 3 vols. New York: Charles Scribner's Sons, 1942-44.

Freytag-Loringhoven, Hugo F .P. J. von. *The Power of Personality in War*. In Art of War Colloquium text. Carlisle Barracks, Pa.: US Army War College, September 1983.

Fuller, J. F. C. *The Conduct of War 1789-1961*. New Brunswick, N.J.: Rutgers University Press, 1961.

Grant, Ulysses S. *Personal Memoirs of US Grant*. 2 vols. New York: Charles L. Webster & Co., 1885.

Hackett, John W. *The Third World War: August 1985*. New York: Macmillan Co., 1978.

Hersey, Paul, and Kenneth H. Blanchard. *Management of Organizational Behavior: Utilizing Human Resources*. Englewood Cliffs, N. J.: Prentice-Hall, 1977.

Hunt, James G., and John D. Blair, eds. *Leadership on the Future Battlefield*. New York: Pergamon-Brassey's, 1985.

Huntington, Samuel P. *The Soldier and the State*. New York: The Belknap Press, 1957.

Janowitz, Morris. *The Professional Soldier: A Social and Political Portrait*. New York: Free Press, 1971.

Johnson, Kermit D. *Ethical Issues of Military Leadership*. Carlisle Barracks, Pa.: US Army War College, 1974.

Jomini, Antoine Henri. *The Art of War*. Trans. G.H. Mendell and W.P. Craighill. 1862. Reprint, Westport Conn.: Greenwood Press, Publishers, n.d.

Larrabee, Eric. *Commander in Chief: Franklin Delano Roosevelt, His Lieutenants, and their War*. New York: Harper & Row, 1987.

Lewis, Lloyd. *Sherman, Fighting Prophet*. New York: Harcourt, Brace & Co., 1932.

Luttwak, Edward N. *The Pentagon and the Art of War*. New York: Simon & Schuster, 1984.

Manstein, Erich von. *Lost Victories*. Edited and translated by Anthony G. Powell. Chicago: Henry Regnery Co., 1958. Reprint, Novato, Calif.: Presidio Press, 1982.

Maslow, A. H. *Motivation and Personality*. New York: Harper, 1954.

McCullough, David. *Truman*. New York: Simon & Schuster, 1992.

Montgomery of Alamein, Field-Marshal Viscount. *A History of Warfare*. Cleveland: World Publishing Co., 1968.

Moskos, Charles C., Jr. *The American Enlisted Man: The Rank and File in Today's Military*. New York: Russell Sage Foundation, 1970.

Musashi, Miyamoto. *A Book of Five Rings*. Woodstock, NY: The Overlook Press, 1982.

Patton, George S., Jr. *War As I Knew It*. Annotated by Paul D. Harkins. Boston: Houghton Mifflin Co., 1947.

Pogue, Forrest D. *George C. Marshall: Ordeal and Hope 1939-1942*. New York: Viking Press, 1966.

Powell, Colin. *My American Journey*. New York: Random House, 1995.

Pratt, Fletcher. *Eleven Generals, Studies in American Command*. New York: William Sloane Associates, 1949.

Ridgway, Matthew B. *Soldier: The Memoirs of Matthew B. Ridgway*. New York: Harper & Brothers, 1956.

Rommel, Erwin. *Attacks*. Vienna, Va.: Athena Press, 1979.

———. *The Rommel Papers*. Translated by Paul Findlay and edited by B. H. Liddell Hart. New York: Harcourt, Brace & Co., 1953.

Ryan, Cornelius. *A Bridge Too Far*. New York: Simon & Schuster, 1974. Reprint, New York: Popular Library, 1977.

Sarkesion, Sam C. *Beyond the Battlefield: The New Military Professionalism*. New York: Pergamon Press, 1981.

Summers, Harry G., Jr. *On Strategy: The Vietnam War in Context*. Carlisle Barracks, Pa.: US Army War College, Strategic Studies Institute, 1982.

Sun Tzu. *The Art of War*. Trans. Samuel B. Griffith. New York: Oxford University Press, 1971.

Van Creveld, Martin L. *Command in War*. Cambridge: Harvard University Press, 1985.

Wavell, Sir Archibald P. *Soldiers and Soldiering*. New York: Avery Publishing Group, 1986.

Weigley, Russell F. *Eisenhower's Lieutenants: The Campaign of France and Germany, 1944-1945*. Bloomington, Ind.: Indiana University Press, 1981.

Williams, T. Harry. *McClellan, Sherman, and Grant*. New Brunswick, N. J.: Rutgers University Press, 1962.

Index

Entries are by paragraph number unless stated otherwise.

W-X-Y-Z